Environmental Change and Human Security: Recognizing and Acting on Hazard Impacts

NATO Science for Peace and Security Series

This Series presents the results of scientific meetings supported under the NATO Programme: Science for Peace and Security (SPS).

The NATO SPS Programme supports meetings in the following Key Priority areas: (1) Defence Against Terrorism; (2) Countering other Threats to Security and (3) NATO, Partner and Mediterranean Dialogue Country Priorities. The types of meeting supported are generally "Advanced Study Institutes" and "Advanced Research Workshops". The NATO SPS Series collects together the results of these meetings. The meetings are coorganized by scientists from NATO countries and scientists from NATO's "Partner" or "Mediterranean Dialogue" countries. The observations and recommendations made at the meetings, as well as the contents of the volumes in the Series, reflect those of participants and contributors only; they should not necessarily be regarded as reflecting NATO views or policy.

Advanced Study Institutes (ASI) are high-level tutorial courses intended to convey the latest developments in a subject to an advanced-level audience

Advanced Research Workshops (ARW) are expert meetings where an intense but informal exchange of views at the frontiers of a subject aims at identifying directions for future action

Following a transformation of the programme in 2006 the Series has been re-named and re-organised. Recent volumes on topics not related to security, which result from meetings supported under the programme earlier, may be found in the NATO Science Series.

The Series is published by IOS Press, Amsterdam, and Springer, Dordrecht, in conjunction with the NATO Public Diplomacy Division.

Sub-Series

A.	Chemistry and Biology	Springer
B.	Physics and Biophysics	Springer
C.	Environmental Security	Springer
D.	Information and Communication Security	IOS Press
E.	Human and Societal Dynamics	IOS Press

http://www.nato.int/science
http://www.springer.com
http://www.iospress.nl

Series C: Environmental Security

Environmental Change and Human Security: Recognizing and Acting on Hazard Impacts

edited by

P. H. Liotta
Pell Center for International Relations and Public Policy
Salve Regina University, Newport, RI, USA

David A. Mouat
Division of Earth and Ecosystem Sciences
Desert Research Institute, Reno, NV, USA

William G. Kepner
U.S. Environmental Protection Agency
Office of Research and Development
Las Vegas, NV, USA

and

Judith M. Lancaster
Division of Earth and Ecosystem Sciences
Desert Research Institute, Reno, NV, USA

Published in cooperation with NATO Public Diplomacy Division

Proceedings of the NATO Advanced Research Workshop on
Environmental Change and Human Security: Recognizing and Acting
on Hazard Impacts
Newport, Rhode Island
4–7, June 2007

Library of Congress Control Number: 2008928520

ISBN 978-1-4020-8550-5 (PB)
ISBN 978-1-4020-8549-9 (HB)
ISBN 978-1-4020-8551-2 (e-book)

Published by Springer,
P.O. Box 17, 3300 AA Dordrecht, The Netherlands.

www.springer.com

Printed on acid-free paper

PREFACE

Environmental and Human Security: Then and Now

ALAN D. HECHT[1] AND P. H. LIOTTA[2]*

[1] *U.S. Environmental Protection Agency Office of Research
and Development*
[2] *Pell Center for International Relations and Public Policy
Salve Regina University*

1. Nontraditional Threats to Security

The events of September 11, 2001 have sharpened the debate over the meaning of being secure. Before 9/11 there were warnings in all parts of the world that social and environmental changes were occurring. While there was prosperity in North America and Western Europe, there was also increasing recognition that local and global effects of ecosystem degradation posed a serious threat. Trekking from Cairo to Cape Town thirty years after living in Africa as a young teacher, for example, travel writer Paul Theroux concluded that development in sub-Saharan Africa had failed to improve the quality of life for 300 million people: "Africa is materially more decrepit than it was when I first knew it—hungrier, poorer, less educated, more pessimistic, more corrupt, and you can't tell the politicians from the witch-doctors" (2002).

While scholars and historians will debate the causes of 9/11 for some time, one message is clear: An often dizzying array of nontraditional threats and complex vulnerabilities define security today. We must understand them, and deal with them, or suffer the consequences. Environmental security has always required attention to nontraditional threats linked closely with social and economic well-being. When three billion people currently live on less than 2 euros a day, 2.6 billion lack proper sanitation, 1.2 billion lack safe drinking water, 924 million are living in slums (a number expected to grow by 27 million per year to 2020), 829 million are chronically undernourished, 790 million lack health services, 191 million are unemployed, and 33.2 million are living with HIV/AIDS, today's world cannot be considered a secure world (UN Population Division; *The Millennium Development*

*Address corresponding to P. H. Liotta, Pell Center for International Relations and Public Policy, 100 Ochre Point Avenue, Newport, Rhode Island 02840, USA; tel +001-4013412371; fax: +001-4013412974; e-mail: peter.liotta@salve.edu

Goals Report 2006; The 2nd UN World Water Development Report; State of the World Cities Report 2006–2007; ILO Global Employment Trends for Women 2004; ILO Global Employment Trends for Women 2007).

Without doubt, the combination of environmental and human challenges has become more complicated in the contemporary landscape. As recently as a decade ago, policy makers recognized the changing nature of security:

The central security challenge of the past half-century—the threat of communist expansion—is gone. The dangers we face today are more diverse. Ethnic conflicts are spreading and rogue states pose a serious danger to regional stability in many corners of the globe. The proliferation of weapons of mass destruction represents a major challenge to our security. Large-scale environmental degradation, exacerbated by rapid population growth, threatens to undermine political stability in many countries and regions. (Clinton, 1996)

2. NATO's Third Dimension

Today's focus on a range of nontraditional security threats coming from environmental, social, or economic stressors was anticipated at least as early as 1969. Speaking at NATO's twentieth anniversary in 1969, U.S. President Richard Nixon challenged the organization to help make "the world fit for man." He called for the creation of a "committee on the challenges of modern society, responsible to the deputy ministers, to explore ways in which the experience and resources of the Western nations could most effectively be marshaled toward improving the quality of life of our peoples." Nixon spoke prophetically about the coming challenges:

The Western nations share common ideals and a common heritage. We are all advanced societies, sharing the benefits and the gathering torments of a rapidly advancing industrial technology. The industrial nations share no challenge more urgent than that of bringing twentieth-century man and his environment to terms with one another—of making the world fit for man, and helping man to learn how to remain in harmony with the rapidly changing world. (Nixon, 1969)

Making the world fit for humans is part of NATO's "third dimension," which goes beyond cooperation in political and defense fields to encourage cooperation related to civil emergency planning and scientific and environmental cooperation. Article 2 of the North Atlantic Treaty specifically addresses this requirement, highlighting NATO's responsibility to promote stability, well-being, and economic collaboration among nations (NATO Treaty, 1949). Today, what we call environmental security and human security are elements of NATO's long-standing focus on nontraditional security threats.

3. NATO and CCMS

The establishment in 1969 of the Committee on the Challenges of Modern Society (CCMS) reflected NATO's concern for nontraditional security threats. Breaking new institutional and political ground in creating the

CCMS, NATO intended "to improve in every practical way the exchange of views and experience among the allied countries in the task of creating a better environment for their society" and to "consider specific problems of the human environment with the deliberate objective of stimulating actions by member governments."

The roadmap for implementing environmental aspects of the CCMS program, nonetheless, was far from clear. In 1969, few European officials or agencies were dealing with environmental matters and the United States would not create the Environmental Protection Agency (EPA) until the following year. In a broad historical review, Harry Blaney noted that the new committee "went to work immediately to establish an innovative approach to international multilateral technology cooperation." CCMS determined that it would develop among interested member and nonmember countries pilot projects aimed at establishing a common body of knowledge, developing mutual understanding of each other's needs and abilities, and finding state-of-the-art solutions to the problems. Filling a vacuum in environmental awareness, CCMS established a regular environmental policy roundtable at its biannual plenary sessions and, in Blaney's words, "urged member governments to get their environmental houses in order" (1973).

Since the origin of CCMS nearly four decades ago, environmental security and human security have grown in importance as elements of national and international security policies. Subsequently, security strategies have focused increasingly on environmental and human security themes. Economic and social concerns, in particular, took center stage:

A world where some live in comfort and plenty, while half of the human race lives on less than $2 USD a day, is neither just nor stable. Including all of the world's poor in an expanding circle of development—and opportunity—is a moral imperative and one of the top priorities of U.S. international policy. (Bush, 2002)

Recent assessments of potential military impacts by climate change, including abrupt climate change scenarios prepared for the Pentagon's Office of Net Assessment, as well as calls by the U.S. Congress for analyses and national intelligence estimates on the impacts of climate change on national security, and European strategies that deal with human vulnerabilities, all contribute to a renewed emphasis and sense of urgency regarding emerging security issues (CNA, 2007; Congressional Record, 2007; Liotta and Shearer, 2007; Solana, 2004). The environmental linkages between water, sanitation, food availability, drought, and poverty have also reemerged as aspects of a strategy to address the root causes of terrorism. The attention to climate change and to terrorism echoes, once again, the recognition of NATO charter Article 2 that security depends on far more than military might. Indeed, changing contemporary conditions make the NATO/CCMS vision today more relevant than ever for global security.

4. Science for Peace and Security

In 2006 CCMS merged with NATO's Science Committee to form the Science for Peace and Security (SPS) program. The broad goals of the forerunner committees, nonetheless, remain strategically relevant to the SPS and show a long-term commitment to nontraditional concerns: to contribute to security, stability, and solidarity among nations by applying the best technical expertise together with collaboration, networking, and capacity building. As the primary committee promoting practical cooperation in civil science and innovation, SPS underscores NATO's mission to link science to society for practical, real-world applications. The charter of SPS specifically illustrates a clear mandate:

The aim of the Science for Peace and Security Programme is to contribute to security, stability and solidarity among nations, by applying the best technical expertise to problem solving. Collaboration, networking and capacity-building are means used to accomplish this end. A further aim is to facilitate continued democratic growth and support economic development in NATO's Partner countries. (NATO SPS, 2007)

Through horizon-scanning agenda that seek to share knowledge, SPS is expanding international dialogue and understanding through broad-based networks of specialists. Scientific cooperation, under SPS guidelines, is therefore an instrument to finding answers to critical questions and a means of connecting nations—and scientists. Indeed, the constructive power of science and technology has in every era had the potential to propel humankind to new levels of well-being. This potential of science and technology is now greater than ever before, but realizing the potential today requires doing things differently. The challenge is to build scientific and technical cooperation in ways that enhance knowledge, spur economic growth, and alleviate poverty. As former U.S. EPA Administrator William Ruckelshaus remarked in 1994:

This is the real lesson of the Soviet collapse. No amount of expenditure on weaponry, nor the most ferocious internal security apparatus in history, could make up for failures in the two areas that will henceforth be regarded as central to the national security of nations. These two areas are economic strength and environmental integrity.

In that same spirit, NATO workshops and pilot studies have demonstrated that scientific cooperation and technology transfer are critical elements in promoting global security.

The *Arab Human Development Report 2002*, a joint United Nations report prepared by experts from twenty-two Arab states, also presents pertinent examples of current challenges that call for scientific and technological partnerships. The *Report* points to serious human resource development issues, such as access to and use of new technologies, and which significantly lags behind other regions: "Only 0.6 percent of the population uses the Internet and the personal computer penetration rate is only 1.2 percent. More generally, investment in research and development does not exceed 0.5 percent of the gross national product, well below the world average" (UNDP, 2002).

Noting that "knowledge is a cornerstone of development," the *Report* warns that "Arab countries face a significant knowledge gap" (UNDP, 2002). The knowledge gap in the Arab world is embedded in deeper fissures in the social system. The *Report* concludes that a lack of political freedom, discrimination against women, and inadequate investments in education drain the potential to achieve real development. While acknowledging the very significant economic and social progress made in the last decades, the *Report* illustrates the challenges facing the Arab world. "There is substantial lag between Arab countries and other regions in terms of participatory governance. ... This freedom deficit undermines human development and is one of the most painful manifestations of lagging political development" (UNDP, 2002).

5. Looking Ahead

In 2001, a RAND Corporation–convened panel produced specific foreign and defense policy recommendations for the incoming Bush administration. The report goes to the heart of the issues of environmental security, human security, nontraditional threats, and new vulnerabilities:

A host of new global challenges may soon require imaginative and sustained responses by the United States and its allies. These nontraditional challenges include uncontrolled migration across borders, international crime, pandemics like AIDS and malaria, and *environmental degradation*. Many of these problems will affect Africa in particular. In few of these cases is there a consensus that they represent serious "security" threats to the United States or its allies. However, *in this era, we in the United States, along with our major industrial-state allies, have the resources and opportunity to ask ourselves whether we want to live in a world where such problems continue to fester, or whether we will try to make a difference.* (Carlucci et al., 1994; emphasis added)

This analysis conveys both a message and mandate for all states and regions: We need to work cooperatively to shape solutions to increasingly complex vulnerabilities and threats.

6. Environmental Change and Human Security at the Pell Center

In the spirit of scientific cooperation and the need to speak across borders and disciplines, an unusual mix of natural and social scientists from nineteen different nation-states—spanning four continents—met at The Pell Center for International Relations and Public Policy, Salve Regina University, in Newport, Rhode Island, for an intense and often exhilarating series of meetings, open fora, and public events—the culmination of which is represented with the publication of this scientific volume.

The Pell Center was established at Salve Regina University by an Act of the United States Congress to pay tribute to Senator Claiborne Pell, whose life and career uniquely exemplify a remarkable mixture of wisdom and courage. The Center perpetuates two of the Senator's principal goals: furthering international dialogue to achieve a more peaceful world and preparing citizens for an informed and active role in local, national, and world affairs. This NATO Advanced Research Workshop on "Environmental Change and Human Security: Recognizing and Acting on Hazard Impacts" has extended and deepened the international presence of The Pell Center and Salve Regina University and demonstrates its own form of unique collaboration: representing sponsorship and support from the NATO SPS Programme, the United Nations Environment Progamme, and the U.S. Environmental Protection Agency.

This scientific volume thus takes both a conceptual and a pragmatic approach to the issues of environmental and human security. Beginning with conceptual approaches to understanding the intersections of risk, uncertainty, and environmental challenges—as well as the methodological challenges to measuring human security—we move immediately to region-specific challenges for environmental and human security in North Africa, the Balkans, and the Middle East. In parallel, we present case studies for human security, which range from examinations of urban challenges, security and sustainability, lost opportunities for human security, and a first-ever presentation (in a NATO scientific volume) on environmental justice and health disparities in American Appalachia. We conclude with means and methods to recognize and act on security hazard impacts, offering case examples and innovative approaches from sub-Saharan Africa, Eastern Europe, and Central Asia. We also offer pathways to the future—in our recommendations for both future research and policy action.

This volume represents far more than just "another book on the shelf." Just as The Pell Center NATO ARW meeting represented an extraordinary endeavor in collegial collaboration, the work presented here is a fluid and evolving process— far more an ongoing exploration than an ending. In the spirit of the NATO SPS Programme, and as an extension of our ongoing spirit of cooperation to address and resolve emerging environmental and human security challenges, this volume's central aim is to focus on security, stability, and solidarity among peoples, states, and regions.

References

Blaney, H. C., 1973, NATO's new challenges to the problems of modern society, *Atlantic Community Quarterly* 11(2):236–247.

Bush, G. W., September, 2002. Expand the circle of development by opening societies and building the infrastructure of democracy, *National Security Strategy of the United States of America,* U.S. Government Printing Office, Washington, DC; http://www.whitehouse. gov/nsc/nss7.html, VII.

Carlucci, F., Hunter, R., and Khalilzad, Z., 1994, Executive briefing: A global agenda for the U.S. president, Washington, D.C.; http://www.rand.org/publications/randreview/issues/ rr.03.01/global.html.

Clinton, W. J., 1996, *A National Security Strategy of Engagement and Enlargement,* U.S. Government Printing Office, Washington, DC; http://www.fas.org/spp/military/docops/ national/1996stra.htm.

CNA Corporation, 2007, *National Security and the Threat of Climate Change,* Washington, DC; http://securityandclimate.cna.org/report.

Congressional Record, 2007, Providing for consideration of H.R. 2082, Intelligence authorization act for fiscal year 2008: H4779–H4786, May 10; http://www.fas.org/ irp/ congress/2007_cr/h051007.html.

International Labor Organization, 2004, *Global Employment Trends for Women*; http://www. cinterfor.org.uy/public/english/region/ampro/cinterfor/temas/gender/doc/trends.htm.

International Labor Organization, 2007, *Global Employment Trends for W*omen; http://www. ilo.org/public/english/region/ampro/cinterfor/temas/gender/news/gl_tren.htm.

Liotta, P. H., and Shearer, A. W., 2007, *Gaia's Revenge: Climate Change and Humanity's Loss,* Praeger, Westport, CT.

Nixon, R., 1969, Address at the Commemorative Session of the North Atlantic Council, April 10; http://www.presidency.ucsb.edu/ws/index.php?pid=1992.

North Atlantic Treaty, 1949, Washington, DC; http://www.nato.int/docu/basictxt/treaty.htm.

North Atlantic Treaty Organization Science for Peace and Security Programme, Introduction; http://www.nato.int/science/about_sps/introduction.htm.

Ruckelshaus, W., 1994, National security and the environment, unpublished address, March 30, United States Military Academy, West Point, NY.

Schwartz, P., and Randall, D., 2004, An abrupt climate change scenario and its implications for United States national security, prepared for the U.S. Department of Defense, Office of Net Assessment by Global Business Network; http://www.environmentaldefense.org/ documents/3566_AbruptClimateChange.pdf.

Solana, Javier, 2004, A human security doctrine for Europe: Presentation of the Barcelona report of the study group on European security; http://www.lse.ac.uk/collections/press AndInformationOffice/newsAndEvents/archives/2004/HumanSec_Doctrine.htm.

Theroux, P., 2002, *Dark Star Safari: Overland from Cairo to Cape Town,* Penguin, New York.

United Nations, 2006, *The Millennium Development Goals Report 2006*; http://mdgs.un.org/ unsd/mdg/Resources/Static/Products/Progress2006/MDGReport2006.pdf.

United Nations, 2006, *The 2nd United Nations World Water Development Report*: Water, a shared responsibility; http://www.unesco.org/water/wwap/wwdr2.

United Nations Development Program and Arab Fund for Economic and Social Development, 2002, Creating opportunities for future generations, *Arab Human Development Rep*ort, United Nations, New York; http://www.nakbaonline.org/download/ UNDP/ EnglishVersion/Ar-Human-Dev-2002.pdf.

UN-Habitat, World Urban Forum, 2006, *State of the World Cities Report* 2006–2007, 19–23 June 2006, Vancouver; http://ww2.unhabitat.org/wuf/2006/documents/HSP.WUF.3INF .3.pdf.

United Nations Population Division, World Population Trends; http://www.un.org/popin/ wdtrends.htm.

ACKNOWLEDGEMENTS

This NATO Advanced Research Workshop, "Environmental Change and Human Security: Recognizing and Acting on Hazard Impacts," held at The Pell Center for International Relations and Public Policy, Salve Regina University, represented an extraordinary collaboration. We wish to acknowledge support and encouragement from the Science for Peace and Security (SPS) Programme of the North Atlantic Treaty Organization, as well as from the United Nations Environment Programme (UNEP) and the United States Environmental Protection Agency (EPA). First, we thank Deniz Beten, Programme Director, Environmental Security of the NATO Public Diplomacy Division for her steady guidance, and Martine Deweer and Elizabeth Cowan for their always immediate help with questions and concerns as we progressed in ARW preparations. Equally, Alan Hecht, Director for Sustainable Development in the EPA's Office of Research and Development, and Cristina Boelcke, Director, Division of Regional Co-operation at UNEP, provided much-needed guidance and support. Joel Scheraga, National Program Director for the Global Change Research Program and the Mercury Research Program in the U.S. Environmental Protection Agency's Office of Research and Development, and David Gallo, Director of Special Projects at Woods Hole Oceanographic Institution, delivered two outstanding public keynote addresses to overwhelming audience response. We gratefully acknowledge and offer our gratitude to Sister M. Therese Antone, RSM, President of Salve Regina, for her vision and leadership and to The Pell Center staff—Associate Director Michele Corbeil-Sperduti, Office Managers Aïda Neary and Teresa Haas, and Work Study student Lauren Brown—for their unwavering commitment and the extraordinary hours put in. Salve Regina Graphic Design Services produced an exceptional product for our program proceedings. We recognize that making our ARW appear flawless required a superhuman effort on their part. Our Technical Editor, Jo-Ann Parks, demonstrated brilliant editorial skills and a great sense of humor. The panel of 22 anonymous reviewers provided invaluable and timely input, and Carol Marsh translated one of the chapters. Finally, we owe thanks to our ARW contributors, whose insight and illuminating analysis make this scientific volume such a significant contribution.

CONTENTS

SECTION III: HUMAN CHALLENGES: CASE STUDIES

SECTION IV: ACTING ON HAZARD IMPACTS: EXAMPLES FROM SUB-SAHARAN AFRICA, EASTERN EUROPE, AND CENTRAL ASIA

SECTION V: ENVIRONMENTAL CHANGE
AND HUMAN IMPACT LINKAGES

LIST OF CONTRIBUTORS

Madelfia A. Abb
Seton Hall University, South Orange, NJ 07079, USA

Mu'taz Al-Alawi
Mu'tah University, P.O. Box 3, Karak 61710, Jordan

Hrabrin Bachev
Institute of Agricultural Economics, 125 Tzarigradsko Shosse Blvd. Block 1, 1113 Sofia, Bulgaria

Nikolai Bobylev
Research Center for Interdisciplinary Environmental Cooperation of the Russian Academy of Sciences, 14, Kutuzova nab., 191187, St. Petersburg, Russia

Deborah J. Chaloud
US Environmental Protection Agency, Office of Research and Development, 944 East Harmon Avenue, Las Vegas, NV 89119, USA

Ryerson Christie
York University Centre for International and Security Studies, 375 York Lanes, 4700 Keele St., Toronto, ON, Canada, M3J 1P3

Liviu-Daniel Galatchi
Ovidius University of Constanta, Mamaia Blvd 124, RO 900527 Constanta-3, Romania

Maciek Hawrylak
Foreign Affairs and International Trade Canada, 125 Sussex Dr., Ottawa, ON, Canada, K1A0G2

Steven R. Hearne
Army Environmental Policy Institute, Suite 1301, 1550 Crystal Drive, Arlington, VI 22202-4144, USA

Sarah Houghton
Foreign Affairs and International Trade Canada, 125 Sussex Dr., Ottawa, ON, Canada, K1A0G2

Laurel J. Hummel
United States Military Academy, West Point, NY 10996, USA

Cindy R. Jebb
United States Military Academy, West Point, NY 10996, USA

William G. Kepner
US Environmental Protection Agency, Office of Research and Development, 944 East Harmon Avenue, Las Vegas, NV 89119, USA

Rashid Khaydarov
Institute of Nuclear Physics, Ulugbek, 100214, Tashkent, Uzbekistan

Renat Khaydarov
Institute of Nuclear Physics, Ulugbek, 100214, Tashkent, Uzbekistan

J. E. Kulenbekov
Kyrgyz Russian Slavic University, Kyrgyz Republic, 44 Kievskaya Str. Bishkek, 720000 Kyrgyzstan

Judith M. Lancaster
Division of Earth and Ecosystem Sciences, Desert Research Institute, 2215 Raggio Parkway, Reno, NV 89512, USA

Robert J. Lawson
Foreign Affairs and International Trade Canada, 125 Sussex Dr., Ottawa, ON, Canada, K1A0G2

V. M. Lelevkin
Kyrgyz Russian Slavic University, Kyrgyz Republic, 44 Kievskaya Str. Bishkek, 720000 Kyrgyzstan

P. H. Liotta
Pell Center for International Relations and Public Policy, Salve Regina University, 100 Ochre Point Avenue, Newport, RI 02840, USA

Gulnar Mailibayeva
Satpayev Kazakh National Technical University, 22 Satpayev st., Almaty 050013, Kazakhstan

Toni Mileski
Skopje University, P.O. Box 567, Skopje 1000, Macedonia

Michele Morrone
School of Health Sciences, Ohio University, Athens, OH 45701, USA

David A. Mouat
Division of Earth and Ecosystem Sciences, Desert Research Institute, 2215 Raggio Parkway, Reno, NV 89512, USA

Mayra Mukusheva
Satpayev Kazakh National Technical University, 22 Satpayev st., Almaty 050013, Kazakhstan

Maliha S. Nash
US Environmental Protection Agency, Office of Research and Development, 944 East Harmon Avenue, Las Vegas, NV, 89119, USA

Gulzhan Ospanova
Satpayev Kazakh National Technical University, 22 Satpayev st., Almaty 050013, Kazakhstan

Taylor Owen
The University of Oxford, Jesus College, Turl Street, Oxford, OX13DW

Luis Rios
United States Military Academy, West Point, NY 10996, USA

Samuel Sarri
College of Southern Nevada, Department of Philosophical and Regional Studies, 333 Pavilion Center Drive, Las Vegas, NV 89144, USA

Allan W. Shearer
Department of Landscape Architecture and Center for Remote Sensing and Spatial Analysis, Rutgers University, Cook - Landscape Architecture, 93 Lipman Drive, New Brunswick, NJ 08901, USA

D. H. Smith
United Nations Development Programme—United Nations Environment Programme Poverty and Environment Initiative, P. O. Box 30552, Nairobi, Kenya

Manat Tlebayev
Satpayev Kazakh National Technical University, 22 Satpayev st., Almaty 050013, Kazakhstan

A. K. Tynybekov
Kyrgyz Russian Slavic University, Kyrgyz Republic, 44 Kievskaya Str. Bishkek, 720000 Kyrgyzstan

Oleg Udovyk
National Institute for Strategic Studies, 3, Dobrohotov St, apt.71, Kyiv 03142, Ukraine

Biljana Vankovska
Skopje University, P.O. Box 567, Skopje 1000, Macedonia

Participants of the NATO Advanced Research Workshop on "Environmental Change and Human Security: Recognizing and Acting on Hazard Impacts" outside the Pell Center for International Relations and Public Policy, Young Building, Salve Regina University, Newport, Rhode Island. (Photographer: Kim Fuller)

INTRODUCTION

Environmental Change and Human Security: Recognizing and Acting on Hazard Impacts

P. H. LIOTTA*

Pell Center for International Relations and Public Policy
Salve Regina University

1. Background and Purpose

The NATO Advanced Research Workshop (ARW), sponsored by the Science for Peace and Security Programme (SPS) on "Environmental Change and Human Security: Recognizing and Acting on Hazard Impacts" was held at The Pell Center for International Relations and Public Policy, Salve Regina University, Newport, Rhode Island, USA. Additional sponsorship for the workshop came from both the United Nations Environment Programme (UNEP) and the United States Environmental Protection Agency (EPA).

This workshop examined how complex environmental issues (such as soil erosion, desertification, water degradation, demographic shifts, food security and agricultural prospects, urbanization trends, hazard-induced migrations) affect human security. Uniquely, this workshop involved the work of natural and social science disciplines to address how best to *mitigate*, *adapt*, or *achieve resilience* in the face of changing environmental conditions.

Scientists from nineteen different European, North American, and Mediterranean Dialogue and Partnership states attended four days of intensive sessions examining the ARW subject topic. The NATO SPS promotion of the ARW, coupled with other media notifications, helped to generate international attention for this critical and important workshop.

2. Defining and Distinguishing Environmental and Human Security

This NATO ARW focused on emerging environmental vulnerabilities that require a broadened and deeper understanding of both traditional and

---

*Address correspondence to P. H. Liotta, Pell Center for International Relations and Public Policy, 100 Ochre Point Avenue, Newport, Rhode Island 02840, USA; e-mail: peter.liotta@salve.edu

1

P.H. Liotta et al. (eds.), Environmental Change and Human Security, 1–5.

nontraditional security issues. Notably, in the opening session, ARW participants recognized that Article 2 of the original North Atlantic Treaty emphasized critical aspects of nontraditional security concerns.

Participants thus focused on specific aspects of environmental change
as it affects regions and communities directly connected to NATO scientific
concerns, while accepting that no single operative definition exists for environmental security. A number of principles regarding environmental security
were commonly accepted, nonetheless, as working constructs for this meeting. Alan Hecht, of the U.S. EPA, noted in the 1990s, for example, that

Environmental Security is a process whereby solutions to environmental problems contribute to national security objectives. It encompasses the idea that cooperation among
nations and regions to solve environmental problems can help advance the goals of
political stability, economic development, and peace. By addressing the environmental
components of potential security "hot spots," threats to international security can be
prevented before they become a threat to political or economic stability or peace. (Cited
in Butts, 2007)

A more recent attempt at defining this sometimes problematic concept
suggests that

Environmental Security refers to the impact of environmental factors on security. These environmental factors include both natural environmental phenomena, such as natural disasters,
and environmental changes caused by human activity such as depletion of natural resources,
loss of biodiversity and climate change. The impact of human activity on the environment
can, however also be positive and depends on the quality of governance as it applies, among
others, to the use and management of natural resources. (Anonymous, 2007)

A further definition, one that emphasizes the need for recognition and
cooperation, suggests that

Environmental security centers on a focus that seeks the best effective response to changing environmental conditions that have the potential to reduce stability, affect peaceful
relationships, and—if left unchecked—could contribute to the outbreak of conflict.
(Liotta, 2003: 72)

Similarly, approaches to defining human security remain varied. ARW participants commonly accepted Owen's working definition of human security,
presented and considered at length in the opening chapter of this volume:
Human security is the protection of the vital core of all human lives from
critical and pervasive threats.

At its most essential, human security is about

• *Protecting* people
• *Providing* peoples the opportunities for progress
• *Promoting* stability, security, sustainability

3. Environmental Change and Human Impact Linkages

The linkages between environmental change and human security, while not always obvious, can nevertheless have profound influence. As Brian Shaw noted in a 2007-released White Paper prepared for the Office of Intelligence, U.S. Department of Energy:

> The relationships between the environment and human security are certainly close and complex. A great deal of human security is tied to individuals' access to natural resources and vulnerabilities to environmental change—and a great deal of environmental change is directly and indirectly affected by human activities and conflicts. ... The central issue is that, unlike war, traditional or man-made disasters, most environmental issues evolve very slowly. It is difficult to determine when a critical environmental change is occurring or when a change in some aspect of the environment will have a significant and destabilizing effect on a nation or a region. (Shaw, 2007: 5–6)

Participating ARW scientists at this Pell Center event commonly recognized that human security and environmental security are not synonymous terms. Environmental change and impact, nonetheless, do directly impact human security outcomes. In the broadest sense, environmental security considers issues of environmental degradation, deprivation, and resource scarcity. By contrast, human security examines the impact of systems and processes on the individual, while recognizing basic concerns for human life and valuing human dignity.

4. ARW Program Format

The overall breakdown of the ARW comprised three separate approaches, which were extended and refined in the preparation of this volume:

- Conceptual approaches to environmental and human security.
- Case study applications and examples, represented by environmental challenges from North America, the Middle East and North Africa (MENA), the Black Sea and Caucasus, the Balkans, and Central Asia.
- Pragmatic approaches to environmental challenges, to include making scientific data relevant and comprehensible to policy and decision makers.

The ARW also included two open fora, composed of experts examining the themes of human security and making the linkages between economics, poverty, and environmental change, and their implications for security.

In accordance with NATO SPS guidelines, the workshop was closed to the public. The Pell Center did, however, present two keynote public addresses in conjunction with the ARW. The public speakers were Joel Scheraga, Director, Global Change Research Program, U.S. Environmental Protection Agency, speaking on "Anticipating and Adapting to the Effects of Climate Change" as well as Dr. David Gallo, Director of Special Communications

Projects, Woods Hole Oceanographic Institution, speaking on "Humanity and the Oceans: Environmental Change and Impact."

5. Related Field Activities

In addition to the ARW activities at The Pell Center, participants engaged in field activities within the Narragansett Bay area, to include symposia and visits to

- University of Rhode Island, Graduate School of Oceanography (GSO), incorporating an extended tour of the GSO oceangoing research vessel, *Endeavor*, which has conducted extensive expeditions in the Mediterranean Sea and the Black Sea.

- National Oceanic and Atmospheric Administration, Northeast Fisheries Science Center, Narragansett Laboratory (NOAA), for extended briefings on Geographic Information Systems (GIS) mapping data of recent ocean phenomena subject to environmental change, new technologies for deep sea research, and the collaborative NOAA and UNEP program for fisheries data arising from the 2002 Johannesburg Agenda and the Global Environment Programme (GEP).

- United States Environmental Protection Agency, Office of Research and Development, National Health and Environmental Effects Research Laboratory, Atlantic Ecology Division (EPA), for research facilities orientation and strategic overview brief from EPA Director Jonathan Garber.

Given the resources and scientific research ongoing in Rhode's Island Narragansett Bay, these field activities provided tremendous benefit. Scientists from the MENA, the Caucasus, and Central Asia were especially enthusiastic about these site visits and were provided ample opportunity for exchange and dialogue.

6. Primary Objectives and Outcomes

The primary objectives of the NATO ARW and this accompanying text represent efforts to

- Foster crossover dialogue with stakeholders and decision makers for policy-relevant applications that relate to environmental security and human adaptability.

- Develop common mechanisms for response to global environmental change while recognizing and acting on region-specific threats, challenges, and vulnerabilities.

- Demonstrate the validity of human welfare assessment under changing socioeconomic and ecological conditions.

As environmental security and human security are linked, interdependent, and necessary conditions for bringing about sustainable development and sustainable societies, this ARW explored these complex linkages in the spirit of collaborative problem solving.

This scientific volume thus seeks, as primary focus, to illustrate and offer concrete solutions to changing variabilities and conditions—and to consider commonalities among scientists from North America, Europe, the Middle East, Central Asia, the Caucasus region, and North Africa in finding solutions. Drawing on previous collaborative efforts, this work contributes to a transatlantic debate on key environmental security challenges in the first quarter of the twenty-first century. It is a debate—and a dialogue—we need to continue.

P.H. Liotta, MFA, Ph.D.
Executive Director
Pell Center for International Relations and Public Policy
Salve Regina University

References

Anonymous, 2007, draft preparatory document for workshop participants, Water scarcity, land degradation, and desertification in the Mediterranean region—environment and security linkage, Museo de las Ciencias Príncipe Felipe, City of Arts and Sciences, Valencia, Spain, 10–11 December 2007.

Butts, K., 2007, Environmental security and the Army war college, PowerPoint presentation for workshop on Teaching Population, Environment, and Security, Woodrow Wilson International Center for Scholars, Washington, DC, May 20.

Liotta, P. H., 2003, *The Uncertain Certainty: Environmental Change, Human Security, and the Future Euro-Mediterranean,* Lexington Books, Lanham, MD.

Shaw, B. R., 2007, From environmental security to environmental intelligence, Paper prepared for the Office of Intelligence, U.S. Department of Energy, Pacific Northwest National Laboratory, February 7, with specific acknowledgment to the Office of Intelligence, U.S. DOE, for release of this report.

SECTION I

APPROACHES TO ENVIRONMENTAL AND HUMAN SECURITY

ZOMBIE CONCEPTS AND BOOMERANG EFFECTS

Uncertainty, Risk, and Security Intersection through the Lens of Environmental Change

P. H. LIOTTA[1]* AND ALLAN W. SHEARER[2]

[1] *Pell Center for International Relations and Public Policy*
Salve Regina University
[2] *Department of Landscape Architecture and*
Center for Remote Sensing and Spatial Analysis
Rutgers University

Abstract: Focusing on the challenges of environmental change and human impact, the authors consider how different mindsets or mental maps lead to alternative risk responses and, consequently, alternative prioritizations of different kinds of security. The uncertainties associated with environmental change are difficult to quantify, yet the impacts may be severe. We argue that we cannot so reduce the uncertainty of the science that can definitively end debate about appropriate policy. Instead, we must learn to integrate uncertainty into decision making processes and consider how our near-term actions enable or constrain future options. Presenting a critical approach to defining human and environmental security, we also distinguish between threats and vulnerabilities and their impact. To examine the relationships between security and risk, we draw on two central metaphors. First, the tenets of traditional security are critiqued vis-à-vis Ulrich Beck's "zombie concepts" of modernism, which emphasize the state and thereby fail to engage the multiple and interdependent processes of change we now face. In this context, we discuss broadly how new solutions beget increased risk and how new knowledge yields greater uncertainty. Second, using P. H. Liotta's "boomerang effect," we look more narrowly at how policies intended to address some specific dimension of security can undermine other dimensions. When these metaphors are considered as a set of related

---

*Address correspondence to P. H. Liotta, Pell Center for International Relations and Public Policy, 100 Ochre Point Avenue, Newport, Rhode Island 02840, USA; e-mail: peter.liotta@salve.edu

P.H. Liotta et al. (eds.), Environmental Change and Human Security, 9–33.

ideas, it becomes apparent that the world is confronted with socially produced and human-centered vulnerabilities. Further, the potential for local
and localized risk has mutated into systemic risk that affects both the
"developing" and "developed" parts of the world. Responses to climate
change, in particular, must therefore accommodate thinking in terms of
multiple facets of security.

Keywords: Risk; uncertainty; environmental change; policy; human security; environmental security

1. Introduction: Setting the Template

The issue of environmental change and its impact on humankind—and
the security that affects humankind—requires, we argue in this chapter
like the authors throughout this volume, a focused, nuanced, and strategic approach. Fundamentally, we must begin with a sense, if not total
understanding, of how the security landscape before us has shifted in the
twenty-first century. While addressing how the security architecture has
changed, we must also consider how the mindsets—or "mental maps"—
of decision makers drive the willingness (or, more appropriately, the
unwillingness) to take on policies that attempt to respond to changing
conditions.

In thinking about the difficulties decision- and policy-makers have in
addressing best choices for security within a changing environment we
recognize the paradox of "manufactured uncertainties," which follows
from the work of Anthony Giddens and others (Giddens, 1990, 1998).
A significant element of Modernism has been the continuing effort to
minimize the uncertainties that stem from natural processes and social
dynamics. The phrase *manufactured uncertainties* can be understood to
capture two aspects of this (largely but not exclusively Western) societal
development. The first might be considered a matter of our attentiveness.
The creation of new knowledge makes us aware of what we were previously
unaware. That is, to use a common truism, "the more we know, the less we
know." More precisely, as we scientifically and technologically advance,
our attitude toward risk can be seen to escalate as we become more and
more aware of new uncertainties. In this way, risk management (for lack
of a better term) becomes increasingly prevalent in our collective thinking
about the world.

The second aspect of manufactured uncertainties concerns how our
interconnected social world operates. The work of civilization to manage

risk has been accomplished by the application of an expanded knowledge base through ever larger and increasingly abstract expert solutions. In doing so, our solutions have transformed what was once local—local in terms of both time and space—and largely idiosyncratic risk into global and systematic risk. In part, this circumstance means that if (or when) dangers overcome protective measures, the impacts can register over a larger area, affect more people, or otherwise be more severe. This point must be emphasized. As noted by John Maynard Keynes in his thinking about the world economy, systematic risk is underscored by "radical uncertainty," under which problems become ill-defined and the possible outcomes of our actions are unknown (Keynes, 1936, 1937). As most commonly understood, and even as most commonly practiced by actuaries, risk can be calculated based on a quantification of uncertainty. Indeed, the conventional notion of risk is dependent on the ability to calculate probabilities. But this concept of risk is only valid under so-called routine conditions, when the law of large numbers and bell curve distributions describe the world. By contrast, probabilities under radical uncertainty are not known, and, as Nassim Nicholas Taleb has recently described, the underlying mathematics is more that of Mandelbrot than of Gauss or Quételet. Under radical uncertainty, new and very surprising events can materialize from nonlinear relationships and feedback loops. Taleb calls these surprises "black swans," after the bird that European scientists said could not exist given their Continental observations, but that they could not deny after visiting Australia (2007).

Thus, the dilemma presented by manufactured uncertainties is cyclic and seemingly feeds on itself: our increasing perception of risk propels us to do more and more to cope with ever new uncertainty, but our modern (read: Modern) answers bring about a kind of uncertainty that our tools cannot ultimately address.

To approach the challenges presented by manufactured uncertainties, systematized risk, and radical uncertainty, we offer an examination of two central metaphors: the "zombie concept" and the "boomerang effect." The zombie concept—sometimes called the "zombie category"—stems from the work of German sociologist Ulrich Beck, whose writings have examined how the concept of risk permeates contemporary life. Important for our topic, Beck argues that we have entered an age of true interlinked, interdependent globalization, involving a multidimensional process of change; however, he cautions that memories of past social structures—such as zombie concepts—linger and continue to inform actions even though these actions no longer respond to or are responsive in our new context. Zombie concepts present two potential problems for decision makers. The first is that more than merely leading to undertakings that are ineffective in the current world, zombie concepts are all too often (if not always) totalizing and they blind decision makers from recognizing complex connections across issues and relationships between solutions. Without such recognition, acting on more holistic—or at least better balanced and more robust—options is impossible. Not killing off the zombie concept of traditional state-centric security—that is not recognizing and engaging the nontraditional security concerns of individuals and regions in a proactive and preemptive manner—may result in the self-fulfilling, but ultimately inadequate, call for military intervention in situations that were provoked by natural disaster, environmental degradation, or pandemic—to name only a few nontraditional security concerns.

The "boomerang effect" encapsulates the paradox of manufactured uncertainties and draws attention to the second potential (and we think) potent consequence of zombie concepts. The failure of a course of action intended to achieve some goal can take several forms. Perhaps benignly, nothing happens and the problem persists as initially identified. More dishearteningly, the failed attempt to achieve success exacerbates some underlying condition, causing the problem to be more acute, more intense, or more frequent. And most dauntingly, the actions produce a new problem. Herein lies the boomerang effect. A boomerang toss that misses its intended target is a failed effort. But the boomerang's design accommodates the risk of a miss by allowing it to come back to the thrower for another pitch. However, the return also makes the person vulnerable to the impact of the throw. That is, the *target-er* becomes the *targeted*. For example, the most obvious boomerang effect of our time is anthropogenic global warming, which is the return of our solution to accommodate a specific lifestyle. As Liotta has noted, in the context of a globalized, interconnected world, the pursuit of the state-security zombie concept can boomerang into nontraditional security problems. For example, military intervention—the most common means to achieve national security ends—can aggravate human security issues and can thereby be a cause rather than a solution to human security predicaments. Such was the case in Kosovo. Such has become the case in Iraq.

The significance of the boomerang effect lies in understanding that the "nontraditional" security issues that have long plagued the so-called developing world could increasingly affect the policy decisions and future choices of powerful states and world leaders. Notably, Ulrich Beck also employed the "boomerang effect" in his 1992 work addressing agricultural policy. Beck suggests that widespread risks "boomerang" individuals producing risks—who will be exposed to these risks (such as pollution and watershed contamination). Risk is fundamentally related to and dependent on knowledge and access to information, which may or may not be linked to economic wealth (1992; Giddens, 1999). Beck's boomerang effect is illustrated most recently with Chinese contamination of (pet and human) food and use of lead paint in toys—leading to embargoes and product recalls. With the production of goods at intended lowest, globalized prices the guarantees of safety are not assured, thus adding to risk and hazard—and economic impact.

The future therefore requires decision makers in both the developing and developed world to focus on a broad—and broadened—understanding of the meaning of security. Focusing on one aspect of security at the expense or detriment of another aspect may well cause us to be "boomeranged" by a poor balancing of ends and means in a radically changed security environment. Before proceeding further, therefore, we consider a basic approach to what "security" as concept and policy choice involves.

2. The Multiple Facets of "New" Security

Although security—as basic concept—is frequently considered in the study and analysis of international relations and strategy, military history, and national policy decisions, its essential meaning might better be widely debated than agreed upon. Commonly considered a basic concept in policy and academic debates, security at the national and subnational level is in reality an ephemeral quantity, its definition in large measure a reflection of the perspectives and physical situations of the student and analyst. Thus academics and analysts raised in a particular school—particularly the "realist school" that emphasizes power relationships between and among states being the most influential, at least in government circles—tend to interpret events with the blinders that the school's focus provides. That realism has stood the test of time accounts for the skepticism that statesmen and many scholars have toward nontraditional, wider definitions that would include many of the issues raised in this book.

While we are sympathetic to realist concerns that using overly broad definitions of security can lead to prescriptions for the misuse of military power, or to the underestimation of the role that military forces should and must play in world affairs, we also recognize that many of the "security" issues we consider here do not—and cannot—fit comfortably within a state-centric, power-driven level of analysis.

To be blunt, there have been specific reasons for those intending to proscribe the terms of the debate on what is or is not a matter of security: doing so both makes the topic accessible for decision makers and provides a basis for determining present and future policy. Most often such decision makers only conceive of security concepts in a power-dominant, state-centric mindset. There is a hazard, nevertheless, of adding the term "security" to either environmental or human-centered concerns. Conflating national security, human security, and environmental security all within a distinct conceptual framework, furthermore, limits effective policy choices in an increasingly complex, uncertain world.

But, couching the emerging "nontraditional" concepts, such as environmental security and human security, solely on the relationships to potential or real threats places the ideas hostage to "traditional" state-centered national security paradigm. We argue that the typology of power is inadequate to represent the often nuanced and always complex dynamics of security concepts.

In the classical sense, security—from the Latin *securitas*—refers to tranquility and freedom from care, or what Cicero termed the absence of anxiety upon which the fulfilled life depends. Thus, in theory, the stable state both provides tranquility for its citizens and extends upward in its relations to influence the security of the overall international system. The overall system promotes security by supporting the stability of states.

Individual security, stemming from the liberal thought of the Enlightenment, was also considered both a unique and collective good. Moreover, despite the abundance of theoretical and conceptual approaches in recent history, the right of states to protect themselves under the rubric of "national security" and through traditional instruments of power (political, economic, and especially military) has never been directly, or sufficiently, challenged. The responsibility, however, for the guarantee of the individual good—under any security rubric—has never been obvious.

In general, one could find little to argue with in these principles. There are problems, nonetheless. On the one hand, all security systems are not equal—or the same. Moreover, all such systems collectively involve codes of values, morality, religion, history, tradition, and even language. Any system that enforces, as it were, human security inevitably collides with conflicting values—which are not synchronous or accepted by all individuals, states, societies, or regions.

On the other hand, in the once widely accepted realist understanding, the state was the sole guarantor of security. For Thomas Hobbes, the classic state-centered realist, an individual's insecurity sprang from a life that was "solitary, poor, nasty, brutish and short" (1985 edition). The state protected the individual from threats, whether these threats came at the hands of a local thief or from an invading army. For this protection, the citizen essentially relinquished individual rights to the state, as the state was the sole protector.

It is significant, nonetheless, that aspects of "nontraditional" security issues that have long plagued the so-called developing world—issues that include environmental degradation, resource scarcity, epidemiology, transnational issues of criminality and terrorism—can increasingly affect the policy decisions and future choices for powerful states and world leaders as well. Indeed, the future may well require decision makers to focus on a broad—and broadening—understanding of the meaning of security.

We thus consider general conceptual approaches to both human and environmental security, along with their often entangled complexities.

2.1. HUMAN SECURITY

As others examine the topic in this volume (especially Owen), human security centers on the individual (rather than the state) and that individual's right to personal safety, basic freedoms, and access to sustainable prosperity. In conceptual terms, human security is both a "system" and a systemic practice that promotes and sustains stability, security, and progressive integration of individuals within their relationships to their states, societies, and regions. In abstract but understandable terms, human security allows individuals the pursuit of life, liberty, and the pursuit of both happiness and justice.

The 1994 United Nations Development Programme (UNDP) report remains the most widely recognized first post-Cold War attempt to recognize a conceptual shift that needed to take place in considering security shifts:

The concept of security has for too long been interpreted narrowly: as security of territory from external aggression, or as protection of national interests in foreign policy or as global security from the threat of nuclear holocaust. It has been related to nation-states more than people. ... Forgotten were the legitimate concerns of ordinary people who sought security in their daily lives. For many of them, security symbolized protection from the threat of disease, hunger, unemployment, crime [or terrorism], social conflict, political repression and environmental hazards. With the dark shadows of the Cold War receding, one can see that many conflicts are within nations rather than between nations. (UNDP, 1994: 22–23)

In 2003, the UN Commission on Human Security expanded this concept to include protection for peoples suffering through violent conflict, for those who are on the move whether out of migration or in refugee status, for those in post-conflict situations, and for protecting and improving conditions of poverty, health, and knowledge (UN, 2003).

With the fall of the Berlin Wall it should have been clear that despite the macro-level stability created by the East-West military balance of the Cold War, individual citizens were not safe. They may not have suffered outright nuclear attack, but remained vulnerable to the effects of environmental degradation, poverty, disease, hunger, violence, and human rights abuses. Yet the protection of the individual was all too often negated by an over-attention to the state. "Traditional security," in the classic Hobbesian sense, thus failed at one of its primary objectives: protecting the individual.

This new type of instability led to the challenging of the notion of traditional security by such concepts as cooperative, comprehensive, societal, collective, international, and human security. The 1994 UNDP report therefore attempted to argue that freedom from chronic threats such as hunger, disease, and repression (which require long-term planning and development investment) as well as the protection from sudden disasters (which require often immediate interventions of support from outside agents) required action. Thus the UNDP offered seven "nontraditional" security components (UNDP, 1994: 22–25):

- Economic security: the threat is poverty; vulnerability to global economic change.

- Food security: the threat is hunger and famine; vulnerability to extreme climate events and agricultural changes.

- Health security: the threat is injury and disease; vulnerability to disease and infection.

- Environmental security: the threat is resource depletion; vulnerability to pollution and environmental degradation.

- Personal security: the threat is violence; vulnerability to conflicts, natural hazards, and long-term encroaching disasters.

- Community security: the threat is loss of the integrity of cultures; vulnerability to cultural globalization.

- Political security: the threat is political repression; vulnerability to conflicts and warfare (Threats and vulnerability assessment: Liotta and Owen, 2006a, b).

In this conceptual approach to "new" security, the overarching focus was on the individual. Thus, the broad conceptualization of security is quite different from a traditional, state-centric view of security.

Human security remains, nonetheless, a contested concept. Other than a common agreement on the focus on the individual, the concept of human security is still a work very much "under construction".[1] In simple terms, the

---

[1] A sufficient review of the literature of human security is not possible in the space available here. Some of the most stimulating pieces on the subject of environmental and human security are found in *Security and Environment in the Mediterranean: Conceptualizing Security and Environmental Conflict*, Han Günter Brauch, P. H. Liotta, Antonio Marquina, Paul Rogers & Mohammed El-Sayed Selim, eds. (Berlin: Springer Books, 2003): Bjorn Møller, Chapter 12, "National, Societal and Human Security: Discussion—Case Study of the Israel-Palestine Conflict"; Nils-Petter Gleditisch, Chapter 26, "Environmental Conflict: Neomalthusians vs. Cornucopians." Other work includes Jorge Nef, *Human Security and Mutual Vulnerability: The Global Economy of Development and Underdevelopment,* 2nd ed. (Ottawa: International Development Research Centre, 1999); Roland Paris, "Human Security: Paradigm Shift or Hot Air?" *International Security* 26, no. 2 (Fall 2001), 87–102; Peter Stoett, *Human and Global Security: An Explanation of Terms* (Toronto: University of Toronto Press, 1999); Caroline Thomas and Peter Wilkin, eds., *Globalization, Human Security, and the African Experience* (Boulder: Lynne Rienner, 1999); Joseph Stiglitz,

United Nations Commission on Human Security defines human security as the protection of "the vital core of all human lives in ways that enhance human freedoms and fulfillment" (2003).

2.2. ENVIRONMENTAL SECURITY

Environmental security emphasizes the sustained viability of the ecosystem, while recognizing that natural processes are the ultimate weapons of mass destruction. In 1566 in Shensi province, for example, tectonic plates shifted and by the time they settled back into place, 800,000 Chinese were dead. Roughly 73,500 years ago, a volcanic eruption in what is today Sumatra was so violent that ash circled the earth for several years, photosynthesis essentially stopped, and the precursors to what is today the human race amounted to only several thousand survivors worldwide (Bissell, 2003: 35).

From an alternative point of view, nonetheless, mankind itself is the ultimate threat to ecosystems. Thus, there now exists a nascent understanding that there is a fundamental linkage between recognizing and acting on environmental effects and impacts. Long-term strategic actions that will mitigate or even prevent serious future negative security outcomes are seriously considered—largely due to the great attention given to climate change, in particular, and to the work of international agencies such as the UN's Intergovernmental Panel on Climate Change (2007). Recognizing and acting on problems of environmental degradation and resource scarcity may come to be a common feature of future security policy. Yet the difficulty in predicting—let alone the ability to determine with *complete certainty*—hard "trigger" events that will promote peace and prevent conflict will continue for some time.

While specific transnational aspects of environmental (and human) security vary on a regional basis—such as the pandemic of HIV in sub-Saharan Africa or in South Asia—a number of features cut across regions in shared commonalities:

Globalization and Its Discontents (New York: W. W. Norton, 2002); Majid Tehranian, ed., *Worlds Apart: Human Security and Global Governance* (London: I.B. Tauris, 1999); Tatsuro Matsumae and L. C. Chen, eds., *Common Security in Asia: New Concept of Human Security* (Tokyo: Tokai University Press, 1995); Yuen Foong Khong, "Human Security: A Shotgun Approach to Alleviating Human Misery?" *Global Governance* 7, no. 3 (July–September 2001). Moreover, a recent issue of *Security Dialogue,* 35(3) 345–371, displayed the rich diversity and division of perspectives on "What is 'Human Security?'" Taylor Owen's concluding essay, "Human Security—Conflict, Critique, and Consensus: Colloquium Remarks and a Proposal for a Threshold-Based Definition" (373–387) is especially useful.

- *Population growth* rates in North Africa may make the region less stable. While admittedly a number of recent United Nations Population Division reports question the ability to accurately predict population growth, such growth rates offer projective measures for assessing regional environmental stability (Liotta, 2003).
- Coupled with population increase is the expectation that *increased urbanization* will be the norm globally. In 1925, for example, 80 percent of human population was located in rural areas; in 1995, only 52 percent remained in rural locales—and the expectation remains that this trend toward urbanization will continue (Espenshade, 1995).
- *Climate change,* regardless of how uncertain the exact change will be, may have significant regional impact. The disparate proportionality of pollution (to include carbon dioxide and greenhouse gas emissions) from Europe, for example, will affect the entire Euro-Mediterranean space (Liotta and Shearer, 2007).
- *Increased desertification and soil erosion and decreased food production, access, and availability* may lead to further demographic shifts to include future environmental refugees.
- *Increased resource scarcity,* most notably water, as well as the potential for dwindling (or less critical) petroleum resources fuel the economy of several states.

While all of the above are specific environmental and human security issues, it remains crucial to recognize that such issues cannot be separated from larger regional partnership interests—and thus not separated from the whole.

2.2.1. Defining Environmental Security

In terms of precise categorization, there are distinct differences between human and environmental security. In the broadest sense, environmental security considers issues of environmental degradation, deprivation, and resource scarcity; by contrast, human security examines direct impact of systems and processes on the individual, while recognizing basic concerns for human life and valuing human dignity. Yet as numerous examples in this text illustrate, complex interactions within various environments often place stress on the security of the individual. Thus, environmental and human security often coexist in complex interdependence best conceptually considered as "extended security."

Integrating environmental issues into security concerns, nonetheless, has its naysayers. Ole Wæver, one of the earliest (along with Barry Buzan) to critique the "new security agenda" expressed skepticism about the ability to influence policy through refocusing the understanding of security:

A security issue demands urgent treatment: it is treated in terms of threat/defence, where the threat is external to ourselves and the defence often a technical fix ... traditionally the state gets a strong say when something is about security. To turn new issues (such as the environment) into "security" issues might therefore mean a short time gain of attention, but comes at a long-term price of less democracy, more technocracy, more state and a metaphorical militarisation of issues. For this reason, environmental activists and not least environmental intellectuals who originally were attracted to the idea of "environmental security" have largely stepped down. ... Security is about survival.... The invocation of security has been the key to legitimizing the use of force, and more generally opening the way for the state to mobilize or to take special force.... Security is the move that takes politics beyond the established rules of the game. (2000)

There are, however, any number of overextending assumptions in the above reference. Above all is the assumption that security is an extreme term that can only be couched in terms of threat, and that the state—as political monolith—can only respond with the use of force.

Yet security is a basis for both policy *response* and long-term *planning*. Further, the use of force—particularly military force—is often an ineffectual and irrelevant response to the "new security agenda." Thus, the argument that "environmental security" is simply a mask for military intervention is an argument that is, at best, thin.

What *is* true is that the understandings of, and definitions for, environmental security range so broadly that its meaning takes on something for everyone—and perhaps, ultimately, nothing for no one. A 1998 study from the U.S. Army Environmental Policy Institute, for example, documented the current confusion among contrasting "versions" of environmental security but was unable to solve the definition problem (Glenn et al., 1998). A further pertinent illustration of this confusion rises out of a senior U.S. military conference that took place in 2000: The Army general responsible for force structure understood that environmental security meant "force protection"; another senior officer seemed convinced that environmental security addressed basic health and safety programs. Not surprisingly, therefore, a review of the literature for defining environmental security shows that more than *twenty* common definitions can be found (King, 2000a, b: 2:1–2). Moreover, a review of the definitions provided by U.S. government agencies for environmental security only complicates further precision (United States Environmental Protection Agency, 1999; Glenn et al., 1998: 19). Perhaps the least useful definition comes from the U.S. Department of Defense, which provides a definition of environmental security that encompasses everything from "explosives safety" to "pest management" (1996).

For the specific application of the term "environmental security" in this scientific volume, the broadest relevant definition should be, and should remain, an understanding that environmental security centers on a focus that seeks the best effective response to changing environmental conditions that have the potential to reduce stability and affect peaceful relationships, and—if left unchecked—could contribute to the outbreak of conflict (Liotta, 2003). The most encompassing definition for environmental security, nonetheless, was written two decades ago, when Norman Myers argued that:

National security is not just about fighting forces and weaponry. It relates to watersheds, croplands, forests, genetic resources, climate and other factors that rarely figure in the minds of military experts and political leaders, but increasingly deserve, in their collectivity, to rank alongside military approaches as crucial in a nation's security. (1986)

In contrast to Wæver's pessimism regarding the true political motives for the environmental security agenda, and in support of Myers's above ideas on the need to rethink and reconceptualize security, one would hope that both military and political leaders have come to recognize the validity of environmental security for strategy and policy initiatives.

3. Recognizing Threats, Distinguishing Vulnerabilities

Environmental and human security remain both evolving and contested concepts. Yet the *vulnerability* aspects that these security issues involve present serious long-term challenges for policy makers. Learning to distinguish between clear threats and complex vulnerabilities may help make these two security concepts both more relevant and more actionable for decisions.

In the past, most discussions of environmental and human security distinction centered on the best approach to dealing with environmental *threats*. Even those who first promulgated the idea of environmental security and then backed away seemed to have done so because of the implication that security contextualization must be couched solely in terms of threat response with use of (almost always, military) force. Few of the definitions, with the possible exception of Myers's all-encompassing approach above, recognize that *vulnerability* can also be a key feature of the security calculus. Although few policy makers today might immediately recognize the difference between the threat and vulnerability, both concepts suggest different realities.

A threat is both an external and internal cause of harm, that is: identifiable, often immediate, and requiring understandable response. Military force, for example, has traditionally been sized against threats: to defend a state against external aggression, to protect vital national interests, and to enhance state security. (The size of the U.S. and USSR nuclear arsenals during the Cold War made perhaps more sense than today because the perceived threat of global holocaust in the context of a bipolar, ideological struggle was far greater then. Force structures were linearly related in the sense that their sizes reflected generally accepted assessments in each country of the other's hostility and military capability. In fact, the U.S. budget process more or less requires the identification of credible threat scenarios and attempts to keep the level of military investments in rough equivalence to the magnitude of the threat scenarios.)

Equally, the ability to apply force—not just the application of it—matters in recognizing threat challenges. Accordingly, force options traditionally have included a range of responses—from deterrence to intervention to preemptive strike. All have been sized, shaped, designed, and budgeted for response to threats. A threat, in short, is *clearly visible or commonly acknowledged.*

Vulnerabilities are less clear and less immediate. Often vulnerabilities are signaled, at best, by suggestive indicators; in some instances—a critical point here—a vulnerability, bound by its often multiple linkages to other complex factors, might not even be recognized. In the broadest understanding, vulnerability may not even be recognized or understood—which can be maddeningly frustrating for decision makers. When it is recognized, a vulnerability often remains only an indicator, often not clearly identifiable, often linked to a complex interdependence among related issues, and does not always suggest a correct or even adequate response.

Essentially, a number of vulnerability issues, if left unchecked over time, can take on significance that could easily match the challenges of the ongoing Arab-Israeli conflict or the concern over proliferation of weapons of mass destruction. These issues include:

- *Global climate change* could wreak havoc on coastal states, particularly those that lack the infrastructure and capacity to rebuild/recover after catastrophe occurs.

- The *rapid spread of disease,* particularly HIV/AIDS, will compound the negative effects of rapid urbanization and destabilizing migrations.

- *Shifts in the balance of threats and vulnerabilities* may require changes in military force structures, missions, and budgets of armed forces.

- The role of *alliances and critical coalition partnerships* in supporting long-term environmental security initiatives *will be more critical than ever.* Accordingly, the notion of "security communities" and cooperative security principles will present opportunities for resolving common/comprehensive security challenges.

- The need *for preventive action*—which in the past has often been referred to as preventive diplomacy or conflict prevention—will become imperative. Investing early may well prevent a number of future long-term contingencies.

To be explicit: Vulnerabilities, if left unchecked over time, become threats.

These kinds of vulnerabilities are, for example, clearly relevant to the Euro-Mediterranean and Eurasian spaces, where we may have moved from a dynamic of the old *security dilemma* to encompass issues in the twenty-first century that will also include a new *human dilemma* for a vast percentage of populations in the Middle East and North Africa, the Caucasus, and Central Asia (Liotta, 2003).

3.1. EXTREME VULNERABILITIES

The *time* element in the perception of vulnerabilities can also further compound the problem—and make vulnerabilities far more controversial and far less pressing than the clear and present dangers of threats. It serves well to recognize that the core identity in a security response to issues involving extended security is that of recognizing a condition of *extreme vulnerability*. Extreme vulnerability can arise from those living under conditions of severe economic deprivation, to victims of natural disasters, and to those who are caught in the midst of war and internal conflicts. Long-term human *development* attempts thus make little to no sense and offer no direct help. The situation here, to be blunt, is not one of sustainability but of rescue.

R. H. Tawney, describing rural China in 1931, recounted the extreme vulnerability among peasants through a powerful image: "There are districts in which the position of the rural population is that of a man standing permanently up to the neck in the water, so that even a ripple is sufficient to drown him" (Scott, 1997). In such instances, the need for intervention is immediate.

3.2. "ENTANGLED" VULNERABILITIES

Most difficult for policy analysts and decision makers, who are often driven by crisis response rather than the needs of long-term strategic planning, are what we call *entangled vulnerabilities*. To be sure, in dealing with such vulnerabilities, the best response is most often unclear. Yet, given the uncertainty, the complexity, and the sheer non-linear unpredictability of such long-term, encroaching vulnerabilities, the frequent—and classic—mistake of the decision maker is to respond with the "gut reaction": the intuitive response to situations of clear ambiguity is, classically, to *do nothing at all*. The more appropriate response is to take an adaptive posture; to avoid the instinct to act purely on gut instinct; and to recognize what variables, indicators, and analogies from past examples might best inform the basis of action (Courtney et al., 1997).

To be clear: avoiding disastrous long-term impacts of entangled vulnerabilities (which can evolve over decades) requires strategic planning, strategic investment, and strategic attention. To date, states and international institutions seem woefully unprepared for such strategic necessities. Moreover, environmental and human security, since they are contentious issues, often fall victim to the *do nothing* response because of their vulnerability-based conditions in which the clearly identifiable cause and the desired prevented effect are often ambiguous.

Some examples of time-dependent, entangled vulnerabilities might help illustrate this decision-making problem. Although now well known that the United States, among so-called developed nations, has spent billions of dollars on *studying* the issue of environmental change—particularly regarding the issues of global warming and greenhouse gases—there still remain linkage problems between cause and effect. If the temperatures were to rise, by some estimates, between 1.4°C and 5.8°C over the course of the twenty-first century, there could be a concomitant rise in sea levels of between 9 and 88 cm.[2] Such a rise in sea level, although not of immediate concern to most nations, would be the single greatest national security issue for island nations such as the Maldives or Tulu; in essence, such a sea level rise would mean the end of both of them (because entire landmasses would be under water).

A second example is equally striking. Because of rising temperatures—although no one precisely knows how high, at what rate, and how much the levels will fluctuate—Canada faces a unique conundrum. Perhaps as early as 2013, the Northwest Passage—which Canada claims as territorial waters—may become navigable. A navigable Northwest passage (which would cut the journey from Europe to Asia by 7,000 km in comparison with transiting the Panama Canal) could lead to a rise in illicit crime, human trafficking, drug smuggling, pollution of the fragile Arctic ecosystem, human disasters at sea, and violations of Canadian sovereign territory (MacDonald et al., 2007).

Given the amount of study that has been devoted to such vulnerability-based security issues, there has been far less attention given to the potential strategic responses. The time, perhaps, for study has passed and the time for action, implementation, and preparatory response may well have arrived. Daniel Esty phrases this dynamic well in describing how the notion of sustainable development is rapidly becoming a "buzzword largely devoid of content" and how new methods and ideas for action need, quickly, to be set in place:

---

[2] These data are based on the Summary for Policymakers of the IPCC (2001: 13–14) that was approved in January 2001 in Shanghai by the IPCC member governments. As such, they do not offer definitive, discrete "proof." Final, convincing, and irrefutable data for these issues do not exist. Equally, the estimate of 5.8°C exceeds the estimates of recent UN and American National Academy of Sciences data. These illustrations are meant to show the nature of security issues that rise out of vulnerabilities rather than out of direct threats. The issues themselves and the best responses to these issues lack the precision and clarity of threats. Intergovernmental Panel on Climate Change, *Climate Change 2001. The Scientific Basis* (Cambridge, UK: Cambridge University Press, 2001).

[The] world needs concrete pollution control and natural-resource management initiatives—for starters, a better global environmental regime, improved data and performance measurement and dissemination of environmental best practices, and a beyond-Kyoto climate change strategy. ... The time for grand vision and flowery rhetoric has passed. The challenges ahead require sharper focus, real commitment, and concrete actions. (Esty, 2000)

Entangled vulnerabilities will not mitigate or replace more traditional hard security dilemmas. Rather, we will see the continued reality of threat-based conditions contend with the rise of various vulnerability-based urgencies. Perhaps paradoxically, entangled vulnerabilities will likely receive the least attention among pressing security issues, even as their interdependent complexities grow increasingly difficult to address over time.

4. Zombie Concepts and Boomerang Effects

To examine the relationships among security, risk, and uncertainty, we draw on two central metaphors. First, the tenets of traditional security are critiqued vis-à-vis Ulrich Beck's "zombie concepts" of modernism which emphasize the state and fail to engage the multiple and interdependent processes of change we now face. In this context, we discuss broadly how new solutions beget increased risk and how new knowledge yields greater uncertainty. Second, using P. H. Liotta's "boomerang effect," we look more narrowly at how policies intended to address some specific dimension of security can undermine other dimensions.

4.1. THE ZOMBIE CONCEPT

As described earlier, zombie concepts linger from a time when the approach to understanding not only security, but also one's place in the global order, rested on an identity that was "nation-state centered" (Urry, 2004: 6). More specifically, zombie concepts are characterized by three primary assumptions, each of which carries a dimension of now obsolete thinking: first, that territory is essential to the nature of society; second, that to understand individuals in sociological terms, we must use preexisting social groupings; and third, that the social organization of the West is the best organization possible (Beck and Willms, 2004: 21–24).

Beck—as well as many others—argues that we have now entered an age of true interlinked, interdependent globalization, involving a multidimensional process of change that has irrevocably transformed world order and the function of states within that order. Yet, like undead zombies, elements of earlier society—and for us, notably elements relating to governance—continue to persist in contemporary life through the mindsets of individuals and institutions. In other words, we are living as if we are in an industrial (or preindustrial) society organized according to nation-states, but we are no longer living in such a world (Urry, 2004: 7–8; Beck, 1999; Beck and Willms, 2004).

Thus, we are beyond the nation-state. We live in the "world risk society," where common threats and common opportunities abound (Beck, 1992, 1998). Yet while the term "globalization" remains vague—something better sensed than perfectly comprehended—there is often attached to the concept and to the process of globalization the recognition of *flows*, that involve not only economic but also cultural, sociological, religious, historical, political, ecological, and economic exchanges.

As we use the term here, globalization involves these disparate and complex flows in the vast networking of financial transactions, rising expectations in standards of living, and increasingly common forms of (largely democratic) state and regional identity and governance. And moreover, this global and interlocked system induces the kind of radical uncertainty that does not present finite and obvious ends. Instead, the dangers or hazards we face should be understood as open-ended events (Beck, 1995). As we and other authors in this volume will continue to emphasize, environmental change and human security—and their shared, interdependent, entangled "effects"—imply that burdens and consequence will be shared as well.

4.2. THE BOOMERANG EFFECT

Zombie concepts constrain the understanding of a problem and thereby limit the possible responses to address it. But more disconcerting is the way zombie concepts veil the collateral consequences of the actions that are pre-scribed. This combination of narrow attention and feedback come together in the boomerang effect.

To best explain the intended significance of this term, we offer a simple, though personal, anecdote. One of the authors (Liotta) grew up in Australia. One weekend in Queensland, after exploring several opal mines, his father elected to "try out" a new killer boomerang. The weapon was heavy enough to be lethal yet light enough to supposedly return when thrown. And sure enough, with the family gathered in an open field, his father heaved a mighty toss and the weapon worked as advertised. To the family's general awe and amazement, the boomerang arced several hundred meters into the air; to their horror, the boomerang then started to return. Without hesitating, his father, two brothers, and he fled to all four quadrants of the compass. His mother, frozen in place, caught the full impact of the massive boomerang. She has never truly forgiven any of them (Liotta, 2002b).

For the authors, the significance of the "boomerang effect" lies in the emerging understanding that aspects of "nontraditional" security issues that have long plagued the so-called developing world could increasingly affect the policy decisions and future choices of powerful states and world leaders as well. As disparate as these "nontraditional" issues may be—whether linked to climate change, resource scarcity, declining productivity, or transnational issues of criminality and terrorism—the "developed" world is now confronted with similar, human-centered vulnerabilities that had often been present previously only in the context of "nontraditional" challenges for developing regions (Liotta, 2002a; Liotta and Shearer, 2007).

In short, we may need to worry less about focusing on protecting the "state" and more on protecting "individual" citizens, which means protection of individual rights and liberties as well as the way of life that most have become accustomed to in the "developed" world. The irony in this claim, of course, is that many proponents for development in some of the poorest states have long argued that the focus on the individual—rather than on sustaining the power base of the state—is the best guarantee for long-term stability, prosperity, and security.

The implications of the changing security landscape for the analyst and policy maker are therefore potentially profound. In essence, we may be witnessing a boomerang effect in which we must focus *both* on aspects of "national security," in which military forces may continue to play a preeminent role, as well as human security, in which "nontraditional" security issues predominate. Thus, we may well witness renewed focus on failed or failing states, epidemiology (such as HIV/AIDS or H5N1), environmental stress, resource scarcity and depletion, drugs, terrorism, small arms, inhumane weapons, and narcotrafficking. Predominant among these issues, of course, is climate change in an apparent time when "technologically advanced society could choose, in essence, to destroy itself, but that is now what we are in the process of doing" (Kolbert, 2006).

5. Respecting Risk, Understanding Uncertainty

Risk is the term always mentioned in security discussions—usually mentioned last, usually in the context of "Yes, it's important"—and almost always ignored in the final decision process. Yet, in weighing the importance of risk, it seems increasingly important to recognize that it is the one factor that cannot be ignored in that it gives form to the understanding of a problem.

At the height of the Cold War, for example, both stability and parity between the United States and the USSR was based on what could well be called a "balance of terror" with nuclear weapons. Accordingly, this basic state of "insecurity" drove the international system toward, rather than away from, stability. This insecurity also influenced the recognition that risk was as much a driving force in the guarantees of basic security as the absence of fear or the desire to be free to make choices on behalf of the collective good. This risk also defined the context in which potential actions were considered. As Günther Anders commented during the Cold War, "it is misleading to say that atomic weapons exist in our political situation. ... As the situation today is determined and defined exclusively by the existence of 'atomic weapons,' we have to state: political actions and developments are taking place within the atomic situation" (Anders, 1962: 494). Risk, in our estimation, is also critical to understanding how—in an age of "new" security—humans will adapt to or suffer in response to the effects of climate change.

Yet as critical as it is to recognize the importance of acknowledging and dealing with risk, risk remains difficult to define in precise terms. Our own approach to defining, and by extension understanding, risk involves an admittedly minimalist conception: *the ability to expose oneself to damage during the process of change and the resilience to be able to sustain oneself during such change.*

While many might well object to this identification as too broad, we do find it useful in that our definition should remind one that risk, if respected and acknowledged, can never be assumed away. One useful example of how risk is dangerously assumed away is taken from contemporary east Africa. During a period when climate change has induced semipermanent drought effects in much of the region, natives have come to depend upon, indeed expect, permanent food support through international agency distribution. While this may seem an odd form of dependency to some, it can also incur fatal consequences when environmental impacts potentially expand beyond the region and food distribution would not, in the future, remain as assured as it is today.[3]

Ulrich Beck classifies risk in a highly provocative context—one related to human action:

---

[3] Based on a joint lecture titled "Water Woes: The Critical Importance of Scarce Resources," presented at The Pell Center for International Relations and Public Policy, Newport, Rhode Island, April 24, 2006, by Dr. Patricia Kameri-Mbote, Africa Policy Scholar, Woodrow Wilson International Center for Scholars, and Program Director, International Environmental Law Research Centre, University of Nairobi, Kenya, and Dr. Geoffrey Dabelko, Director, Environmental Change and Security Program and Coordinator, Global Health Initiative, Woodrow Wilson International Center for Scholars, Washington, DC.

Danger is what we face in epochs when threats can't be interpreted as resulting from human action. Rather than being experienced as decision-dependent, they are interpreted as being unleashed by natural catastrophe or as punishment from the gods. They are experienced as collective destiny. Risk, by contrast, marks the beginning of a civilization that seeks to make the unforeseeable consequences of its own decisions foreseeable, and to subdue their unwanted side effects through conscious preventative [sic] action and institutional arrangements. (Beck and Willms, 2004: 110–111)

While slightly oxymoronic to claim that it is essential to "understand" uncertainty—particularly radical uncertainty—it remains likely that it is essential to be able to appreciate its complexity and to consider its consequences and potential impact.

Admittedly, levels of uncertainty when considering the catastrophic risk of global climate change can prove maddeningly frustrating, as the following graphics adapted from the Union of Concerned Scientists review of the Intergovernmental Panel on Climate Change Special Report on Emission Scenarios illustrate:

Figure 1 shows that while there may be a suggested leveling of temperature increase by the late twenty-first century, rapid increases in the early decades

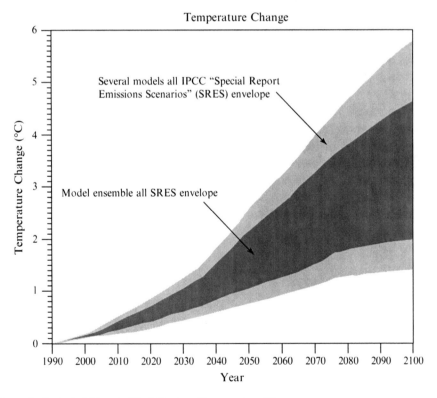

Figure 1. Projected Twenty-First-Century Temperature Rise.

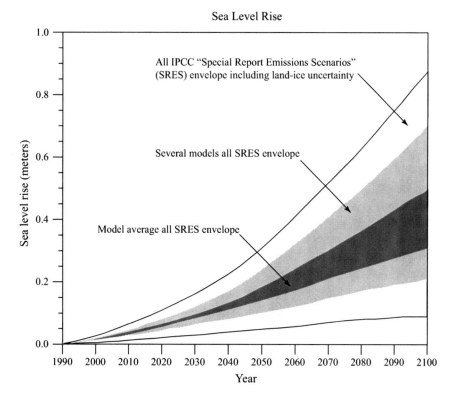

Figure 2. Projected Twenty-First-Century Sea Level Rise.

cannot allow for complacency. Figure 2 suggests a similar dynamic. The disturbing implication of this second graphic lies in continued sea-level rise decades after "heat-trapping gases are stabilized in some scenarios and the upward tempera-ture trend levels off" (Union of Concerned Scientists, 2007).

Both graphics, of course, can prove maddeningly frustrating (for decision makers) in their variability. That which proves "good science" by allowing for uncertainty tends to go beyond the pale of allowable boundaries for many who might seek to implement actions to mitigate, adapt, or enhance resiliency for those human victims who must suffer the consequence of such potentially immense change.

6. The More We Know, the Less: Addressing Uncertainty and Shaping Policy

When these metaphors of zombie concepts and boomerang effects are con-sidered as interrelated in their complexity and meaning for policy choices, it becomes more apparent how states and regions are increasingly confronted with socially produced and human-centered vulnerabilities. Further, the potential for local and localized risk has mutated into *systemic risk* that affects both the developing and developed world. Responses to environmental change and human impact must therefore accommodate thinking in terms of these multiple facets of security.

In the mid-1990s, Emma Rothschild usefully depicted how security has changed horizontally, vertically, and on multiple axes. Beginning with the state, she described security as brought down to the individual, up to the international system or supranational physical environment, across (broadening) from a focus on the military to include the environment, society, and economy; and finally argued that the responsibility to ensure security diffused in all directions to include local governments, international agreements, NGOs, public opinion, forces of nature, and the financial market. Although not an explicit definition, this conceptualization provides an example both of how narrow the traditional paradigm has been and how complex the expansion of the concept can become (1995).

Essentially, states and regions, in a globalized context, can no longer afford to solely emphasize national security issues without recognizing that other values, norms, and expectations—which are often abstract—also influence both choice and outcome. Societies, whether in the emerging world or in the "developed world" (admittedly a rather arrogant term), are increasingly witnessing an unfolding tension: citizens increasingly hold their states accountable for multiple kinds of projection with multiple kinds of security.

Within the debate on security, there emerged an increased focus on the rights of the individual. This debate has led to intriguing possibilities and, most definitely, uncertain outcomes. It remains unclear, however, whether an ethical and collective policy to support human security will be the focus of most states in the future—or whether any such policy could logically be de-linked from a systematic association with power and powerful states.

Security, per se, no longer represents merely security of territory from external aggression or protection of national interests in foreign policy. "Extended security" must also represent protection from the threat of disease, hunger, unemployment, crime, terrorism, social conflict, political repression, and environmental hazards.

In such an environment, to avoid the boomerang effect, decision makers and policy analysts will, again, need to focus on multiple aspects, axes, and levels of security. Appreciating the complex interdependence and dynamic interplay among a host of dependent and independent variables that cannot fully be predicted or understood will only help decision and policy makers better address "entangled" vulnerabilities.

In its way, this complex interaction approaches what was once known as chaos theory. And yet, as disparate as these nontraditional security aspects indeed are, they all will (in one form or another and in multiple geopolitical contexts) increasingly have influence on future strategic relationships and decisions. The issue truly is not one of "hard" traditional security (often based on state-to-state power relationships) or "soft" nontraditional security (that can involve multiple transnational aspects). The future will require decision and policy makers in both the developing and developed world to focus on broad—and broadened—understandings of the meaning of security. Focusing on one aspect of security at the expense or detriment of another aspect, nevertheless, may well cause us to be "boomeranged" by a poor balancing of ends and means in a radically changed security environment.

6.1. IN LIEU OF CLOSURE

The uncertainties associated with environmental change's human impact are difficult to quantify, yet the impacts may be severe. We argue that we cannot so reduce the uncertainty of the science that can definitively end debate about appropriate policy. Instead, we must learn to integrate uncertainty into decision making processes and consider how our near-term actions enable or constrain future options.

We live in what Ulrich Beck terms the "Global Risk Society," an age typified by zombie concepts and boomerang effects. Undeniably, Beck suggests that with vast leaps in technological advance and interlinked communications, we have only increased the potential for local, and localized, risk to mutate into *systemic risk*; both invoke and involve radical uncertainty. The crucial error, therefore, is not to believe that there is simply data or information missing that could fill the uncertainty "gap"; to the contrary, manufactured uncertainties occur when increased knowledge fuels the production of ever more increased uncertainty (Beck, 1999: 112; Beck and Willms, 2004: 127; Beck et al., 1994).

Regarding the uncertainties of environmental change and their potential human impact, the implications are astonishing. We do not suffer from scanty information; instead we endure too much scientifically conclusive data. We cannot so reduce our uncertainty that we can definitively resolve contemporary problems in the way early modernists aspired to do. We can only hope to increase our adaptability to uncertainty with the insecurities that come with our increased knowledge.

References

Anders, G., 1962, Theses for the atomic age, *Massachusetts Review* **3**(3):493–505.

Beck, U., 1992, *Risk Society: Towards a New Modernity*, trans. M. Ritter, Sage, London.

Beck, U., 1995, *Ecological Politics in the Age of Risk*, Polity, Cambridge.

Beck, U., 1998, *The Politics of Risk Society*, Institute for Public Policy Research, London, pp. 587–595.

Beck, U., 1999, *World Risk Society*, Polity, Cambridge, pp. 112–116.

Beck, U., and Willms, J., 2004, *Conversations with Ulrich Beck*, trans. M. Pollak, Polity, Cambridge.

Beck, U., Giddens, A., and Lash, S., 1994, *Reflexive Modernization: Politics, Tradition, and Aesthetics in the Modern Social Order*, Polity, Cambridge.

Bissell, T., 2003, A comet's tale: On the science of Apocalypse, *Harper's* **306**(1833):33–47.

Courtney, H., Kirkland, J., and Viguerie, P., 1997, Strategy under uncertainty, *Harvard Business Review* **75**(November):66–79.

Espenshade, E., 1995, *Goode's World Atlas*, Rand McNally, New York, p. 27.

Esty, D., 2000, A term's limits, *Foreign Policy*, September–October.

Giddens, A., 1990, *The Consequences of Modernity*, Stanford University Press, Stanford, CA.

Giddens, A., 1998, *Beyond Left and Right*, Polity, Cambridge.

Giddens, A., 1999, Risk and responsibility, *Modern Law Review* **62**(1):1–10.

Glenn, J., et al., 1998, *Defining Environmental Security: Implications for the U.S. Army*, U.S. Army Environmental Policy Institute, Atlanta.

Hobbes, T., 1985, *The Leviathan*, edited and with an introduction by C. B. MacPherson, Penguin Books, New York.

Intergovernmental Panel on Climate Change (IPCC), 2001, *Climate Change 2001: The Scientific Basis*, Cambridge University Press, Cambridge.

Intergovernmental Panel on Climate Change (IPCC), 2007, *Climate Change 2007: Impacts, Adaptation and Vulnerability, Working Group II Contribution to the Fourth Assessment Report of the IPCC*, Cambridge University Press, Cambridge.

Keynes, J. M., 1936, *The General Theory of Employment, Interest and Money*. 1964 reprint, Harcourt Brace, New York.

Keynes, J. M., 1937, The general theory of employment, *Quarterly Journal of Economics* **51**:209–223.

King, W. C., 2000a, Remarks from the U.S. Army Senior Environmental Leadership Conference, Washington, DC, March.

King, W. C., 2000b, Understanding environmental security: A military perspective, unpublished advanced research project written while a West Point Fellow serving at the U.S. Naval War College.

Kolbert, E., 2006, *Field Notes from a Catastrophe: Man, Nature, and Climate Change*, Bloomsbury, London, p. 187.

Liotta, P. H., 2002a, The boomerang effect: The convergence of national and human security, *Security Dialogue* **12**(33):473–488.

Liotta, P. H., 2002b, The boomerang returns, *Security Dialogue* **12**(33):495–498.

Liotta, P. H., 2003, *The Uncertain Certainty: Environmental Change, Human Security, and the Future Euro-Mediterranean*, Lexington Books, Lanham, MD, p. 72.

Liotta, P. H., and Owen, T., 2006a, Sense and symbolism: Europe takes on human security, *Parameters* **36**(3):85–102.

Liotta, P. H., and Owen, T., 2006b, Why human security? *Whitehead Journal of Diplomacy and International Relations* **7**(1):1–18.

Liotta, P. H., and Shearer, A. W., 2007, *Gaia's Revenge: Climate Change and Humanity's Loss*, Praeger, Greenwood, CT, pp. 38–40; 43; 56; 62; 128.

MacDonald, B. et al., eds., 2007, *Defence Requirements for Canada's Arctic*, Vimy Paper 2007, The Conference of Defence Associations Institute (July 26, 2007); http://www.cda-cdai.

ca/Focus%20Briefs/4-07%20Arctic%20Patrol%20Ships%20and%20Canada's%20Arctic%20Sovereignty%20.pdf.

Myers, N., 1986, The environmental dimension to security issues, *The Environmentalist* 6(4):251–257.

Rothschild, E., 1995, What is security? The quest for world order, *Dædulus: The Journal of the American Academy of Arts and Sciences* **124**(3).

Scott, J. C., 1997, *The Moral Economy of the Peasant: Rebellion and Subsistence in Southeast Asia,* Yale University Press, New Haven, CT, p. 1.

Taleb, N. N., 2007, *The Black Swan: The Impact of the Highly Improbable,* Random House, New York.

Union of Concerned Scientists, 2007, Science of global warming: Future projections of climate change, July 27; http://www.ucsusa.org/global_warming/science/projections-of-climate-change.html.

United Nations Commission on Human Security, 2003, *Protecting and Empowering People,* http://www.humansecurity-chs.org/finalreport/index.html.

United Nations Development Programme (UNDP), 1994, *Human Development Report: Annual Report,* Oxford University Press, New York, pp. 22–25.

United States Department of Defense, 1996, *Environmental Security,* Directive 4715.1, U.S. Government Printing Office, Washington, DC.

United States Environmental Protection Agency, 1999, *Environmental Security,* 160-F99-01, U.S. Government Printing Office, Washington, DC.

Urry, J., 2004, Thinking society anew, Introduction in: *Conversations with Ulrich Beck,* U. Beck and J. Willms, trans. M. Pollak, Polity, London, pp. 6–8.

Wæver, O., 2000, The traditional and new security agenda: Influence for the third world, paper presented at Universidad Torcutato Di Tella, Buenos Aires, 11 September, quoted in Scheetz, T., 2002, The limits to environmental security as a role for the armed forces, paper presented at the Center for Hemispheric Defense Studies symposium, Brasilia, 8 August.

MEASURING HUMAN SECURITY

Methodological Challenges and the Importance
of Geographically Referenced Determinants

TAYLOR OWEN*

The University of Oxford, Jesus College

Abstract: Human security is a new and contested concept. Although it is gaining legitimacy in many academic and policy communities, many argue that it has no single accepted definition, no universal foreign policy mandate, and no consensus-commanding analytic framework for its measurement. For others this is of little concern; that "human security" was the coalescing force behind the International Convention to Ban Landmines and the International Criminal Court is enough to prove that it is both representative of popular sentiment and legitimate as a tool of international policy making. Increasingly central in this debate over the utility of human security is the feasibility of its measurement. This paper first argues that measuring human security—despite its critics' concerns—is a worthy academic exercise with significant policy relevant applications. Then, by analyzing the four existing methodologies for measuring human security, it is argued that all fall victim to a paradox of human security. After introducing the notions of space and scale to the measurement of human security, a new methodology is proposed for mapping and spatially analyzing threats at a subnational level.

Keywords: Human security; measuring human security; hotspot analysis; human security thresholds; Cambodia

1. Why Measure Human Security?

Up until the end of the Cold War, what we now refer to as "traditional security" or "national security" dominated the field of international relations. In this view, the state was responsible for the preservation of territorial

---

*Address correspondence to Taylor Owen, The University of Oxford, Jesus College, Turl Street, Oxford, OX13DW; e-mail: taylor.owen@jesus.ox.ac.uk

P.H. Liotta et al. (eds.), Environmental Change and Human Security, 35–64.

integrity, domestic order, and international affairs. "Security" generally meant the protection of the state from external attack, nuclear proliferation, international espionage, and internal rebellion. In response, the security and defense infrastructure was tailored to address threats through military buildup, nuclear stockpiling, and foreign intelligence.

While the security policy and apparatus of many countries remains focused on preserving national sovereignty, today's reality is that most deaths are not a result of interstate war. Instead, disease, violence, natural disasters, and civil conflict are the leading causes of preventable premature mortality as we enter the twenty-first century (see Table 1). This rapid evolution in primary threats has prompted a substantial shift in relevant security issues and thinking. This includes a widening of the potential array of security threats as well as a broadening of the security mandate from a narrowly focused national perspective to one focusing on human and community-based requirements.

We can expect the broadening of security threats to continue as the complex adaptive systems from which people and communities are threatened—global environmental change, disease spread, impoverishment, violence from nongovernment forces—continue to evolve at a rapid pace. These changes, coupled with a security infrastructure that is slow to change, national in scope, and hierarchical, suggest that the gap between security threats and the capacity to ameliorate these threats will grow even larger in the years to come. Clearly, there is an urgent need to address this disconnect between the theory and mechanisms of the traditional security paradigm and the majority of harms that people and communities face.

One alternative to traditional security, human security, shifts the focus of the concept from the state to the individual and community. Whereas a security threat was once something that threatened the integrity of the state, under the *human security* rubric, it is the set of harms that threatens

TABLE 1. A Global Death Registry: 2000[1].

Cause of death	Global death totals
Interstate war	10,000
Internal and internationalized internal conflict	90,000
Disasters	65,000
Homicide	730,000
Communicable disease	18,000,000

[1] Disasters, EM-DAT, 2000; War, homicide, and suicide, Krug et al., 2002; Communicable disease, World Health Organization, 2001.

the integrity of the individual and community. This shift in the referent of security provides an expanded set of threats, resulting in a more inclusive and comprehensive security paradigm. However, the concept of human security poses some difficult analytic and policy problems, namely, how does one distinguish and prioritize threats if all possible harms to individuals are deemed security concerns? *If traditional security is overly restrictive, human security risks being expansive and vague.*

While the validity of the normative interpretation of human security is relatively uncontested, at least among proponents of the concept, its analytic utility is fiercely debated. It is one thing to say that individuals are at risk from a much wider array of threats than the current security paradigm addresses; it is quite another to identify, measure, and assess these many possible harms. Central to this debate are the parameters with which one selects human security threats. If, for example, a broad definition of human security is used, all threats that could potentially harm an individual must be included. A global assessment using this criterion is impossible. Quite simply, people can be harmed by such a vast array of threats that complete coverage is conceptually, practically, and analytically impossible. Practitioners have circumvented this reality using two qualifiers in their measurement attempts—researcher and data defined threat identification. Both arbitrarily limit the included threats to those falling under categories such as violence, human right abuses, or health. More importantly, both marginalize the very core principle of the human security concept, that actual insecurity must drive our response mechanisms.

The question of whether or not human security should be measured is contested. Critics raise a number of objections.

First, measuring implies a predetermined definition. What is included in the measurement necessarily provides a de facto list of what is and is not a human insecurity. For those who are hesitant to limit human security to one definition, this is problematic.

Second, the term "measuring" in itself implies a degree of certainty that the existing data do not warrant. Moreover, "objective" and "subjective" measures may be contradictory. Subjectively, opinion polls indicate that people in developing countries fear violence more than disease. Objectively, however, mortality statistics tell us that by far the greatest threat is disease. How such contradictory assessments might be combined into a broad measure is not clear. A final consideration is that organizations often become defined by their measuring methodology. For example, regardless of other extensive and varied academic work, the International Institute for Sustainable Development (IISD) and the UNDP's Human Development Report (HDR) are arguably known best by their measuring indices. Although this is not necessarily bad, it can distract from other important work being done, especially if early flawed research results are released.

Despite the above concerns, the measurement of human security should be undertaken for the following four reasons. First, measurement helps define the often ambiguous concept of human security. Second, measurement can reveal patterns that would not otherwise be observed. A substantial measurement exercise will inevitably help identify and locate human insecurities. Studying the relationships between multiple insecurities could reveal chains of causality and cumulative impacts not currently recognized, articulated, or quantified.[2] Third, in the positivistic social sciences—with the goal of determining causal and correlative relationships—measurement is seen as essential. Fourth, measurement provides "objective" evidence of trends that can be of great value in policymaking and political debate and can influence public and media perceptions of the issue in question.

2. Review of Proposed Human Security Measures

Four broad frameworks have been proposed for measuring human security.[3] They can be distinguished from one another in three ways: how they define the concept of human security; what they propose to measure; and the methodology they use to aggregate and analyze the data.

2.1. GENERALIZED POVERTY

Definition Used. Gary King and Christopher Murray (2000) define human insecurity as a state of "generalized poverty." This state exists when a human being ranks below a predetermined threshold in any of a number of domains of well-being.

King and Murray argue that there are important qualitative differences in well-being or life experience, above and below a certain level across a wide variety of domains of measurement. If any individual or group falls below a particular threshold on any one of the domains in question, it is said to be in a state of generalized poverty (Table 2). Implicit in this definition is that no one indicator, if below the threshold, is any more significant than another. This avoids the difficulties of attempting to weight different measures when creating a composite index and leaves the establishment of thresholds as the only subjective aspect of the process.

---

[2] A good example is the UN skepticism that the levels of warfare had declined in the 1990s, until this downward trend was proven wrong by data from the Universities of Maryland and Uppsala, and in the 2005 Human Security Report.

[3] These four were chosen because they make a clear reference to measuring "human security." There are many measuring methodologies that analyze similar variables and could be considered measurements of human security type threats. It would be useful for a more extensive study to look at these other methodologies.

TABLE 2. Generalized Poverty.

Domain	Indicator
Income	GNP per capita converted to purchasing power parity
Health	Quality of health scale
Education	Literacy rate or average years of schooling
Political freedom	Freedom house measure of societal freedom
Democracy	Fraction of adults able to participate in elections

What is Measured. King and Murray (2000) define the domains of well-being as income, health, education, political freedom, and democracy, closely paralleling the United Nations Development Program (UNDP) definition.[4] For each domain, a single dichotomous indicator is chosen on which individuals can score 1 or 0. Interestingly, these indicators do not include a measure of violence, although the domains chosen were all deemed to be "things worth fighting for" (King and Murray, 2000: 14–15).

Methodology. Three methods have been proposed to advance King and Murray's theory of generalized poverty.

The first measure is the Years of Individual Human Security (YIHS). This represents the expected years that an individual will spend outside of a "state of generalized poverty." The latter exists if an individual scores zero on any one of the above indicators.

The use of language becomes a little unusual in this measurement scheme. For example, to score zero on "political freedom" would put an individual in a state of "generalized poverty" even though that individual might be quite affluent. "Generalized poverty," in other words, does not necessarily equate with poverty in the traditionally understood sense of the term.

The second measure is "Individual Human Security" (IHS). This represents the *proportion* of an individual's lifespan that she or he could expect to spend outside of a state of generalized poverty. This ensures a more accurate control for age.

Finally, a method of aggregating the YIHS for a particular population—within a nation-state, for example—has been proposed, "The Population Years of Human Security" (PYHS). This may be useful in the study of development and security policies (King and Murray 2000: 13–14).

---

[4] The UNDP definition of human security was introduced in the 1994 Human Development Report (UNDP, 1994). Including seven categories (economic, food, health, community, environmental, political, and personal), it is considered the broadest possible conceptualization of human security. It is often criticized for closely resembling what are usually considered "development" rather than "security" concerns.

2.2. THE HUMAN SECURITY AUDIT

Definition Used. For Kanti Bajpai (2000), human security is defined as the protection from direct and indirect threats to the personal safety and well-being of the individual. This is derived from commonalities between the Canadian and UNDP conceptions of human security.

What is Measured

Methodology. The first stage of Bajpai's methodology is to measure the *potential* threat to the individual. This is done by collecting quantitative data for an array of potential direct and indirect threats (see Table 3). Although Bajpai claims that the data for this stage are abundant, there may be some problems with continuity and accuracy, given the broad range of indicators. Much of the data he requires is either aggregated from sparse and questionable data sources or simply are not available for the developing world. Moreover, judgments about *potential* threats—as against actual physical harm—are necessarily conjectural and are unlikely to command consensus.

Second, Bajpai seeks to measure the capacity of the individual to cope with potential threats. This is done through a qualitative assessment of people's or governments' capacities. Possible indicators might include government antiracism policies, as opposed to incidents of racist abuse (Bajpai, 2000: 53–56).

Although the threat vs. capacity methodology is interesting, it has some practical limitations. The method by which indicators would be aggregated is not explained and the weighting between the threats and the capacities would have to be entirely subjective. In addition, there is no assurance that the capacities will be directly relevant to the threats posed. If they are not, then any process of aggregation will have little meaning. For example, a country

TABLE 3. Human Security Audit.

Direct threats	
Local	Violent crime, abuse of women/children
Regional	Terrorism, genocide, government repression
National	Societal violence, international war, banditry, ethnic violence
International	Interstate war, weapons of mass destruction, landmines
Indirect threats	
Societal	Lack of basic needs
	Disease
	Employment levels
	Population growth or decline
	Natural disasters
Global level	Population movement
	Environmental degradation
	Unequal consumption

may have a robust social welfare system (a strong indicator of capacity to deal with an economic threat), and consequently score a high capacity ranking. This same country, however, may have no effective disaster response program and be highly vulnerable to the threat of major natural disasters.

2.3. THE GLOBAL ENVIRONMENTAL CHANGE AND HUMAN SECURITY PROJECT (GECHS) INDEX OF HUMAN INSECURITY (IHI)

Definition Used. The GECHS definition of human security (Lonergan et al., 2000) builds on the premise that certain environmental and social conditions, when coupled with increasingly vulnerable societies, may lead to insecurity. Security in this context is only achieved when individuals have the option, physically and politically, to end or adapt to threats to their environmental, social, or human rights. This methodology attempts to measure a broad range of human security threats with focus on the environmental components.

The GECHS definition of human security puts significant focus on a cumulative causal relationship between the environment and personal safety. They also point out potential indirect threats stemming from either an environmental condition or the human response to that condition.

What is Measured. Social, environmental, economic, and institutional domains of security are the focus of the GECHS Index of Human Insecurity (IHI). Four indicators for each domain (see Table 4) are then selected using the following criteria: relevance to the selection framework; existence of a

TABLE 4.   Index of Human Insecurity.

Social	Urban population growth
	Young male population
	Maternal mortality ratio
	Life expectancy
Environmental	Net energy imports
	Soil degradation
	Safe water
	Arable land
Economic	Real GDP per capita
	GNP per capita growth
	Adult literacy rate
	Value of imports and exports of goods and services
Institutional	Public expenditures on defense versus education
	Gross domestic fixed investment
	Degree of democratization
	Human Freedom Index

theoretical or empirical relationship; general availability of the data; data commensurability; and adequacy of the spatial coverage.

There is, however, no mention of a prioritization of these factors. If their relative priority is based on an entirely subjective reasoning, the soundness of the methodology should be questioned (Lonergan et al., 2000).

Methodology. Once indicators are selected, the methodology involves three stages. First, data are collected for all indicators at the national level with as detailed a time series (1970–1995) as available. Missing data are estimated using linear regression. Second, data are standardized into a common scale.[5] Finally, the index is calculated using cluster analysis, assigning a degree of severity (insecurity) between one and ten to each indicator for each country (Lonergan et al., 2000).

The IHI is the only one of the four proposed indices that has been actualized using real data. Interestingly, there is a strong linear relationship between the IHI and the Human Development Index (HDI) (Figure 1). This is perhaps not surprising since both are seeking in effect to measure human development.

But the GECHS approach raises several questions. First, what is the difference between development and security as defined by GECHS? This

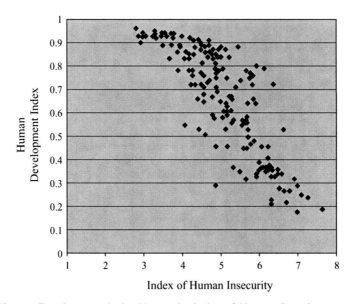

Figure 1. Human Development Index Versus the Index of Human Security.
Source: Lonergan et al. (2000), The Index of Human Insecurity, *Aviso Bulletin*, 6.

---

[5] The details of this step are not clear and are crucial to the validity of the final measurement.

question is at the center of the definitional debate about human security. If the list of harms under human security becomes too exhaustive, it starts to include many of those found under the concept of development. This has the potential to render both its analytic and policy utility redundant. In this regard, it is also possible that the results of the IHI, being so similar to the HDI, could be used to strengthen the case for a narrower definition of human security such as the one used in the Human Security Report (see section 2.4). If broad measurements are redundant, narrow definitions will likely prevail.

Second, what differentiates the IHI from the HDI? Is it telling us anything new and if not, why use it? It could also be asked whether either index tells us anything more than the single indicator of infant mortality, which is often thought to be a good proxy for the HDI.[6]

2.4. THE HUMAN SECURITY REPORT

Definition Used. The *Human Security Report*, at the Program for Human Security at the University of British Columbia, uses the most restrictive measure of human insecurity, limiting it for stated pragmatic and methodological reasons to deaths caused by armed conflict and criminal violence. This is a much narrower spectrum of threats than those embraced by any of the other measuring methodologies (see Appendix). The Report does not propose mapping injuries from war and criminal violence due to lack of data, but does suggest that numbers of deaths would be a good proxy (HPHPCR, 2001).

What is Measured. The Human Security Centre remains uncommitted on whether or not to include a composite Human Insecurity Index in the Report. If so, the most likely indicators would be armed conflict deaths and homicides. The convention already used for deaths from disease, natural disasters, and criminal violence—i.e., number of deaths per 100,000 per year in a particular country—would be used. The data from each would be aggregated on a national and possibly a regional level.

Like all others, this index confronts a number of difficulties. Currently no data are collected on the absolute numbers of conflict deaths per year; they are only collected from the armed conflicts that exceed a certain death threshold (HPHPCR, 2001: 2–5).

---

[6] The authors present the following three defenses of the GECHS index. The first is that the HDI index is intended to achieve a stable indicator and therefore might not catch the degree of detail present in the sixteen indicators of the IHI. Second, the IHI has a much stronger theoretical link to the concept of human security and development. And third, the IHI attempts to deal with some issues of perception by incorporating some qualitative data. Again, these three statements would have to be further defended, and hopefully will be in a forthcoming GECHS publication.

APPENDIX. Chart Proposed Methodologies and Indicators Used.

		Proposed methodologies		
The UNDP's Broad Spectrum of Human Security Indicators	Generalized poverty index	Human security audit	Index of human insecurity	Human security report
Economic security	Income (GNP per capita in int'l purchasing power parity)	Employment levels	Real GDP per capita	
Assured basic income (employment)		Unequal consumption	GNP per capita growth	
Social safety net			Adult literacy rate	
Income levels			Import and exports	
Food security		Access to food		
Food production				
Food availability				
Calorie intake				
Health security	Quality of Health Scale	Access to primary health care	Maternal mortality ratio	
Disease		Disease	Life expectancy	
Access to health services				
Maternal infant mortality				
Environmental security		Environmental degradation	Net energy imports	
Water (pollution and scarcity)			Soil degradation	
Land use		Natural disasters	Safe water	
Air quality			Arable land	
Natural disasters				

Personal security Security from physical Violence (war, crime, rape, child abuse, drug abuse, accidents)		Violent crime, Child/women abuse Terrorism, Genocide, Government repression Societal violence, Intrastate war, Banditry, Ethnic violence Interstate war, WMD, Landmines		Battle-related deaths in armed conflict Homicides
Political security Democratic freedom Strength of civil society Priority on military spending	Political freedom (Freedom House) Democracy (participation rate of eligible adults)		Public expenditures on health vs. education Degree of Democracy Human freedom index Gross domestic fixed investment	
Community security Demographics Stability	Education (literacy rate) Population movement Population stability		Urban population growth	Young male population

TABLE 5. The Human Security Report.

Violence	Battle-related deaths in armed conflict
	Homicides

Second, the data are subject to a variety of biases. Although there is no reason to assume that they are all skewed, the data are almost certainly not accurate enough to have any confidence in individual country rankings—a difference of one or two places on the individual country ranking having little substantive meaning. This problem is similar to that faced by the Human Development Index of the Human Development Report. Ranking countries in quintiles may, however, be possible.

Third, while criminal violence data (unlike armed conflict data) are collected by governments, they too are subject to inaccuracies, often being both under-recorded and under-reported. National collection of data is often subject to political bias. Moreover, data is often published several years after it has been collected—making it difficult to use in an annual index (HPHPCR, 2002: 2–5).

While the Gary/Murray and GECHS indices do not include violence, the Human Security Report Index would not include deaths caused by disease, starvation, or natural disasters. The Human Security Report argues that there are important relationships between underdevelopment and human insecurity (narrowly defined), but that to explore them each must be treated separately for the purpose of analysis (Table 5).

Methodology. Since there is a common scale for both indicators—number of deaths—aggregation of the two measures is unproblematic. When death toll data from armed conflicts are collected (they are not collected at the moment), it will be possible to measure battle-related deaths per 100,000 of population, allowing for more meaningful national comparisons.

3. Methodological Challenges

This brief review of attempts to create human security indices reveals how contested—not to say confused—contemporary treatments of human security are. The four methodologies are based on quite different conceptions of human security. Whereas GECHS stresses environmental variables and King and Murray stress developmental issues, neither addresses violence. The Human Security Report Index, on the other hand, would neither include environmental nor development factors in its measure of human security.

The attached chart displays the indicators used by each and demonstrates how much of the broad definition each methodology incorporates (see Appendix).

All of the approaches included in Appendix claim to measure "human security" but it is clear that they are in fact emphasizing different aspects of the concept. For example, the proposed Human Security Report Index focuses exclusively on personal security and does not consider other dimensions such as economic security, food security, and so on. Most of the other approaches employ a broader notion of human security and include various combinations of at least four of the seven dimensions associated with the UNDP definition. Characteristics shared by these broader approaches include:

- They are based on a comprehensive set of indicators that are derived from the human security literature.

- National data for each indicator are derived from readily available data sources such as the World Bank and the United Nations (UN).

- All indicators are applied in each nation and assumed to be of equal importance.

This systemic, international approach provides a basis for comparing national human security estimates but ignores the reality that not all of the factors represented by the indicators are relevant in each country and that most human security threats are regionally dispersed *within*, not simply between, countries. For example, the employment of a standard indicator set across all nations may overemphasize those factors that are not relevant in a particular country as well as underrate the importance of factors that are actually driving human security.

In addition, national-level assessments may submerge regional differences and thereby not provide timely assessments of when factors are converging and leading to rapid deterioration in human security. The collective implications of these limitations could include misleading signals to the policy community with respect to where and when interventions are most needed.

Indicator selection is also problematic in all of these proposed methodologies. One way to overcome an unmanageable list of possible human security threats is to simply list which threats will and will not be included in the research design. This "laundry list" method is subject to the political, institutional, and cultural biases of the research designer. Methodologies using this approach will inevitably leave out numerous causes of insecurity. For example, a violence-based measurement methodology does not account for the 18,000,000 annual deaths from communicable disease.

The second way around the broad nature of human security is to let data availability drive the assessment parameters. One could compile all available datasets depicting conceivable threats to human security. An assumption with this approach is that if a harm is serious enough, someone will most likely measure it. The problem with this is that it takes a considerable institutional capacity to compile a global scale data set. There are few institutions capable of doing this, and their mandates almost certainly dictate the types

TABLE 6. Measuring Problems and Proposed Solutions.

Problems	Solutions
Data availability	Only measure regionally relevant threats; look subnationally for data
Data integrity	Accept subjectivity, mitigated by local knowledge and disciplinary experts
Data aggregation	Use space as a common denominator; use Geographic Information Systems (GIS)

of threats they will prioritize. In addition, the very point of human security is to shift our attention to threats usually not considered, and most likely not measured on a global scale. An unbalanced focus on economic data is also sure to occur in a data-driven threat assessment.

What also becomes clear if the chart in Appendix is paralleled with the feasibility of each methodology is that when attempts are made to broaden an index by including more indicators, issues of data availability, integrity, and aggregation become increasingly problematic (Collier and Hoeffler, 2001; Brauer, 2001). This results in a difficult paradox:

The more conceptually accurate (i.e., broad) a methodology attempts to be (closer to representing all possible threats), the less practically and analytically feasible it becomes.

As a methodology expands its conceptualization of human security closer to the original broad UNDP definition, it becomes increasingly difficult to both aggregate and differentiate between each method's autonomous variables. In addition, particularly on a global scale, the data simply are unlikely to be available to fill out a "laundry list" of threats for every country. This leads to either significant gaps when comparing one country to another or to the use of old, problematic, and unreliable data.

This paradox has resulted in methodologies being either narrow and feasible or broad and impossible to implement and/or inaccurate.

With this in mind, the new measuring methodology introduced below incorporated the following three solutions to the measurement paradox (Table 6).

Absolutely integral to this re-conceptualization of human security measurement is the notion of space—all the above indices rely solely on national level data. By shifting our perception of space, and measuring insecurity at the local level, I will argue that a much more meaningful representation may be achieved and the problems of the measurement paradox overcome.

4. The Space and Scale of Vulnerability

In order to overcome the methodological challenges outlined above, it is first useful to consider the importance of scale in assessing vulnerability. I will first introduce the idea of spaces of vulnerability, and then provide some practical examples of how different scalar interpretations can lead to quite different perceptions of whom are insecure and from what. It will be argued that local vulnerability must be viewed from a multidimensional perspective.

The parallel fields of hazard identification and risk assessment offer a constructive precedent for measuring human security. Rooted in human ecology, hazard research originated as a means of analyzing environmental extremes, namely, floods (Barrows, 1923). As with human security, researchers were looking at a both conceptual and practical problem; conceptually, the societal relationship to flood location (Kates and Burton, 1986), and practically, how this translated into better flood management policy (White, 1945). In this early research, hazards were firmly rooted in the analysis of physical events. It became clear, however, that a hazard could also be a socially constructed situation (Cutter et al., 2000), an event not rooted in the physical environment, but as a consequence of human actions, such as technology, or technological failure (Kates and Burton, 1986). This shift led to a contextualization of hazard research. A hazard was now seen as very much rooted in a social, political, temporal, organizational, and spatial context (Cutter et al., 2000). Perhaps more importantly for this paper is that this shift led to the development of methodologies incorporating both empirical and social analysis (Palm, 1990).

The study of this interplay between hazard and societal context is the focus of the study of risk and of vulnerability. Risk assessments are grounded in the mathematical study of probability and confidence intervals (Covello and Mumpower, 1994). Although based on the precedent of mathematical inquiry, the first risk assessments per se attempted to link an accident, originally seen as random, with a probability of occurrence, moving the interplay from pure chance to having a degree of risk (Kasperson et al., 1988).

Whereas the study of hazards focuses on the event, and the study of risk focuses on the probability of interplay between the event and its impact, vulnerability addresses the nature of the impact itself. The study of vulnerability is based on a combination of exposure and lack of resilience (Cutter et al., 2000).

What is particularly important to this discussion of scale is that this interplay is location specific. What one is exposed to, as well as to some extent one's ability to mitigate the threat, is spatially determined (Wilhite and Easterling, 1987; Blaikie and Brookfield, 1987; Kasperson et al., 1988).

This has led to what Cutter and Solecki call a hazard-of-place (Cutter and Solecki, 1989). The interplay of social, political, and economic factors—interacting separately, in combination with one another, and with the physical environment—creates a mosaic of risks and hazards that affect people and the places they inhabit (Cutter et al., 2000: 716).[7]

On a more practical level, looking at space, or location, as a common attribute provides a new variable with which human security can be assessed. By providing the means to visualize and spatially analyze data, Geographic Information Systems (GIS) allow a new means of interpreting data that would otherwise only be seen as complex tables and graphs. While this can make GIS a very effective tool for advocacy (Michel, 1991), visualizing different scalar interpretations of data can also tell us a great deal about the nature of the measure itself. This is particularly true for concepts that have complex scalar constructions, such as human security.

The following examples serve to show how varying scales can dramatically alter our perception of data. In some cases, a very local resolution is the most appropriate, whereas in others, a coarser interpretation actually provides a better reflection of the particular phenomena. While this is a relatively intuitive claim, it is very important to keep in mind as we seek to shift security to the lowest common denominators, the individual.

In Figure 2, map 1 depicts the 2004 U.S. election results at the state level. This map is widely used to demonstrate the red state–blue state divide that supposedly bifurcates American politics. This perceived divide has significant political ramifications as it is widely believed that the country is more polarized than ever before.

There are numerous possible scalar interpretations of this same data, however, that provide a significantly different picture of American political division. Map 2, for example, shows the same data at the county level. Already, by simply displaying the data at a new scale, we see far more nuance. Map 3 adds further refinement by ranking each county based not on final vote results, but rather on percentage of votes for either party, resulting in what is popularly known as the "Purple America" map. However, as scale is more than simply resolution, this map can be weighted by other variables such as population. The result, map 4, now provides a dramatically different picture of American political division. Most notably, there is very little red.

---

[7]Curiously, while recognized, this mosaic has only once, and quite recently, been quantitatively assessed. Cutter et al. (2000) provide a practical methodology, the hazard-of-place model of vulnerability, for combining varying biophysical and societal vulnerabilities and have conducted a case study for a county in South Carolina. Their methodology, however, being somewhat narrow in both its theoretical base and spatial focus (only conducted for one county), is only a very early trial of what is sure to be a growing field.

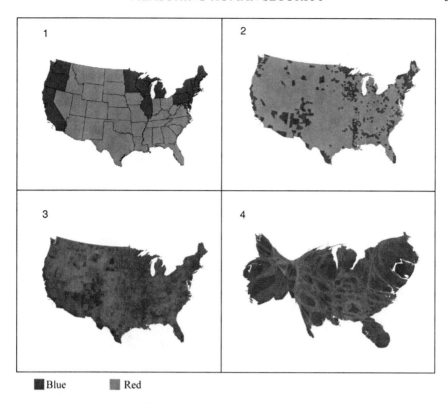

Figure 2. Maps of 2004 U.S. Election Results.

A second example, somewhat more relevant to the discussion scales of human security, involves the interpretation of landmine data in Cambodia. Taken at a national level, Cambodia is one of the most contaminated countries in the world. There are approximately 3,000 landmine fields that are estimated to take over 100 years to de-mine. However, viewing landmine data in various socio-spatial scales demonstrates a far more nuanced picture. In Figure 3, map 1 shows the communes with landmine contamination. While certainly not affecting the entire country, in this map, it appears as if a large percentage of the country is afflicted by mine fields. Map 2, however, shows where the landmine injuries occurred in 2000. These injuries are a function of a highly localized phenomenon—the presence of landmine fields near impoverished villages. Map 3, landmine polygons, and map 4, poverty severity, display what turns out to be the most appropriate scalar construction to represent the human security threat of landmines in Cambodia. In this case, it is a highly spatially specific map of the landmine fields themselves, combined with the poverty severity that forces people to farm on known contaminated land (Owen and Benini, 2004).

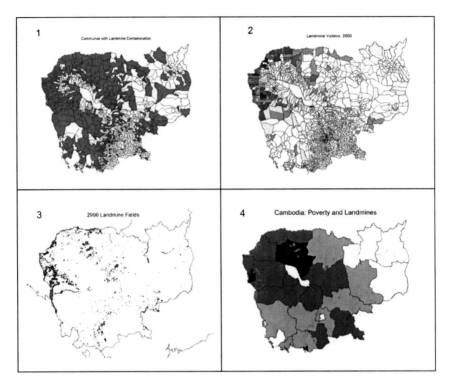

Figure 3. Maps of Landmine Data in Cambodia.

What is important about these two examples of various scalar interpretations is that the scale in which data is viewed can have a significant impact on what we see. This needs to be kept in mind when we view security threats at the local level. Many of these threats will have wide-ranging causes and consequences that extend far beyond the spatial sphere. Mitigating mechanisms will therefore require a broader analysis than one based purely on localized vulnerabilities.

5. Human Security Mapping

A central challenge in the measurement of human security is how can a measure stay true to the broad nature of human security—viz., not leave out any serious threat harming individuals—but also limit or refine our included threats to a manageable and measurable list?

In addition to the usual requirements of providing more reliable data in an efficient and timely manner, advancing the policy relevance of human security measures requires addressing two fundamental limitations associated with previous approaches. The new method must

- Move away from a standardized set of input indicators to a set that focuses only on those threats that are of critical importance in a specific region or nation and

- Re-orient the scale to the subnational level to capture and monitor locally dispersed threats to human security

5.1. METHODOLOGY

Human Security is the protection of the vital core[8] of all human lives from critical and pervasive economic, environmental, health, food, political, and personal threats.

The above definition articulates a general conceptualization of human security. By itself, however, it does little to indicate what the threats are, who they are affecting, and where they are of concern. Without this qualifying information, critics rightly point out that the broad concept of human security has little theoretical grounding or policy relevance (Paris, 2000; Krause, 2000; Foong Khong, 2001).

With this in mind, I have developed a measuring methodology that facilitates the collection, organization, and spatial analysis of human security information using a Geographic Information System (GIS).

This methodology takes a subnational approach to data collection and analysis. The method isolates "hotspots" of aggregated insecurity and determines spatial correlations between rarely compared security threats.

This methodology engages the paradox of measuring human security (that the broader the conceptualization used, the greater the difficulty of establishing a consistent metric), by addressing the problems associated with aggregating differing data types. By building the methodology around the spatial reference of data a common denominator is created—space—allowing for direct aggregation and analysis without creating a subjective nominal scale.

Also critical is the methodology's focus on identifying local hotspots rather than creating a national index. Although many global indices point policy makers to underdeveloped or insecure countries,[9] none isolate specific regional vulnerability within these countries. This coarse resolution arguably contributes to poor, nationally ubiquitous development policy and a lack of meaningful correlation analysis between harms.

Critics of human security rightly point out that it would be impossible to collect data for all imaginable human security threats for every location in

---

[8] For a discussion of the concept of "vital core" see CHS, 2003.
[9] For example, the Human Development Index and The Human Poverty Index.

the world. Past measures have responded by limiting the list of considered threats. This limiting is often driven by data availability (requiring global coverage) and inevitably leaves out important harms.

This new methodology takes a different approach. By recognizing that there is no difference in a death from a gun or a disease, it lets a threshold of severity guide the list of included threats in any one region being assessed (Owen, 2004).

5.1.1. Stage One: Threat Assessment

By shifting the focus of vulnerability from the state to the individual, human security attempts to incorporate what are traditionally considered development concerns into the realm of security studies—focusing on threats such as natural disasters, communicable diseases, dire poverty, human trafficking, or landmines rather than nuclear or interstate war. This shift in focus, however, presents a potentially unmanageable mandate. Critics have pointed out that if all potential harms are included, the concept is elusive and analytically indeterminate (Mack, 2002). Although this may be true if attempting to measure all possible harms in all places in the world, by controlling the location of the study, the list of relevant harms is limited significantly and the use of a GIS makes the spatially referenced data amenable to analysis.

The first stage of the methodology, therefore, seeks to determine from grounded empirical and qualitative research what specific threats affect a particular country or region. This can be achieved in a number of ways. Ideally, regional experts in each of the six categories[10] of security would be interviewed and asked whether there are any issues that would qualify as human security threats in their region—threats that present a critical and pervasive vulnerability to the vital core. By way of illustration, for the UNDP's Environmental Security category, this could be an extreme flood, and for the Personal Security category, this could be a high risk from landmines.

The most important point about this stage of the methodology is that it has reduced a seemingly endless list of threats (anything that can seriously harm an individual) down to only those that in practice affect a particular country or region. By shifting scales from the national to a local focus, human security becomes a manageable concept, going from hundreds of threats down to a handful.

Figure 4 is a visualization of stage one. For each of the six human security subsystems, boxes a, b, and c represent threats chosen by regional and thematic experts. The vertical lines separating the subsystems demonstrate the isolation of the raw data within disciplines.

---

[10] The threats falling under Community Security, the seventh category of the UNDP conceptualization (UNDP, 1994), were deemed insufficiently harmful to cross the threshold of insecurity.

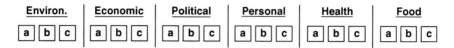

Figure 4. Stage One Diagram: Threat Assessment.

5.1.2. Stage 2: Data Collection and Organization

With the human security threats affecting a country or region determined and classified, data detailing them must be collected. These data can be both quantitative and qualitative, but all must have a spatial dimension. Ideally, the data sets collected will detail the indicator that best represents each specific threat. The indicators are chosen and the data collected using local researchers, the NGO community, government ministries, and international organizations. There will of course be some overlap with the experts consulted in stage one. A key to this stage is data availability. It is argued that the challenge is best addressed by looking at the subnational level, by using disciplinary experts, and by focusing only on relevant threats—those that surpass the human security threshold.

The concept of subsidiarity[11] is particularly important to the feasibility of the data collection process. Although information on all threats will not be available for all areas, the data are likely to cluster in the scale and regions in which they are relevant. For instance, if a flood affects only region x of a country, there will not necessarily be relevant hydrologic data for region y.

Once data sets detailing each threat are collected, they are organized in a Geographic Information System (GIS) by their spatial reference. This reference can be either a political boundary, a coordinate, or a grid space. What is important is that there is a link between specific threat severity and location, or space.[12]

At this point, we can now determine the level of threat for any point in our study region for any of the initial threats.

Table 7 depicts an example of the Stage Two data organization. The spatial reference does not have to be the same for all threat data sets. In the final table, however, all data will be disaggregated down to match the set with the finest resolution.

---

[11] The notion that problems be addressed at the relevant or most appropriate scale. Commonly used in relation to central political administrations only performing functions that are subsidiary to essential local functions. This is particularly important in this methodology, as data for different threats must be represented using different scales and data types. This is the benefit of using spatial data and also the reason why we need to break data into classes before aggregating together.

[12] This is ideally a standardized GIS code linked to a particular boundary. This is often not available, however, and the data will often need to be recoded. A feasible but timely process.

TABLE 7. Example of Stage Two Table

Province name	Economic			Health		
	Threat A	Threat B	Threat C	Threat A	Threat B	Threat C
1						
2						
3						
4						
5						
...						

5.1.3. Stage Three: Data Visualization and Analysis

The final stage of the methodology is to map and analyze the spatially referenced threat data. For each of the determined threats, we now have data sets detailing the location and severity of the threat within the country. As all information is linked to a spatial reference (e.g., province, city, coordinate, etc.), each threat can be mapped using a Geographic Information System. This process involves three steps: base map creation, hotspot analysis, and functional analysis (Figure 5).

Base Maps. Base maps are created in the GIS by linking threat data sets to digital boundary maps using their like spatial reference.[13] Once this is done, each threat can be mapped. These base maps are called layers and will be the foundation for the subsequent spatial analysis.

Although at this point each data set represents the total range of severity for each threat, this methodology is designed to isolate where each threat is most severe. This is done by first classifying the data based on their natural breaks.[14] This process produces a map for each threat showing where the threat severity is "high," "medium," or "low."

Hotspots. Hotspots are regions of aggregated human insecurities. They are places that experience multiple "high" level human security threats. Although a country as a whole may experience many different threats, these threats are

---

[13] The databases are connected using the "join" function in ArcGIS.

[14] There are many other ways that data can be classified. For a detailed description see Slocum (1999). We chose natural breaks because we are only looking for the range in the data where the threat is most severe, the entire range of the data being relatively unimportant to us. It is also conceptually elegant. As our primary collaborators are development workers and policy makers, there is no need to overcomplicate what is a simple stage of our methodology.

Province Boundary Shapefile

GIS Code	Province	Area	Perimeter
1	Banteay Meanchey	6054204712	535684.9
2	Otdar Meanchey	6681936370	556668.9
3	Siemreap	10863661754	638385.1
4	Preah Vihear	13655706145	713819.8
5	Stueng Treng	12076709457	702785.5
6	Battambang	11803125598	684479.7
7	Pailin	1102792730	171511.7

Joined with Domestic Violence Data Table

GIS Code	Province	Domestic Violence Rate
1	Banteay Meanchey	12.3
2	Battambang	8.9
3	Kampong Cham	16.8
4	Kampong Chhnang	19.8
5	Kampong Speu	2.7
6	Kampong Thom	13.4
7	Kampot	11.1

Results in Combined Table (Linked by GIS Code)

GIS Code	Province	Area	Perimeter	Domestic Violence Rate
1	Banteay Meanchey	6054204712	535684.9	12.3
2	Battambang	6681936370	556668.9	8.9
3	Kampong Cham	10863661754	638385.1	16.8
4	Kampong Chhnang	13655706145	713819.8	19.8
5	Kampong Speu	12076709457	702785.5	2.7
6	Kampong Thom	11803125598	684479.7	13.4
7	Kampot	1102792730	171511.7	11.1

Which Can Then Be Thematically Mapped

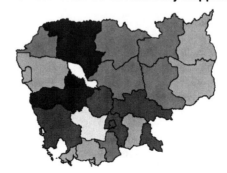

Figure 5. Diagram of Spatial Joining Process.

often regionally dispersed—different areas afflicted by different harms to different degrees of severity. In some locations, however, these threats overlap. Presumably, a person in a region suffering from five high-level threats will be less secure than someone in a region with only two threats.[15]

Hotspots are found by first separating only the regions with "high" levels of insecurity in each of our threat severity maps. All of these maps can then be overlaid[16] to show the regions subject to multiple high levels of human insecurity—how many "high" rankings a spatial unit has received.

In Table 8 it is clear that province 5, with six high-security threats, is less secure than, say, province 2, with only one high-security threat.

Human security hotspot analysis is useful for a number of reasons. First, conceptually, hotspots demonstrate the necessity of using a broad conception of human security. They clearly show that people remain insecure while not at war, and that within their borders may be suffering from a much wider range of possible threats than the traditional human security paradigm suggests.

Second, spatially aggregating varying data sets facilitates a degree of interdisciplinary analysis that is rarely achieved. By way of illustration, although many people know where floods are harming people, and many people know where poverty is worst, few people know both. Also, difficulties of data aggregation and cross-discipline communication often hinder

TABLE 8. Example of Stage 3.

Economic		Health		Hotspot count		
Province name	Threat A	Threat B	Threat C	Threat A	Threat B	Threat C
1	1	1	2	2	2	0	3
2	1	1	1	2	0	1	1
3	2	0	0	2	0	1	2
4	0	0	0	2	0	2	2
5	2	2	2	2	2	2	6
...						

[15] It is of course possible that an indicator for one system will also be relevant to another. This will result in a degree of spatial statistical interdependence between variables. However, as the hotspot analysis only includes the worst of the threats at their highest level, we feel that there is enough independence in each indicator to warrant separate categorization. However, in subsequent work, we have statistically analyzed the relations between the threats. As this work requires much more detailed data rather than generalities regarding threat "degree," we keep the data sets in their cardinal format and build explanatory multivariate regression models.

[16] For a description of "overlaying" see Bernhardsen (1999).

well-meaning broad analysis. By limiting subjective decisions on the relative severity of various threats to the early discipline-specific first stage (threats assessment), the analysis limits subjective data aggregation.

Third, hotspot analysis has practical utility for development and humanitarian relief efforts. The logistical benefits of knowing exactly what harms are affecting which region of a country are clearly evident. In addition, having all the information in a GIS allows for easy access to large data sets that generally do not get shared, let alone used widely within the development community.
Results. Figure 6 provides an example of part of the above methodology as it was implemented through a trial in Cambodia.[17] It outlines the threat assessment, data collection, and spatial analysis stages for two of the six human security subsystems—economic security and health security.[18] The process is divided into seven stages that are briefly outlined below and in Figure 6.

1. Human security is first conceptually divided into six subsystems following the United Nations Development Program's (UNDP's) categorization (UNDP, 1994).

2. The relevant Cambodian threats for each of the threat subsystems were determined. Sixty interviews were conducted in Phnom Penh with experts spanning the six human security categories. They were asked what, if any, concern within their area of expertise would qualify as a threat to human security. For example, for economic security, poverty was the threat and for health security, HIV/AIDS, malaria, dengue fever, and tuberculosis were the threat.

    In order to simplify the demonstration of the methodology, from this point on we will follow poverty, dengue fever, and tuberculosis from data collection to final correlation mapping. Again, in the full Cambodia study, thirteen threats were identified and measured.[19]

3. Indicators that best represented each of these threats were then determined by consulting disciplinary experts, and the spatial data necessary to measure them were collected. For poverty the indicator chosen by the Cambodian experts was the percentage of population below the poverty line calculated as monetary equivalent of calorie intake collected at the commune level. For tuberculosis the indicator was the number of cases per 100,000 collected at the regional district level. And for dengue fever the indicator was the number of cases per 100,000 collected at the provincial

---

[17] This study was conducted from November 2002 to January 2003, based at the Cambodian Development Research Institute (CDRI) in Phnom Penh.

[18] In the full study data detailing thirteen threats were collected and analyzed. For the details of this analysis and its conclusions see Owen and Le Billon (2004).

[19] Poverty, starvation, flooding, droughts, malaria, TB, dengue fever, HIV/AIDS, domestic violence, gun injuries, landmines, human rights violations, village-level violence.

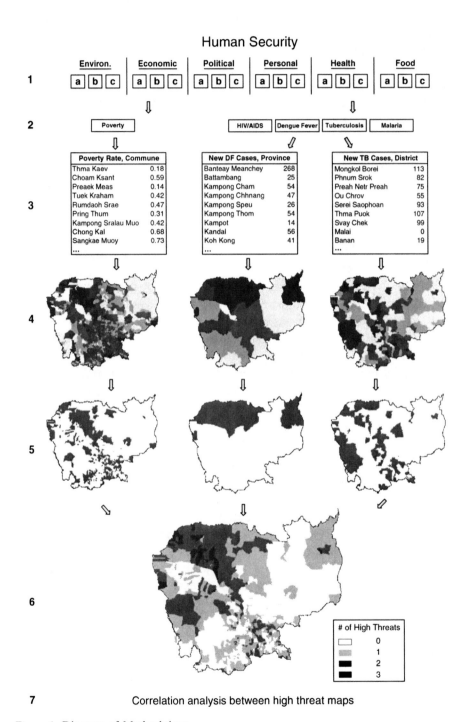

Figure 6. Diagram of Methodology.

level. These data were then joined to the spatial databases of the boundary GIS files, enabling them to be mapped.

4. The data sets were classified by their natural breaks in order to define high, medium, and low threat areas.

5. High-threat areas were isolated and mapped on their own. This is important, as human security involves a threshold of severity. Human security is threatened when the worst of all the possible threats are at their highest levels in a particular location.

6. All three high threat maps can now be overlaid to find hotspots of aggregated insecurity. In Figure 6, the scale of this hotspot map ranges from 0 to 3 high threats in any one commune.

7. Finally, spatial functional analysis can be conducted between pairs of human security threats. Taking the three examples from Figure 6, the national probability of being exposed to a high poverty, TB, and dengue fever threat was calculated. For example, out of a possible 1,626 communes, 451 have a high poverty threat (more than 65 percent of the population below the poverty line), meaning all else being equal, there is a 27 percent probability of a commune having a high poverty threat.

The threat determination and data collection stages have shown that data are available to document a wider range of indicators at a finer resolution than has been suggested in some of the academic literature (Collier and Hoeffler, 2001; Brauer, 2001), and also that, given a degree of institutional collaboration, meaningful collection and organization of broad ranging indicators is possible. Support for a broader definition of human security and the possibility of its rigorous analysis is thereby provided.

For development practitioners in Cambodia, this methodology has clearly pointed to regions that are subjected to the aggregated impact of multiple human insecurities. In the example given here, there are regions that suffer from poverty, TB, and dengue fever, but also regions that suffer from two or all three of these threats simultaneously. Regions with three human insecurities are clearly less secure than those with two, one, or no high-level human security threats. Policy and decision makers should find this added spatial resolution of assistance especially for coordination of work between NGOs with separate but ultimately overlapping interests.

6. Conclusions

Considering the multi-scaled characteristics of human security threats, a critical first step in determining the utility of the concept is to incorporate this referential shift into measuring methodologies. The implications of this shift must be understood if the concept is going to effectively drive policy and gain its rightful place in the international security discourse.

With this in mind, human security hotspot analysis is useful for a number of reasons. First, conceptually, hotspots demonstrate the policy utility of human security. They clearly show that people remain insecure while not at war (countering traditional security), and that within their borders they are suffering from a wide range of possible threats.

Second, the practical utility of knowing exactly which harms are affecting which regions of a country are clearly evident for the policy and development community. In addition, having such information in a GIS allows for easy access to vast amounts of data that may be of tremendous value to, and may not otherwise be shared by, policy makers, practitioners, and humanitarian operations.

Third, hotspots are an effective means of presenting large amounts of information to the public and to the policy-making community. This process is replicable in any region or country and can provide varying levels of detail. The human insecurity of a region can be displayed as one final map for the media, as a summary report for the policy community, or as its base data for academics.

The following policy recommendations emerge from early trials of this measuring methodology.

Capture the relevant threats only. For policy to effectively target the most vulnerable, clear information on only what is harming people is essential. Measures that include data on all possible threats to all people in all places divert attention away from the acute harms needing immediate action.

Draw on local knowledge. A vast amount of valuable expertise and threat information exists at the local level. Local researchers and development workers, not only international theoretical frameworks, should be consulted as to what the relevant threats are.

Use subnational data where available. Once the focus is shifted from world coverage to specific regions, a vast amount of local, subnational data becomes available. These data should be used at its most detailed level in order to capture the critical spatial variation within each threat. Aggregation to the national level will likely distort the validity of threat data.

Human security should inform humanitarian assistance. Development and humanitarian assistance policy should target a more specific level than the state, be tailored to locally relevant threats, and be more reflexive to rapid changes in insecurity levels and causes. A local-level human security assessment team, dispatched prior to international operations, would greatly increase the targeting of specific vulnerabilities and assist collaboration between actors in these complex environments. Ongoing monitoring and updating of the process would allow for real-time anticipation of rising insecurity levels.

As the nature of insecurity evolves, so too must our security mechanisms. This means addressing issues such as poverty, communicable disease, environmental disasters, and the proliferation of small arms with the same vigor we once used to counter the nuclear threat and to secure the Cold War balance of power. To do this, however, we need sophisticated mechanisms for clearly identifying the complex nature of vulnerability. Using the concept of human security and modern GIS mapping technologies, the proposed methodology makes the most of existing local knowledge and points policy makers to the many and complex regionally relevant threats to people's security.

References

Bajpai, K., 2000, *Human Security: Concept and Measurement*, Kroc Institute Occasional Paper 19.

Barrows, H., 1923, Geography as human ecology, *Ann. Assoc. Am. Geogr.* **13**:1–14.

Bernhardsen, T., 1999, *Geographic Information Systems: An Introduction* (2nd Edition), Wiley, Chichester.

Blaikie, P., and Brookfield, H., 1987, *Land Degradation and Society*, Methuen, London.

Brauer, J., 2001, War and Nature: The problem of data and data collection. Uppsala Conference on Conflict Data, June; http://www.pcr.uu.se/ident.html.

Collier, P., and Hoeffler, A., 2001, Data issues in the study of conflict, Uppsala Conference on Conflict Data, June; http://www.pcr.uu.se/ident.html.

Commission on Human Security (CHS), 2003, *Human Security: Now*, Final Report to the Human Security Commission, 2002–2003, New York.

Covello, V., and Mumpower, J., 1994, Risk analysis and risk management: An historical perspective, in: *Environment, Risks and Hazards*, S. L. Cutter, ed., Prentice-Hall, Upper Saddle River, NJ.

Cutter, S. L., and Solecki, W. D., 1989, The natural pattern of airborne toxic releases, *Prof. Geogr.* **41**(2):149–161.

Cutter, S. L., Mitchell, J. T., and Scott, M. S., 2000, Revealing the vulnerability of people and places: A case study of Georgetown County, South Carolina. *Ann. Assoc. Am. Geogr.* **90**(4):713–737.

EM-DAT, 2000, EM-DAT: OFDA/CRED International Disasters Data Base.

Foong Khong, Y., 2001, Human security: A shotgun approach to alleviating human misery? *Glob. Gov.* **7**(3):231–236.

HPHPCR, 2001, *Proposal for the Creation of a Human Security Report*, Harvard Program on Humanitarian Policy and Conflict Research.

Kasperson, R. E., Renn, O., Slovic, P., Brown, H. S., Emel, J., Goble, R., Kasperson, J. X., and Ratick, S., 1988. The social amplification of risk: A conceptual framework, *Risk Anal.* **8**(2):177–187.

Kates, R., and Burton, I., 1986, *Geography, Resources and Environment*, vol. 1, University of Chicago Press, Chicago, IL.

King, G., and Murray, C., 2000, *Rethinking Human Security*, Harvard University Program on Humanitarian Policy and Conflict Research.

Krause, K., 2000, *Une Approche Critique de la Securite Humaine*, Program D'Etude Strategique et se Securite International, Institute Universaire de Haute Etude International, Geneve, Septembre.

Krug, E., Etienne, G., Dahlberg, L., Mercy, J., Zwi, A., and Lozano, R., eds., 2002, *World Report on Violence and Health*, World Health Organization, Geneva, p. 4.

Lonergan, S., Gustavson, K., and Carter, B., 2000, *The Development of an Index of Human Insecurity*, Global Environmental Change and Human Security Project, International Human Dimensions Programme on Global Environmental Change, Research Report.

Mack, A., 2002, *Feasibility of Creating an Annual Human Security Report*, Program on Humanitarian Policy and Conflict Research, Harvard University, February.

Michel, D., 1991, *Image and Cognition*, Harvester Wheatsheaf, New York.

Owen, T., 2004, Human security—Conflict, critique and consensus: Colloquium remarks and a proposal for a threshold-based definition, *Secur. Dialogue* **35**(3):373–387.

Owen, T., and Benini, A., 2004, Human security in Cambodia: A statistical analysis of large-sample subnational vulnerability data, Report written for the Centre for the Study of Civil War, International Peace Research Institute, Oslo.

Owen, T., and Le Billon, P., 2004, *Security Mapping: Measuring Human Security in Cambodia*, Liu Institute for Global Issues Report, UBC.

Palm, R., 1990, *Natural Hazards: An Integrated Framework for Research and Planning*, Johns Hopkins University Press, Baltimore, MD.

Paris, R., 2000, Human security: paradigm shift or hot air? *Int. Security* **26**(2):87–102.

Slocum, T., 1999, *Thematic Cartography and Visualization*, Prentice Hall, Upper Saddle River, NJ.

United Nations Development Program (UNDP), 1994, *New Dimensions of Human Security: Human Development Report*, Oxford University Press, New York, pp. 22–25.

White, G. F., 1945, *Human Adjustments to Floods*, Department of Geography Research Paper 29, University of Chicago Press, Chicago, IL.

Wilhite, D., and Easterling, W., eds., 1987, *Planning for Drought: Towards a Reduction of Social Vulnerability*, Westview, Boulder, CO.

World Health Organization, 2001, *World Health Report 2001*, WHO, Geneva, p. 114.

SECTION II

ENVIRONMENTAL CHALLENGES: EXAMPLES FROM NORTH AFRICA, THE BALKANS, AND THE MIDDLE EAST

DRYLANDS IN CRISIS

Environmental Change and Human Response

DAVID A. MOUAT[1]* AND JUDITH M. LANCASTER[2]

[1,2] *Division of Earth and Ecosystem Sciences, Desert Research Institute, 2215 Raggio Parkway, Reno, NV 89512*

Abstract: Dryland regions occupy 41 percent of the earth's land surface. They occur on every continent, are occupied by over two billion people, and are—in many cases—experiencing a crisis. The combination of inherently fragile ecosystems and human population pressure can result in some environmental changes, such as soil erosion and rangeland and agricultural degradation, that are often easy to detect and measure. One of the most important issues in dealing with these changes and responses is to arrive at possible solutions that can result in ameliorative changes before the crises become irreversible. The addition of climate change exacerbates the situation for dryland landscapes, with predicted changes to hydrological regimes and vegetation composition and cover that would have serious implications for human populations. Under such conditions, environmental security can become threatened, which reinforces the desirability for dryland peoples to develop coping mechanisms and alternative lifestyles. The human response to these changes has been characterized as "adapt, migrate, or die." We add a fourth human response, "revolt" to the possibilities. The technique of alternative futures analysis offers a strategy for modeling the environmental effects of human decisions regarding land use, which if made in a timely fashion, might prevent undesirable outcomes.

Alternative futures analyses can involve communities and policy makers in planning for viable economic and environmental strategies. Contextual concepts, examples, and a discussion of impacts are presented. We attempt to illustrate that environmental and human security are inextricably interlinked;

---

* Address correspondence to David Mouat, Division of Earth and Ecosystem Sciences, Desert Research Institute, 2215 Raggio Parkway, Reno, NV 89512, USA; e-mail: david.mouat@dri.edu

P.H. Liotta et al. (eds.), Environmental Change and Human Security, 67–80.

that if we wish to protect human values, we must ensure environmental security and sustainability.

Keywords: Desertification; climate change; human response; alternative futures analysis; sustainability

1. Introduction

The sustainability of dryland systems—both natural and human—is dependent upon adaptability: not only to unreliable and irregular rainfall, but also to fluctuations in resource availability. Economic and political factors have changed the patterns and strategies of human occupation of dryland regions, where previous lifestyles of transhumance or nomadism and minimal involvement in a market economy permitted human occupation of arid and semiarid lands even in times of drought, without causing ecosystem stress. Population growth and the trend toward permanent occupation of settlements are linked to the imposition of taxes, reliance on cash crops, and export of local resources and raw materials, which together with climate variability have resulted in overcultivation, overgrazing, and deforestation in many areas of the world (Middleton and Thomas, 1997). These areas are highly susceptible to desertification, and economic sustainability is a major social and environmental issue and potential driver for change. Partly as a result of human actions, there are an increasing number of disasters facing dryland people (MA, 2005a).

Drylands are inherently fragile and it is likely that human-induced change, particularly temperature increases, will have escalating effects upon them within a relatively short timeframe (Williams, 2001). For some areas, such as sub-Saharan Africa, there is a long history of environmental and social crisis caused, in large part, by climate, and exacerbated by improper human use. In the late 1960s and early 1970s a drought captured world attention and resulted in the death of over 250,000 people (Reynolds and Stafford Smith, 2002) and the migration of undoubtedly many more. In many areas, loss of productivity results from the exploitation of dryland resources, about 135 million people are at risk of becoming refugees as a consequence of desertification (INCCCD, 1994), and the livelihoods of an additional one billion are also likely to be affected. The addition of climate change exacerbates the situation, with predicted changes to hydrological regimes and vegetation composition and cover that would have serious implication for human populations (Williams, 2001).

2. Background

Drylands are arid, semiarid, and dry subhumid areas of the world where mean annual rainfall is less than 1,000 mm and with evapotranspiration rates that exceed the precipitation (Williams, 2001). Two billion people live in drylands, which occupy 41 percent of the earth's land surface and occur on every continent. Drylands are at risk of desertification, which is defined as

Land degradation in arid, semi-arid and dry sub-humid areas resulting from various factors, including climatic variations and human activities. (UNCED, 1992)

The components of this definition are important; both climate and human activity play a role, and therefore changes in either are likely to have an effect. The impacts may be positive or negative—in both social and environmental contexts. The societal impacts of desertification include migration and refugees (MA, 2005a), which contribute to rural poverty and may create inequalities and tensions in host communities, and prove a challenge for local and regional decision makers and government agencies. Land degradation and the associated loss of ecosystem services result in migration (MA, 2005a). In addition, according to UN reports (INCCCD, 1994), half the world's fifty armed conflicts were associated with dryland areas suffering from the effects of desertification. Desertification is both a cause and consequence of poverty (Glantz, 1994; Reynolds and Stafford Smith, 2002); it results in reduced crop yields, increased distance to potable water and fuelwood, poor animal and human health, hunger, hopelessness, and desperation.

The environmental processes associated with desertification are insidious and cumulative and are undoubtedly exacerbated by the pressures of modern society, increases in population, and social or political inequalities. Environmental processes are interlinked and involve significant feedback mechanisms. For example, dryland soils are usually characterized by low levels of biological activity, organic matter, and aggregate stability—which can easily result in a breakdown or decline of soil structure, accelerated soil erosion, reduction in moisture retention, and increase in surface runoff (Williams and Balling, 1996). Reduction in plant cover may result, which will exacerbate the processes of soil deterioration, leading to a change of scale in the spatial distribution of soil resources (de Soyza et al., 1998), an increasingly "patchy" landscape, and a decline in sustainability that is difficult or impossible to reverse (Schlesinger et al., 1990).

However, desertification can be prevented and people can live "within" the confines of the areas affected—reducing poverty, migration, and social problems as well as maintaining (or even improving) ecosystem sustainability (MA, 2005a). Social participation is an essential component of strategies to abate, mitigate, or prevent desertification (Navone and Abraham, 2006), and many countries have

implemented programs to address the land and water management issues
that are the driving forces behind desertification processes. Increased integration
of land and water management, and of pastoral and agricultural land uses,
may provide viable solutions if approached in a proactive manner, including
capacity building and institutional support at all levels (MA, 2005b).

3. Environmental Change

Climate has been considered a factor influencing desertification since the
1977 United Nations Conference on Desertification (UNCOD, 1978), and
over the past thirty years the importance of climate change in all ecosystems
has become increasingly recognized. Strategies to mitigate desertification
often involve understanding how climate varies, thereby developing mechanisms to allow human use systems to adapt to these variations. The potential
impact of climate change leads to the need to assess and develop long-term
warning systems and techniques as well as social "buffering" strategies and
alternative methods for generating income and providing sustainability.

There have been significant changes in climate during the twentieth century,
with a global increase in temperature of about 0.6°C (MA, 2005b), which is
largely attributable to increased atmospheric CO_2 and is a trend that is likely
to continue—primarily as a result of industrialization (IPCC, 2000). Long-
term temperature and precipitation trends resulting from climate change are
possible to predict at the global scale and can be spatially plotted (IPCC,
2001). However, the global climate change models vary sufficiently in their
predictions to cause considerable uncertainty at the regional level. Drylands
and desertified areas typically have more extreme climates as well as being
among the most variable (Herrmann and Hutchinson, 2006), with effects
upon the hydrological cycle having more drastic implications in dryland areas
than in other regions (Lioubimtseva and Adams, 2004). Under global change
scenarios, droughts are predicted to become more frequent in desert margin
areas, and flood intensity will also likely increase; both conditions will contribute to drier soils and less infiltration (Herrmann and Hutchinson, 2006).

The processes of desertification affect climate at varying degrees. Reduction
in vegetation cover may lead to increased runoff and increased potential evapotranspiration rates. This in turn may lead to increases in surface and near-
surface temperatures, lowered soil and atmospheric moisture levels, and higher
near-surface wind speeds (Williams, 2001). Changes in both soil temperature
and relative humidity affect atmospheric movement, which in turn affects precipitation patterns. Human actions in dryland areas, such as overgrazing, also
contribute to changes in surface and atmospheric conditions, and biomass
burning increases atmospheric CO_2 and causes a loss of fixed nitrogen in the

system (Williams and Balling, 1996). Plants sequester carbon, and therefore reduced vegetation diminishes carbon reserves and increases CO_2 emissions. Hence, reduced vegetation contributes to global warming (MA, 2005b).

Recent studies show that climate change will especially increase the risk of desertification in Africa (de Wit and Stankiewicz, 2006), which is already severely affected. Climate change models for Africa indicate that temperatures are likely to increase 0.2–0.5°C per decade (de Wit and Stankiewicz, 2006), which will contribute to increasing rainfall variability by more than one standard deviation from normal in many areas (IPCC, 2001), and will, in turn, significantly decrease perennial surface runoff. In the Western Cape of South Africa, for example, up to half the present perennial water supply is likely to be lost—even based on a relatively optimistic climate model used by de Wit and Stankiewicz (2006). The Intergovernmental Panel on Climate Change (IPCC, 2001) found that decreases in runoff totaled about 17 percent during the 1990s, which indicates that the trend has most likely already started.

Global climate change models, upon which these studies are based, operate at scales that do not permit a local level of analysis, and given the great spatial variability of dryland ecosystems there will be some areas that are less affected than others, even within the same ecosystem and climate regime (IUCC, 1993). This internal variability will impose additional social pressures upon dryland peoples, adding a further element of uncertainty to society, sustainability, and human and environmental security. However, there are indications that "traditional" lifestyles permitting people to prosper in regions susceptible to desertification are changing—and not necessarily for the worse. San (bushmen) members of a nature conservancy in Namibia that caters to tourists use natural products from the local environment for artwork that serves the dual purpose of income generation and maintaining cultural traditions (Seely, 2006).

There are times when change appears more attractive and people are more willing to consider alternatives; ideally, this is before options become limited as a result of worsening conditions of any kind. Human response to the "tipping point"—or point of no return—is critical, and may have considerable repercussions.

4. Human Response

Dryland peoples have developed coping mechanisms for surviving fluctuations in ecosystem services, which Bradley and Grainger (2004) have studied under three categories: *anticipatory strategies*, *response strategies*, and *survival strategies*. Of the two groups in Senegal studied by Bradley and Grainger, those who were more resilient had higher tolerance of "abnormal" conditions and a greater diversity of survival strategies—to which the minimal advance of desertification in the region was attributed. However, if dryland conditions get "worse," possibly as a result of climate change, environmental security may become threatened (MA, 2005b) and people may run out of options for maintaining a sustainable existence. Thus, a lack of options as an outcome of living in an unsustainable environment is inextricable interlinked to human security.

The strong links, both cause and effect, between desertification and poverty are contributory factors in the social and political marginalization of dryland peoples (MA, 2005b), which is likely to further reduce options in times of environmental or social change and stress. The worst-case options include taking no action (which may ultimately result in starvation), moving (often to urban centers, and not always a negative option), or revolting. In many cases, people in entire regions move to cities or neighboring towns (Glantz, 1994), where resources may be no more abundant, and where they are outsiders with different customs and beliefs.

Environmental refugees are typically poor and need shelter, food, and medical care. Additionally, they may bring different customs, religions, agricultural practices, and diseases with them, and many governments, at national or local levels, simply do not have the resources to deal with this situation (Tickell, 2003). An influx of refugees into an area already experiencing environmental stress (such as desertification) and depletion of resources is, according to Liotta (2003), likely to result in "ethnic" clashes and infrastructure collapse. Adaptation is an infinitely preferable option, but may not be viable if decisions are put off until it is too late.

At regional or national levels, countries may initiate policies to alleviate disasters such as drought, which would, to a large extent, minimize migration and its associated problems. Long-term efforts to promote sustainable development, such as Agenda 21, which arose from the UN Convention on Development in 1992 (the "Earth Summit"), have in some ways revolutionized policy decisions made at national levels and include valuable aspects of capacity building. Despite the fact that capacity building is typically aimed at the community level, the decisions are made at the top. Experience has shown that "top-down" solutions are generally less effective than those originating at community level ("bottom-up"), strongly advocated by the UN Convention to Combat Desertification (UNCCD).

Identification of the timeliest period for intervention and action is important, and a proactive approach will increase the chances of people adapting, thereby reducing the likelihood of migration, revolt, and death. In this context, we see "adaptation" as a peaceful process, whereas the term "revolt" is used to indicate an organized opposition to authority which may involve bloodshed. "Eco-tipping point" is a term coined by Marten (2005) referring to the element of the natural-human system that can activate a chain of events leading to transformation from unsustainable to sustainable (or vice versa). Marten cites examples of bottom-up decisions on topics as diverse as fisheries in the Philippines and a New York City community garden to make his point that once started, community action can create a chain of events and positive spin-offs beyond the immediate target of the proposed activity. The apparently simple steps of taking action and making decisions are critical components of response, and part of a philosophy that suggests examining future scenarios and their spatial manifestations: alternative futures.

5. Alternative Futures Analysis

Alternative futures analysis relies upon developing an understanding of the critical uncertainties that shape a region and the people or stakeholders of that region (much of the following is adapted from the work of Shearer; see, especially, Shearer et al., 2006). The stakeholders include not only those people who live and work in the area but also policymakers, scientists, NGOs, media, and others who have a vested interest in the future of the region. The process that leads to the development of these alternatives empowers people; acts as a forum for sharing problems, ideas, and concerns; and may defuse tensions and conflicts caused by climatic variability, diminishing resources, an influx of refugees, or other issues that face dryland areas.

An alternative futures analysis uses geographical information systems (GIS) as a tool to conduct simulations using a series of process models, each of which is grounded in theory relevant to its own particular discipline. The models used vary with each analysis. For example, in some studies fire ecology may be sufficiently important to demand a model, whereas in others fire may be part of a vegetation change model. The methodology was, in part, developed by Carl Steinitz, Harvard Graduate School of Design (1990, 1993), and considers principles of landscape architecture. Models (and design) are developed to answer questions addressing the landscape in terms of present condition, structure, and function and the projected effects of change in landscape components. These spatially explicit futures models (and design) address human-environment interactions through empirical representation of human-induced change across time; past patterns and correlates of change are then used to "project" alternative future scenarios. Especially important to the future scenario modeling effort is participation by stakeholders, to identify their priorities and indicate viable alternative land-use options. This process leads to the sharing of differing perceptions and views, and is effective in defusing potential conflict situations.

An alternative futures analysis usually will not be effective under a serious crisis situation because people are often too preoccupied to focus and develop the necessary level of attention and interest. However, if conducted in a timely manner, the intervention is positive and is likely to result in increased communication among all involved, the illustration of possible conflicts, their causes and potential solutions, and a regional analysis framework that produces quantified results, maps, and figures showing status and impacts of possible change—which could be used to address policy and decision makers, or by people in those positions to make their point at higher administrative levels.

5.1. METHODS

The analysis employs spatially explicit futures models which address human-environment interactions through empirical representation of human-induced change across time. It may also employ "design" to spatially represent assumptions of development and change. Past patterns and correlates of change are then used to "project" alternative future scenarios and illustrate their potential effects. Scenarios are developed for each study based upon locally relevant framing issues and critical uncertainties developed in stakeholder workshops, questionnaires, and informal discussion sessions. Examples of possible framing issues, which may be relevant in a dryland area and affect land use decisions include:

- Should a group of farmers experiment with new types of drought-tolerant agricultural crop varieties?
- Should the area of currently cultivated land be extended?
- Should local governing bodies impose control over water use until there is significant rainfall?

These "shoulds" refer to factors over which the public, land owners/managers, and policy makers have some control and are the framework for the development of the critical uncertainties—the important issues facing a region, whose outcomes are unknown (Shearer et al., 2006). Examples of these might include:

- Will there be enough water to irrigate this year?
- Will partnerships with neighboring communities provide a "safety net" to permit pursuing a crop diversification strategy?
- Will the drought in a neighboring province cause an influx of refugees?

In the example being developed here, where people are considering crop diversification to provide an alternative livelihood and a potential "drought buffering" mechanism, the stakeholder workshop process might also address such questions as "Will the area under cultivation need to be increased?" and "Which part of the valley is most suitable for increased irrigation?" and will result in the development of scenarios and, in turn, the identification of the who, what, where, and why of actions based upon changing agricultural strategies. Each scenario is fully described in several pages of text (in some cultures oral role-playing or graphics may be more appropriate), and this exercise in itself requires that decision makers and those who will implement the decisions consider the specifics and implications involved in each suite of actions.

Scenarios may be thought of as the means to the end, and alternative futures as the end itself (Shearer et al., 2006). Futures used for analysis are not necessarily those that are most likely to occur, but rather encompass a broad range of possibilities—some of which may not be "desirable" for all parties involved in the process. In the example discussed in the previous paragraph, the resultant futures may be two to four different land use strategies, including the status quo "let's not do anything different at all."

The analysis component of the alternative futures study investigates the local and regional consequences of the futures upon environmental parameters (and societal values) such as water, vegetation change, and (for example) insect populations through a series of process models based partially on the scenario descriptions, each of which will have a greater or lesser impact upon the other models in the system. These impact assessments illustrate the potential changes caused by the future, which, when considered across all variables, shows the overall significance of individual futures.

5.2. RESULTS

Alternative futures analyses have been conducted in many parts of the world. Examples in North America address issues such as agricultural policy (the Iowa Corn Belt), recreational opportunity (Monroe County, Pennsylvania), urban development and ecological preservation (southwest California), water (the Upper San Pedro River Basin in Arizona and Sonora) and diverse land use and planning options (southern Rocky Mountains, Alberta; Willamette River Basin, Oregon; and others).

For example, the Upper San Pedro River Basin of southern Arizona (U.S.) and northern Sonora (Mexico) is a unique ecosystem with the highest avifaunal biodiversity in North America. Competing land uses include development, ranching, and open space. As a result of rapid development and increased ground water pumping, the system is at very high risk to the certainty of conversion of land from one use to another (ranching to housing development, for example). As a result of stakeholder workshops, a number of possible scenarios for how the region might change were consolidated to three:

- "Open" (essentially a development scenario)
- "Plans" (essentially a "business as usual" scenario)
- "Constrained" (essentially a conservation scenario)

The spatial allocation of these scenarios into alternative future patterns of land use ("alternative futures") as they might appear in the year 2025 revealed interesting impacts on the values of the region's stakeholders. For example, all of the alternative futures resulted in lowered groundwater levels and increased drying of the river (Steinitz et al., 2003), and all impact the ecological sustainability of the region. Figure 1 illustrates the impacts that the three spatial development allocations of the scenarios—the "alternative futures"—might have on ground water change. Note that all three result in lowered ground water levels. This region was historically dominated by grass-land ecosystems, which have gradually been replaced by shrub-dominated communities and fragmented by development (Mouat and Lancaster, 1996). Although ranching is still important in the region, ecotourism, mining, and residential development have increased, and result in a range of perceptions among local stakeholders concerning priorities for the region's future.

The previous example discusses changes in land use as a function of future development patterns with resultant impacts on groundwater. In a hypothetical study of a part of the Shiyang He watershed in Gansu Province of northwest China conducted by Zhou and Mouat (2007), the framing issues relate to water shortage, water quality, loss of productivity, and soil

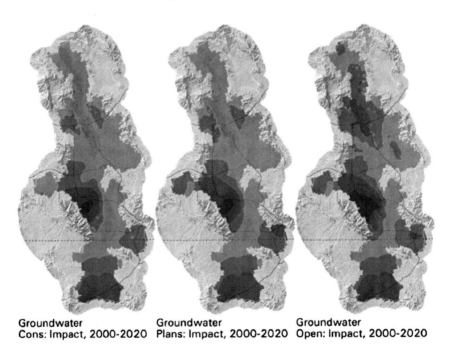

Groundwater   Groundwater   Groundwater
Cons: Impact, 2000-2020   Plans: Impact, 2000-2020   Open: Impact, 2000-2020

Figure 1. Impact of Three Different Development Scenarios (Constrained, Plans, and Open) on Groundwater in the San Pedro River Basin. Darker Tones Indicate the Greatest Drop in Groundwater Level.

salinization. Currently, a large agricultural area lies just to the north of the Qilian Shan (mountains), with the small city of Wuwei located in the western part of the agricultural region. Downstream from Wuwei, further from the snow-covered Qilian Shan, is a reservoir, and to its north is a degrading agricultural area lying to the north of the town of Minqin. That agricultural region is becoming increasingly salinized as a result of an already saline soil and poor irrigation techniques.

In several areas of northwestern China, desertification has been reduced over the past five years due to the application of new policies designed to change human activity and protect the environment. The "Grain for Green" and "Grazing Prohibition" policies are improving the environment, but other challenges remain, such as how to improve farmers' income and develop the economy. These questions were driving forces behind Zhou and Mouat's (2007) study, which identified five critical uncertainties facing the Minqin region:

- Will there be sufficient water for agricultural and domestic use?
- What effect will climate change have?
- Is soil salinization likely to increase?
- What changes to the economy will occur, possibly as a result of Government policy?
- Are land use patterns going to change?

The present land use status and three potential alternative futures were developed by Zhou and Mouat (2007) from these framing issues and critical uncertainties (A, B, C, and D in Figure 2), as follows:

A. Maintain the status quo (the future will be similar to the present)

B. High-tech development in Wuwei

C. Increased population and water use (throughout the basin)

D. Climate change (reduced snowfall and snow cover in the Qilian Shan)

Figure 2A illustrates the present landscape.

Figure 2B illustrates the first alternative future: The scenario leading to this future posits that as a result of a policy decision to diversify the economy of the region, Wuwei develops one or more high tech industries that attract people to the city. The city expands into the agricultural area, at the same time as the rural population declines as a result of migration into town for better paying jobs. A policy to import high water use commodities initiated by the administration in Wuwei, combined with mandated water conservation, results in lowered water use for that area, and agriculture downstream, around Minqin, improves. This future is positive for the environment, the economy, and the people.

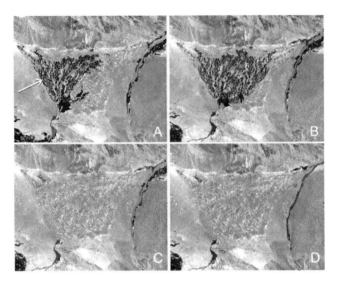

Figure 2. Alternative Futures for the Minqin Area. Area of Irrigated Agriculture, Which Is the Variable Affected by Policies and Human Decision Making, Indicated by Arrow in A.

Figure 2C illustrates the second alternative future: The scenario leading to this future posits that no new government policies are implemented and the population increases throughout the Basin. As a result of higher population, it is anticipated that water use and salinization would both increase. However, agricultural productivity would decrease, resulting in lower incomes for the farmers. This future is negative for all concerned, and desertification would increase in extent and intensity.

Figure 2D illustrates the third alternative future: The scenario leading to this future posits that climate change results in the same amount of annual precipitation. However, with (predicted) warmer winters and warmer nighttime temperatures, the snow in the Qilian Shan melts completely by spring and there is a significant decrease in summer runoff. This places a stress on the system and results in farmers overdrawing the groundwater, hence exacerbating an already salinized system. All of the downstream irrigation in the vicinity of Minqin is abandoned, which results in rural migration into Wuwei and the creation of new urban centers. Depending on people's response, this future could ultimately result in a positive or a negative outlook for the population based upon possible policy interventions and the people's economic choices.

These hypothetical but entirely plausible scenarios illustrating the introduction of new policies, human behavior, and land use demonstrate how an alternative futures analysis might be employed to address land and water management issues and increase understanding of the interaction between socioeconomic and biophysical processes. Both are vital components of dealing with the impact of climate and ecosystem change in the world's drylands (MA, 2005b), and necessary for maintaining sustainability and viable livelihoods.

6. Conclusions

The challenges of environmental change, including desertification, are increasingly recognized and becoming better understood. Over-legislation, particularly of the "top-down" variety, is not likely to contribute as much as would the implementation of policies that take advantage of local knowledge and experience, improve communication, and reduce redundancy. We are increasingly part of a global society, and the impacts of climate change and desertification extend beyond the geographical area in which they originate (MA, 2005a), emphasizing the urgent need for integrated and concerted action to identify those areas and peoples at greatest risk, and those countries that are unable to address these issues unaided. The challenge of connecting science and policy with the real world is still present, but hopefully will be eroded as human responses become increasingly assertive and directed. Alternative futures analysis is an example of proactive community involvement and decision making that will provide information, illustrate the consequences of human responses, and assist people to develop options and alternatives for maintaining dryland sustainability.

References

Bradley, D., and Grainger, A., 2004, Social resilience as a controlling influence on desertification in Senegal, *Land Degrad. Dev.* **15**:451–470.

de Soyza, A. G., Whitford, W. G., Herrick, J. E., Van Zee, J. W., and Havstad, K. M., 1998, Early warning indicators of desertification: Examples of tests in the Chihuahuan Desert, *J. Arid Environ.* **39**(2):101–112.

de Wit, M., and Stankiewicz, J., 2006, Changes in surface water supply across Africa with predicted climate change, *Science* **311**:1917–1921.

Glantz, M. H. ed., 1994, *Drought Follows the Plow: Cultivating Marginal Areas,* Cambridge University Press, Cambridge.

Herrmann, M., and Hutchinson, C. F., 2006, Desert outlook and options for action, in: *Global Deserts Outlook*, E. Ezcurra, ed., UNEP, Nairobi, pp. 111–139.

Information Unit on Climate Change (IUCC), 1993, Desertification and climate change, UNEP, Châtelaine.

Intergovernmental Negotiating Committee for a Convention to Combat Desertification (INCCCD), 1994, *The Almeria Statement on Desertification and Migration, 11 February, 1994.* Châtelaine.

Intergovernmental Panel on Climate Change (IPCC), 2000, *IPCC Special Report on Emission Scenarios*, Intergovernmental Panel on Climate Change, Geneva.

Intergovernmental Panel on Climate Change (IPCC), 2001, *Climate Change: Impacts, Adaptation and Vulnerability*, World Meteorological Organization, Geneva.

Liotta, P. H., 2003, *The Uncertain Certainty: Human Security, Environmental Change and the Future Euro-Mediterranean*, Lexington Books, Lanham, MD, p. 157.

Lioubimtseva, E., and Adams, J. M., 2004, Possible implications of increased carbon dioxide levels and climate change for desert ecosystems, *Environ. Manage.* **33**:S388–S404.

Marten, G., 2005, Environmental tipping points: A new paradigm for restoring ecological security, *J. Policy Studies* (Japan) **20**:75–87.

Middleton, N., and Thomas, D., eds., 1997, *World Atlas of Desertification*, 2nd. ed., Arnold, London, p. 182.

Millennium Ecosystem Assessment (MA), 2005a, Chapter 22: Dryland systems, in: *Ecosystems and Human Well-Being: Current Status and Trends,* World Resources Institute, Washington, pp. 623–662.

Millennium Ecosystem Assessment (MA), 2005b, *Ecosystems and Human Well-Being: Desertification Synthesis*, World Resources Institute, Washington, p. 25.

Mouat, D., and Lancaster, J., 1996, Use of remote sensing and GIS to identify vegetation change in the Upper San Pedro River watershed, Arizona, *Geocarto Int.* **11**(2):55–67.

Navone, S., and Abraham, E., 2006, State and trends of the world's deserts, in: *Global Deserts Outlook*, E. Ezcurra, ed., UNEP, Nairobi, pp. 73–88.

Reynolds, J. F., and Stafford Smith, D. M., 2002, Do humans cause deserts? in: *Global Desertification: Do Humans Cause Deserts?* J. F. Reynolds and D. M. Stafford Smith, eds., Dahlem University Press, Berlin, pp. 1–21.

Schlesinger, W. H., Reynolds, J. H., Cunningham, G. L., Huenneke, L. F., Jarrell, W. M., Virginia, R. A., and Whitford, W. G., 1990, Biological feedbacks in global desertification, *Science* **247**:1043–1048.

Shearer, A. W., Mouat, D. A., Bassett, S. D., Binford, M. W., Johnson, C. W., and Saarinen, J. A., 2006, Examining development-related uncertainties for environmental management: Strategic planning scenarios in Southern California, *Landscape Urban Plan* **77**:359–381.

Seely, M., 2006, People and deserts, in: *Global Deserts Outlook,* E. Ezcurra, ed., UNEP, Nairobi, pp. 27–45.

Steinitz, C., 1990, A framework for the theory applicable to the education of landscape architects (and other design professionals), *Landscape J.* **9**(2):136–143.

Steinitz, C., 1993, A framework for theory and practice in landscape planning, *GIS Europe* **2**(6):42–45.

Steinitz, C., Arias, H., Bassett, S., Flaxman, M., Goode, T., Maddock, T., III, Mouat, D., Peiser, R., and Shearer, A., 2003, *Alternative Futures for Changing Landscapes: The Upper San Pedro River Basin in Arizona and Sonora*, Island, Washington, DC, p. 202.

Tickell, Sir C., 2003, Risks of conflict: Population and resource pressures, in: *Security and Environment in the Mediterranean—Conceptualizing Security and Environmental Conflicts*, H. G. Brauch, P. H. Liotta, A. Marquina, P. F. Rogers, and M. El-S. Selim, eds., Springer, Berlin, pp. 13–18.

United Nations Conference on Desertification (UNCOD), 1978, *Round-up, Plan of Action and Resolutions*. New York: United Nations.

United Nations Conference on Environment and Development (UNCED), 1992, *Earth Summit Agenda 21: Programme of Action for Sustainable Development*, United Nations Department of Public Information, New York.

United Nations Convention to Combat Desertification (UNCCD); www.unccd.int, accessed May 2007.

Williams, M. A. J., 2001, Interactions of desertification and climate: Present understanding and future research imperatives, *Arid Land Newslerr.* **49**.

Williams, M. A. J., and Balling, R. C., Jr., 1996, *Interactions of Desertification and Climate,* Arnold, London.

Zhou, L., and Mouat, D. 2007. "Whose decisions impact land use change: The people or the government? Thoughts from Northwest China," forthcoming.

DESERTIFICATION IN JORDAN

A Security Issue

MU'TAZ AL-ALAWI*

Mu'tah University, Jordan

Abstract: The existing ecosystems in Jordan are fragile and prone to deterioration. The leading causes of land degradation in Jordan are improper farming practices (such as failure to use contour plowing or overcultivation of the land), overgrazing, the conversion of rangelands to croplands in the marginal areas, where rainfall is not enough to support crops in the long term, and uncontrolled expansion of urban and rural settlement at the cost of cultivable land. Implementation of soil conservation and erosion control measures such as contour plowing, terracing, and stonewall construction on farmer's fields helped in curbing accelerated erosion and in protecting the potential agricultural lands.

Keywords: Arid climate; desertification; Jordan; land degradation; soil erosion

1. Introduction

Desertification is considered one of the most serious environmental problems on a global scale. It affects the drylands of the five continents and has strong relations with climatic change and loss of biodiversity.

Desertification has become the most commonly used term for an insidious environmental problem producing many disasters affecting human lives and civilization all over the world. In 1977, a United Nations Conference on Desertification (UNCOD) was convened in Nairobi, Kenya, to produce an effective, comprehensive, and coordinated program for addressing the problem of land degradation (UNEP, 1992). The UNCOD recommended the United Nations Plan of Action to Combat Desertification (UNPACD). However, the implementation of the Plan was severely hampered by limited

---

*Address correspondence to Mu'taz Al-Alawi, Mu'taz University, P.O. Box 3, Karak 61710, Jordan; e-mail:alawi1979@yahoo.com

P.H. Liotta et al. (eds.), Environmental Change and Human Security, 81–102.
© 2008 *Springer Science + Business Media B.V.*

resources. The United Nations Environment Programme (UNEP) concluded in 1991 that the problem of land degradation in arid, semiarid, and dry subhumid areas was intensified, although there were "local examples of success." In 1992, UNEP produced a World Atlas of Desertification (UNEP, 1992). Many studies over the preceding 20 years indicated that the problem of desertification continued to worsen. The studies have also indicated that irrational cultivation, overgrazing, deforestation, and poor irrigation practices are the main factors for land degradation process.

The question of how to tackle desertification was a major concern for the United Nations Conference on Environment and Development (UNCED), known as "the Earth Summit," which was held in Rio de Janeiro in 1992. The Conference adopted a new, integrated approach to the problem, emphasizing action to promote sustainable development at the community level. It also called on the United Nations General Assembly to establish an Intergovernmental Negotiating Committee (INCD) to prepare a Convention to Combat Desertification for countries experiencing serious drought or desertification. In December 1992, the General Assembly of the United Nations agreed and adopted resolution 47/188. The Committee was established in early 1993 and held five preparatory sessions before adopting the Convention on 17 June 1994 in Paris. The Convention was then opened for signature on 14–15 October 1994. The Convention entered into force on 26 December 1996, 90 days after it had been ratified by 50 countries. As of August 2005, 191 countries have acceded/ratified the Convention. The Hashemite Kingdom of Jordan was among those countries, ratifying the Convention on 21 October 1996.

2. Desertification

The concept of desertification emerged during colonial rule in West Africa out of concerns about signs of desiccation and the creeping of the Sahara desert into the Sahel (Herrmann and Hutchinson, 2005). The term "desertification" was first used to describe the change of productive land into desert as a result of human activity in the tropical forest zone of Africa. Renewed attention was drawn to the desertification concept when a series of drought years began in the late 1960s that contributed to famine conditions in several African countries in the Sahel and was exacerbated by political instability and unrest.

As many of the definitions of desertification have been imprecise, many statistics and maps are "guesstimates" and lack reasonable accuracy. However, with regard to the land degradation process, desertification is not limited to the African continent. Rough estimates suggest that one-third of the world's total land surface is threatened by desertification (Barrow, 1994).

2.1. GLOBAL EXTENT OF DESERTIFICATION

Spatial distribution of desertification was identified and documented in several maps (e.g. Dregne, 1983; UNEP, 1997; Kharin et al., 1999). The global extent of the problem of desertification was expanded in the revised second edition of the World Atlas of Desertification (UNEP, 1997). The atlas included related environmental issues, including concerns about surrounding biodiversity, climate change, and the availability of water. Maps in this atlas show land that is lost, or is in the process of being lost. About 130 million hectares can no longer be used for food production, which is about the land area of France, Italy, and Spain combined.

Social and economic conditions (poverty and food security) also have a major impact on the process and control of desertification. Over one billion people are at risk as they face malnutrition or worse through decreasing productivity of the soil on which they depend for food. The atlas also included estimates of population in the areas at risk and some facts on the impacts of desertification on migration and refugees. The United Nation Convention to Combat Desertification (UNCCD) confirmed these impacts and indicated that desertification is a worldwide problem affecting directly 250 million people and a third of the earth's land surface (UNCCD, 2005).

Global, continental, and national maps of desertification or the desertification process are useful to inform people about the general status of desertification and to call attention to the presence or absence of the problem. These maps, however, are indicative only and their accuracy is relatively low, as their scale is very small and their mapped area is very large. Setting the priorities and planning to combat desertification require more detailed maps covering medium to very large scales. The former scale (1:50.000) enables prioritization at the level of region while the latter scale (>1:1.000) provides a planning tool at the level of local community and village. Many affected countries, including Jordan, still lack these maps.

2.2. DEFINITION

Despite the different debates in definition and causes of desertification, the term "desertification" was used to indicate an irreversible land degradation process that is mainly accelerated by human factors. The most widely accepted definition of desertification comes from the UNEP (1992) as:

Desertification, as defined in Chapter 12 of "Agenda 21" and in the International Convention on Desertification, is "the degradation of land in arid, semi-arid and sub humid dry areas caused by climatic changes and human activities." It is accompanied by the reduction in the natural potential of the land, the depletion of surface and ground-water resources and has negative repercussions on the living conditions and the economic development of the people affected by it. On the other hand, "land degradation" means reduction or loss, in arid, semi-arid and dry sub-humid areas, of the biological or economic productivity and complexity of rain-fed cropland, irrigated cropland, or range, pasture, forest and woodlands resulting from land uses or from a process or combination of processes, including processes arising from human activities and habitation patterns. (UNCCD, 1992)

This definition is also used as the basis of the UNCCD. As included in the definition, the problem of desertification mainly occurs in dry and arid areas. Several definitions of aridity have been presented. The aridity index is defined as the ratio of P/PE, where P is the mean annual precipitation and PE is the mean annual potential evapotranspiration. Accordingly, the degree of aridity can be classified as in Table 1.

It is important to understand that desertification, though it produces conditions similar to deserts, is not occurring in hyperarid deserts. In fact, the "growth" of deserts along their edges is a natural condition, often followed by periods of desert shrinkage. The size of all deserts will vary from year to year based on short-term weather and precipitation patterns. This process of desert area oscillation is not the basis of the more damaging sequence of desertification. As the United States Geological Survey (Walker, 1998) explains, "the presence of a nearby desert has no direct relationship to desertification."

In conclusion, desertification is a deterioration process in which ecosystems lose the ability to survive, leading to the deterioration of land and its production and leading to the vanishing of the land's economic output.

2.3. INDICATORS

An identification indicator of desertification is the first step prior to implementation of monitoring schemes and identification of possible changes in status and rate. These indicators are dependent on the target or component of environment of interest, scale of monitoring, and the ecosystem involved.

TABLE 1. Aridity Index.

Climate zone	P/PE ratio
1. Hyperarid (deserts)	<0.05
2. Arid	0.05. 0.20
3. Semiarid	0.21. 0.50
4. Dry subhumid	0.51. 0.65
5. Humid	>0.65

Some indicators, such as extent and distribution of wind erosion, might be useful at a regional level, but have little value at the level of a village. On the other hand, the extent of gullies and their density is time and cost consuming at the country level. Selection and aggregation of indicators require the following:

1. Representation of the target and its performance

2. Sensitivity to change and

3. Capability of measurement, analysis, interpretation, and comparison

Data collection may involve a large number of indicators that should be aggregated into fewer indices. Hammond et al. (1995) called for aggregation of much primary data into 20 indicators that could be combined to produce a few aggregated ones. Regardless of the desertification process, the indicators can be divided into three main groups and listed as:

1. Physical indicators:

– Decrease in soil depth

– Decrease in soil organic matter and fertility

– Soil crusting and compaction

– Appearance/increase in frequency/severity of dust and sand dunes

– Salinization/alkalinization

– Decline in ground water quality and quantity

– Increased seasonality of springs and small streams

– Alteration in relative reflectance of land (Albedo change)

2. Biological indicators:

Vegetation

– Decrease in vegetation cover and aboveground biomass

– Decrease in yield

– Alteration of key species (biodiversity)

– Deterioration and reduction of seed bank.

Animal

– Alteration in key species distribution and frequency

– Change in population of domestic animals

– Change in herd composition

– Change in livestock yield

3. Social/economic indicators:

– Change in land use/water use
– Changes in settlement patterns and increase of abandonment
– Change in population and demographic structure

3. Biophysical Characterization of Jordan

Jordan is located about 80 km east of the Mediterranean Sea between 29°
11' to 33° 22' north, and 34° 19' to 39° 18' east. The area of land mass is
approximately 88,778 km² (DOS, 2003), while the area of bodies of water
is approximately 482 km², including the Dead Sea and the Gulf of Aqaba.
Altitude ranges from less than −400 m (below mean sea level) at the surface
of the Dead Sea up to the 1,750 m of Jabel Rum. The climate varies from dry
subhumid Mediterranean in the northwest of the country with rainfall of
about 630 mm to desert conditions with less than 50 mm over a distance
of only 100 km.

3.1. CLIMATE

More than 80 percent of the country's area is arid and receives less than 200 mm
annual rainfall, with precipitation patterns being latitude, longitude, and
altitude dependent. Rainfall decreases from north to south, west to east, and
from higher to lower altitudes. Where the ground rises to form the highlands
east of the Jordan Valley, precipitation increases from less than 300 mm in the
south to more than 500 mm in the north. The Jordan Valley forms a narrow
climatic zone that annually receives up to 300 mm of rain in the north and less
than 120 mm at the northern edge of the Dead Sea, the lowest point on earth.

Farther inland from the western highlands forms a considerable part of
the country that is known as the "Badia." The name Badia is an Arabic word
describing the land where Bedouins live and practice seasonal browsing.
The area includes all lands receiving annual rainfall of 50–200 mm and has
general characteristics of seasonal contrasts in temperature with high variations in rainfall within and among years (Dutton et al., 1998).

The major characteristic of the country's climate is the contrast between
hot, dry, uniform summers and cool variable winters. The rainy season is
between October and May with 80 percent of the annual rainfall occurring
between December and March (JMD, 2005).

The country has a long summer, extending from mid-May to the end of
September and reaching a peak during August, with daytime temperatures
frequently exceeding 36°C and averaging more than 32°C.

4. Matrix of Critical Causes Leading to Desertification in Jordan

Since desertification is a long-term process—long in terms of its development and impact—it is difficult to pinpoint cause–effect relationships. It
is also difficult for the general public to understand it, other than as an
existing fact of life. The public usually perceives only the consequences
of desertification, such as famine and dying herds and people. Because of
its long-term nature and association with other more obvious or pressing
problems, decision makers have great difficulty addressing it as a problem.
A matrix of critical issues leading to desertification in Jordan (Appendix)
has been developed.

5. Desertification in Jordan

5.1. OVERVIEW

Most of Jordan's arid and semiarid areas have suffered desertification.
Although the rate of desertification was not identified, several surveys and
studies at the country level indicated that Jordan's land is at the threat of
high rate of desertification. The process has been accelerated by unsupervised management and land use practices of overgrazing, cultivation, and
plowing of marginal soils and wood removal in the high rainfall zones.
The regions of irrigated highlands and the Jordan Valley were also affected
by aspects of salinization and alkalinization of soil. In addition to human-
induced factors, climatic factors of unpredictable rainfall and periodic
droughts are contributing to the problem. According to Al-Hadidi (1996),
the transition zone (between arid areas in the east and subhumid areas in the
west) has suffered from a high risk of desertification and is expected to lose
its productivity over time.

The arid and semiarid lands in Jordan are sensitive to human interference
that resulted in a severe depletion of its natural resources and in different
forms of land degradation due to multiple interactions of socioeconomic
factors. Further, degradation will continue if human activities are not carefully managed. Almost 90 percent of the land area of Jordan receives less
than 200 mm of rainfall annually. This is reflected in poor structural stability
of soils and the subsequent vulnerability to excessive erosion following shal-
low rainstorm events. Such a fragile ecosystem has also been manifested by
non-sustainable land use patterns and poor vegetative cover of the rangeland
and forests. Therefore, most of the economic activities take place on the
remaining 10 percent of the land area, and the competition between different
user groups for these lands is therefore intense.

5.2. CURRENT STATUS AND FORMS OF DESERTIFICATION IN JORDAN

Although modern observation facilities using satellite imagery and comput-
ers to analyze data are widely used at present, there are still many uncertainties at the global, regional, and national levels on the causes, extent, and
seriousness of desertification. These uncertainties make it difficult to plan
properly. Available information indicates that most of the land resources in
Jordan are either desertified or vulnerable to desertification (Table 2), thus
affecting food security and development in the region.

5.2.1. The Main Forms of Desertification Prevailing in Jordan Are

1. Wind erosion: This is considered the most common environmental problem in Jordan leading to
- Loss by removal of fertile top soils
- Encroachment and accumulation of sands on productive rangelands and
  agricultural land, urban areas, and civil construction (infrastructure)

The area affected by wind erosion in Jordan was estimated at 3,237,000 ha
(Table 3). This area is increasing gradually because of cultivation of marginal
lands and movement of sand dunes toward agricultural and rangelands.

2. Water erosion: This process leads to
- Loss by removal of fertile top soils
- Accumulation of eroded material behind dams (siltation) in irrigation
  networks and on productive rangelands and agricultural lands

The area affected by water erosion was about 332,000 ha in 1992 (Table 3).

3. Salinization and water-logging: This process leads to
- Accumulation of salts in irrigated arable lands
- Increased annual loss of large areas of valuable croplands due to rising
  water table

TABLE 2. Areas Desertified and Vulnerable to Desertification in Jordan.

Country area (km²)	Desertified area		Areas vulnerable to desertification	
	(km²)	(%)	(km²)	(%)
89,206	71,000	79.59	10,000	11.21

Source: Shakhatra (1987). The figures in this table are in line with the estimates of
ACSAD in 1996 for some Arab countries and reported by Abdelgawad (1997).

TABLE 3. Land Degradation Types and Causes in Jordan.

Forms of desertification	Area (1,000 ha)
1. Chemical degradation	367
2. Physical degradation	–
3. Water erosion	332
4. Wind erosion	3,237
5. Other inundated lands	287

Source: FAO (1992).

– Increased water consumption for leaching excess salts putting additional pressure on the limited water resources

4. Loss of Nutrients: Decline of soil fertility and productivity due to present policies of agricultural intensification and poor management practices

5. Soil and water pollution

– Increased soil and water pollution with agricultural chemicals caused by inappropriate soil management and extensive use of fertilizers and pesticides in irrigated areas

– Accumulation of pollutants on the soil surface layer of agricultural areas close to industrial complexes

– Accumulation of polluting materials in both land and water resources due to hostilities (wars) in the region during the last 5 decades

5.3. CAUSES OF DESERTIFICATION

The complexity of the causes of desertification and the diversity of its effects make it difficult to evaluate accurately its magnitude. Estimates of the areas lost to or threatened by desertification are a matter of controversy because of the complexity and variety of the forms of desertification, and also due to the different notions of irreversibility in terms of the timescales considered. The following are the major causes for desertification in Jordan.

5.3.1. Population Growth and Urbanization

The population in Jordan is characterized by a high birthrate. The Statistical Year Book of Jordan (DOS, 2002) showed that the population of Jordan in 1994 was 4.139 million. The figure increased to 5.35 million in 2004. The crude birth rate for the years 1994 and 2001 were 32 and 28 per thousand, respectively. The rate of natural population increase dropped in the same period from 2.7 to 2.3 percent. The same period witnessed a similar drop in the population growth rate from 3.3 to 2.8 percent, but obviously the percentages of actual growth rate were higher than the corresponding figures of natural growth rate due to influx of refugees (major waves in 1948 and 1967), as well as Jordanian expatriates during the 1991 Gulf War, and non-Jordanian cheap labor.

The increase in the population in Jordan from about 1,237,000 in 1950 to about 6,330,000 in 2000 (Figure 1) has put considerable pressure on limited land and water resources of the region. This, along with changing lifestyles and consumption patterns and increasing food demand combined with aridity of the environment have hastened the rate of land degradation.

Population growth and other demographic changes have led to losses of land to urbanization and the diminishing of per capita share of cultivated land in most of the countries of the Arab Region. Land misuse and degradation are widespread and are proceeding at accelerated rates prompted by the ever-increasing demand for food. The failure of resource management policies is aggravated by overgrazing, overexploitation, and overcultivation of marginal lands; deforestation; and improper technologies employed.

The population of Jordan is highly urban. In 1951, only 35 percent of Jordan's population lived in urban areas (Figure 2). By 2000, the figure had reached 74 percent, which is largely a result of internal rural-to-urban migration, combined with the influx of refugees and migrants, mainly from Palestine and Iraq. The urban population within Amman, Irbid, and Zarqa governorates now account for 3.378 million people, equaling 63 percent of the total population of Jordan. Urbanization has increased since 2000, and it

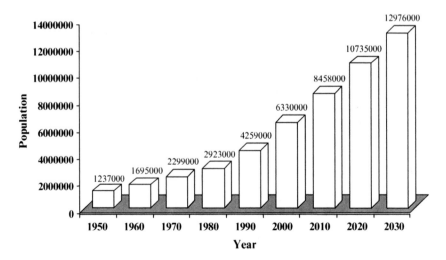

Figure 1. Total Population in Jordan.
Source: UN (1997) Population Division.

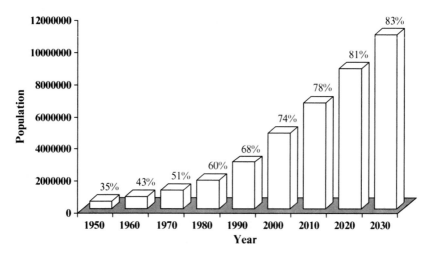

Figure 2. Urbanization in Jordan and Percent Urbanization in Jordan.
Source: UN (1997) Population Division.

is predicted to increase to 78 percent, 81 percent, and 83 percent in the years 2010, 2020 and 2030, respectively.

Urban areas consume natural resources from both near and distant sources. Cities encroach onto agricultural lands, where the urban fringes and peripheries grow faster than the cities, and spontaneous or squatter settlements even more quickly. This uncontrolled growth leads to rapidly increasing amounts of wastes (solid and liquid), causing pollution of land and water resources and aggravating the desertification problem. Intensification of agricultural production to meet urban demands for food leads to heavy, concentrated use of fertilizers and pesticides, causing pollution of land and water resources. This, along with the fast-growing industrialization near urban centers, adds to the active desertification process (UNEP, 1999).

5.3.2. *Water Demand and Desertification*

Desertification directly reduces freshwater reserves. It has a direct impact on river flow rates and the level of groundwater tables. The reduction of river flow rates and the lowering of groundwater levels leads to the silting up of estuaries, the encroachment of salt water into aquifers, and the pollution of water by suspended particles and salinization, which in turn reduces the biodiversity of fresh and brackish water and fishing catches, interferes with the operation of reservoirs and irrigation channels, and increases coastal erosion, adversely affecting human and animal health. Lastly, desertification leads to an accelerated and often unbridled exploitation of underground fossil water reserves and their gradual depletion (Koohafkan, 1996).

The dominant environmental challenge facing Jordan is the scarcity of the Kingdom's water resources in an arid land with unpredictable rainfall and an expanding population. Rainfall is confined largely to the winter season and ranges from around 660 mm in the northwest of the country to less than 130 mm in the extreme east. Major surface water sources are the Yarmouk and Zarqa rivers, and the associated side wadi, all flowing westward into the River Jordan and the Dead Sea. While high evaporation rates result in relatively low annual stream flows, the high infiltration rates common in Jordan result in high rates of groundwater recharge.

Jordan is a semiarid country with very limited freshwater resources. The availability of water is classified as very low on the Water Stress Index, which indicates the degree of water shortage. Water Stress Index is the value of annual rainfall divided by the total population (m^3/capita/year). Countries with less than 1,700 m^3/capita/year are considered to have "existing stress," while countries with less than 1,000 m^3/capita/year have "scarcity," and countries with less than 500 m^3/capita/year are regarded as having "absolute scarcity" according to this index. With 167 m^3/capita/year, Jordan falls into the category of "absolute scarcity." The scarcity of water in Jordan is the single most important constraint to the country's growth and development because water is not only a factor for food production but a very crucial factor of health, survival, and social and economical development. As a result of scarcity, the demands and uses of water far exceed renewable supply. The deficit is made up by the unsustainable use of groundwater through overdrawing of highland aquifers, resulting in lowered water tables in many basins and declining water quality in others. In addition, the deficit is overcome by supply rationing to the domestic and the agricultural sectors. In 2002 the total use of water in Jordan was 809.8 million cubic meters (MCM) or 159 m^3/capita/year with the total population at 5.1 million people. This usage included 88.8 MCM of nonrenewable groundwater (groundwater mining) and 72.4 MCM of treated wastewater.

Table 4 shows the most recent statistical data on water use in Jordan. Water uses vary from year to year depending upon the available surface water supply, which is decreasing due to upstream uses and climatic fluctuations.

TABLE 4. Sources of Sectoral Water Use in Jordan in 2002 (In Million Cubic Meters per Year).

Sector	Surface water	Groundwater	Treated wastewater	Total use
Municipal	50.540	198.685	–	249.225
Industrial	1.863	34.971	–	36.384
Irrigation	156.950	287.556	72.365	516.871
Livestock	6.000	0.835	0.000	6.835
Total	215.353	522.047	72.365	809.765

According to available water supplies, a total of 517 MCM in 2003 were used in agriculture (representing about 63.8 percent of the total water use); the domestic sector consumed 249 MCM (30.7 percent); industry share was only 36 MCM (4.5 percent).

The Kingdom of Jordan is facing an unremitting imbalance between the total sectoral water demands and the available supply of freshwater. By 2020, the total demand for water is expected to increase to 1,685 MCM because of large increases in population, improvements in living standards, and growth in economic activity. While new sources of water supply are expected to increase the available water from the current level of 850 MCM per year to 1,289 MCM per year by 2020, a shortfall of 396 MCM representing 24 percent of total demand will remain and will have to be managed through appropriate demand-reduction programs.

If rapid population growth and fast urbanization continue, the per capita availability of water is likely to be reduced in Jordan. Water shortages will certainly speed up the rate of desertification. Unless appropriate alternative water policies are empowered and corrective measures taken soon, the impacts of desertification will become catastrophic in Jordan.

5.3.3. Food Security and Desertification

By impoverishing the natural potential of ecosystems, desertification also reduces agricultural yields making them more unpredictable. Therefore, it affects the food security of the people living in these areas. The people develop a survival strategy to meet their most urgent requirements, and this, in turn, helps to aggravate desertification and hold up development. The most immediate and frequent consequence of these survival attitudes is the increased overexploitation of accessible natural resources. Lastly, desertification heightens considerably the effects of climatic crises (droughts) and political crises (wars), generally leading to migration, causing suffering and even death to hundreds of thousands of people worldwide. These consequences, in turn, weaken the economies of the countries affected by desertification, particularly when they have no other resources than their agriculture.

5.3.4. Irrational Cultivation and Irrigation

Extension of rain-fed cultivation to the low rainfall zones to meet the increasing demand for food resulted in accelerated land degradation in Jordan, as many of the barley-cultivated areas had low suitability to land utilization type. Irrigated cultivation, on the other hand, had little significance until the beginning of the 1960s. The irrigated areas were scattered around the water sources and along the streams. The expansion of irrigated farming in the eastern parts of the country, where numerous irrigation projects have been initiated, resulted in more land under irrigation and the risk of salinization. Originally, these projects aimed to assist in settling the Bedu and reduce the problem of food security. The increased investment in irrigation projects by the private sector has resulted in intensive irrigation in the low rainfall zones. This has increased stress on water resources, especially in the Badia (arid eastern parts), and resulted in overexploitation of groundwater.

5.3.5. Pollution

Thousands of tons of poisonous chemicals (pesticides) were believed to be used in agriculture in Jordan since the early 1960s. The most prominent pesticides are DDT (used until the 1960s), thallium (used largely until the 1970s), floroacetamids, azodrin, and other chemicals (Disi and Oran, 1995; Hatough-Bouran and Disi, 1995). The unsupervised use of pesticides and chemical fertilizers is certainly expected to result in polluting soils, water resources, aquatic fauna, vegetation, birds and mammals. Eventually, negative impacts on human health and environment are expected.

5.3.6. Soil Salinization

Salinity of soils in Jordan results from salts that originally exist in the soils, salts that are added to the soil because of improper irrigation practices, and those that result from evaporation of the internal close-to-surface water. The geological salt-rich residues and extensive land use nowadays play an important role in salinization. Results of the available studies indicate that soil salinity in the Jordan Valley varies between slight and medium. The area of saline lands was about 420 ha in the northern Ghors, about 800 ha in the middle Ghors, and about 125 ha in the southern Ghors. Gypsum in the Jordan Valley is concentrated in the area between the west borders of South Shounah and the borders of the Dead Sea. Salinity in the marginal lands and the Badia, on the other hand, increases with decreasing rainfall. Soil studies that were conducted by the Ministry of Agriculture in 1990 indicated that salts close to the soil surface were widespread in the Badia.

By 2025, when the population is projected to reach about 10 million, and the percentage of the population with sewerage services will have increased from the current 55 percent to over 65 percent, about 280 MCM per annum of wastewater is expected to be generated. The average salinity of municipal water supply is 580 parts per million (ppm) (range 430–730 ppm of Total Dissolved Solids [TDS]), and average domestic water consumption is low (some $50 m^3$ per capita per year countrywide), resulting in a higher than normal salinity in the wastewater. The primary wastewater treatment technology in Jordan is stabilization ponds, resulting in considerable losses by evaporation and, as a result, increased salinity. The combination of these factors has resulted in a salinity average of 1.180 ppm, but as high as 1.800 ppm of TDS in the effluent of Khirbit as-Samra, affecting crops and irrigated areas reusing the effluent. The above factors have to be effectively addressed to deal with the constraints on the reuse of appropriately treated wastewater. The government's new plans are to convert the stabilization ponds into mechanical treatment plants and to expand other plants.

5.3.7. Soil Erosion in the Highland Zone

Long-term records of the Highland weather stations repeatedly show rainstorm events of more than 100 mm, most likely from high rainfall intensity events. Such intense rainstorms usually generate extensive soil erosion, especially on steep and long slopes. Moreover, man induced destruction of natural forests by deforestation and expansion of farmland (land cultivation) and grazing into these forests over the past centuries.

5.3.8. Overgrazing of the Rangeland in the Steppe Zone by Sheep and Goats

All areas of Jordan named the Badia, which receive less than 200 mm of rainfall per annum, are officially designated as rangeland (pastureland). Productivity of rangelands varies from one region to another. Chronologically, interest in rangeland assessment, rehabilitation, and development in Jordan arose as early as 1950s (HTS, 1956). Many studies and research showed low levels of rangeland productivity that tended to decrease with time. This was mainly attributed to overgrazing of natural vegetation, which accelerated degradation of rangelands in the low rainfall zones. At the same time, the number of grazing animals is constantly growing and results in more pressure on the limited resources of rangelands. Prolonged heavy grazing has changed rangeland quantitatively and qualitatively. Quantitatively, it results in fewer and smaller plants and low vegetative cover. Qualitatively, it results in a decrease in the most palatable and nutritious plants relative to unpalatable plants and those lacking nutrients.

According to Abu-Irmaileh (1994), productivity of the grazed semiarid areas ranged from 11 percent to 33 percent of the amount of vegetation produced by adjacent protected areas. Hatough et al. (1986) found that grazing reduced productivity, cover, and diversity of shrubs while protection resulted in a "highly productive growth of many palatable plants such as *Erucaria boviana* and species of *Avena, Lolium, Phalarris, Bromus, Stips, Salsola, Atriplex, Erodium* and others."

Deteriorating rangeland quality impacted wildlife in desert habitats. The damage occurred from

- Direct competition between sheep and goats with other herbivores (such as gazelles, seed-eating birds, and rodents)
- Loss of food sources for herbivores
- Loss of food sources for predators (the herbivores whose populations declined)
- Transmittal of diseases from domestic animals to wildlife
- Loss of structural components of habitat (bushes and trees)

These structural and biological changes were supported by the large number of bird and plant species inside the natural reserves compared with the outside open rangeland.

5.3.9. Wind and Water Erosion in the Desert Region

Such a process is age-old. Desert soil has a phenomenally poor surface structural stability that manifests in terms of surface crust formation and, subsequently, very low, steady infiltration rate that usually does not exceed 5.0 mm h^{-1} (Abu-Sharar 1993; Abu-Sharar and Salameh, 1995). The fragile soil structure is subject to extensive wind and water erosion, especially when rainfall intensity exceeds infiltration rate. In addition, dramatic changes brought about by irrigated agriculture and subsequent desertification trends on the improved lands remain of concern.

5.3.10. Off-Road Driving

Trampling of vegetation in transitional and fragile ecosystems by off-road vehicles is leading to destruction of vegetation cover and soil erosion.

5.3.11. Deterioration of Forests

Reports indicate that woodlands covered vast areas of Jordan in the past. Most of this plant cover was removed during the third and second centuries B.C. mainly for land conversion to agriculture and by urban encroachment. In recent times (until the 1950s), Jordan was covered with large woodland areas. The area of woodlands in Jordan 100 years ago is estimated at twice today's area (Tillawi, 1995). The area has decreased to half a million dunums (1 ha = 10 dunums) in the 1940s and 400,000 during the 1950s (Figure 3).

The forests in Jordan have been subjected to severe pressure from grazing and wood collection, as well as from grain cultivation in recent years. Overgrazing by goats has been prevalent for centuries. Goats not only eat the leaves and shoots of trees, but also destroy seedlings and ground vegetation, thus preventing regeneration (Atkinson and Beaumont, 1971; Beaumont, 1985). Woodlands of mid-mountainous Jordan were reduced considerably and even parts of the woodlands in the capital were totally wiped out by increased urbanization.

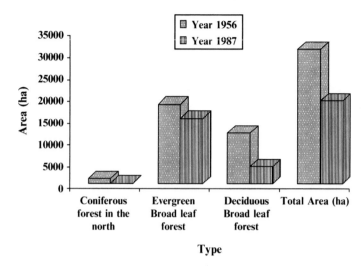

Figure 3. Decline in Northern Woodland Area Between 1956 and 1987.

5.3.12. Lack of Environmentally Friendly National Land Use Management and Policy

Jordan has a vast heritage, with rich environmental and cultural assets. These assets, however, have been under-protected in the past, due to the lack of a clear conservation policy. These assets have also been mismanaged economically, without a policy for using them for the economic benefit of local communities while preserving the value of that heritage.

For example, Jordan has only limited forest resources, with less than 1% of the country classified as forest. These forests face various threats, including woodcutting, overgrazing, quarrying, and the expansion of agriculture. These sites have also been witness to an onslaught of development that must be regulated in order to conserve them for generations to come.

The past several years have seen an increase in construction, development activity, and tourism in many of Jordan's environmentally and culturally sensitive areas. No comprehensive land-use policy has yet been developed for these "special places," resulting in their sub-optimal use. In accordance with the Kingdom's economic development agenda, it is now imperative to develop a comprehensive land use plan that will balance the protection and conservation of these areas with the promotion of their economic potential by ensuring the inoffensive and productive use of the sites.

6. Efforts to Combat Desertification in Jordan

Field observations indicate a gradual increase in land degradation in Jordan. Major reasons for such a phenomenon are argued to be encroaching urbanization into traditional agricultural land, persistent drought, water shortage, deforestation, losses in land productivity, decreasing agricultural feasibility of traditional crops, decreasing availability of cheap labor, abandoning agriculture to other professions, discouraging environment of investment in the agricultural sector, increasing prices of agricultural inputs, deficiencies in "know-how" of soil-water management in arid areas, and introduction of new competitive commercial crops in irrigated agriculture (Abu-Shriha, 2003).

6.1. PRACTICES EMPLOYED IN FIGHTING DESERTIFICATION IN JORDAN

1. Strips: alternating strips of cropped and fallow area, where the fallow strips act as miniature rainwater catchments. The ratio of cropped to catchment area varies from 1:1 to 1:3 depending on slope, soil type, and rainfall. This method can harvest enough rainwater to double crop yields.

2. Contour ridges: parallel stone ridges are built 5–10 m apart to stop runoff water (and the soil it carries) from damaging downstream areas. Each ridge collects runoff water from the area immediately upstream/uphill, and the water is channeled to a small plantation of fodder shrubs. With a combination of well-designed ridges and drought tolerant shrubs project communities were able to meet a large proportion of their fodder requirements.

3. Control of rills and gullies: a combination of vegetative cover (to slow runoff) and physical structures (to stop and divert runoff water) is a cheap, effective way to prevent rills and gullies from expanding.

4. Water reservoirs: small ponds are easy to build, even on slopes, by selecting a suitable depression and sealing off the lower end with a masonry wall. On a slightly larger scale, a low-cost earthen dam can meet most of a community's needs for domestic purposes, supplemental irrigation, or livestock. Reservoir size can be matched to runoff conditions (total amount, flow rates) and the labor and material available for construction.

7. Jordan's Strategy to Combat Desertification

Jordan has long prioritized its most pressing problems as being scarce water resources and land degradation. Accordingly, all relevant institutions address these issues, especially the Ministry of Water and Irrigation. Jordan is one of the thirty original supporters of the World Conservation Strategy. In October 1996, Jordan ratified the Convention to Combat Desertification. The National Strategy for Agricultural Development (ad hoc Committee of the Irrigated Agriculture, 2001) for the decade 2000–2010 stressed sustainable agriculture and protection of natural and biological resources. Finally, with the advent of the twenty-first century, Jordan prepared its National Agenda 21. The document outlines several key areas related directly to natural resources and dryland issues and promotes the participatory approach at all levels to ensure success and sustainability. The Agenda also reflects the integrated approach to environment and development and converges with objectives of poverty alleviation and sustainable human development.

References

Abdelgawad, G., 1997, Deterioration of soil and desertification in the Arab countries, in: *Agriculture and Water*, Arab Center for Studies of Arid Zones and Drylands (ACSAD) (in Arabic), Damascus, Syria, Vol. 17, pp. 28–55.

Abu-Irmaileh, B. E., 1994, Al-Mowaqqar, a model for arid rangelands in Jordan, *J. Arid Environ.* **28**:155–162.

Abu-Sharar, T. M., 1993, Effects of sewage sludge treatments on aggregate slaking, clay dispersion and hydraulic conductivity of a semiarid soil sample, *Geoderma* **59**:327–343.

Abu-Sharar, T. M., and Salameh, A. S., 1995, Reductions in hydraulic conductivity and infiltration rate in relation to aggregate stability and irrigation water turbidity, *Agric. Water Manage.* **29**:53–62.

Abu-Shriha, N. I., 2003, Assessment of local and traditional knowledge in combating desertification in Jordan. A document presented to National Action Plan (NAP) to Combat Desertification Project JOR/00/001/A/99, Ministry of Environment, Amman-Jordan.

Al-Hadidi, L. M., 1996, Evaluation of desertification risk in Jordan using some climatic parameters, Unpublished MSc thesis, University of Jordan, Amman-Jordan.

Atkinson, K., and Beaumont, P., 1971, The forests of Jordan, *Econ. Bot.* **25**(3):302–311.

Barrow, C. J., 1994, *Land Degradation: Development and Breakdown of Terrestrial Environments,* Cambridge University Press, Cambridge.

Beaumont, P., 1985, Man induced erosion in northern Jordan, in: *Studies in the History and Archaeology of Jordan,* A. Hadidi, ed., Department of Antiquities, Amman-Jordan, Vol. I, pp. 291–296.

Disi, A. M., and Oran, R., 1995, *Animal Diversity in Jordan,* 3rd scientific week, Higher Council for Science and Technology (HCST), Amman-Jordan, Vol. V, pp. 26–60.

DOS (Department of Statistics, Jordan), 2002, *Statistical Yearbook.* Amman-Jordan.

DOS (Department of Statistics, Jordan), 2003, *Jordan in Figures: Annual Report*, Department of Statistics, Amman, Jordan.

Dregne, H. E., 1983, *Desertification of Arid Lands.* Harwood Academic, New York.

Dutton, R. W., Clarke, J. I., and Battikhi, A. M., 1998, *Arid Land Resources and Their Management: Jordan's Desert Margin,* KPI: London and New York.

FAO (Food and Agriculture Organization of the United Nations), 1992, Inventory and assessment of land resources for near east and Africa Region. Paper for the 11th session of the Regional Commission on land and Water use in the Near East, FAO, Tunis, Tunisia.

Hammond, A., Adriaanse, A., Rodenburg, Bryant, D., and Woodward, R., 1995, *Environmental Indicators,* World Resources Institute, Washington, D.C.

Hatough, A., Al-Eisawi, D. M., and Disi, A. M., 1986, The effect of conservation on wildlife in Jordan, *Environ. Conserv.* **13**(4):331–335.

Hatough-Bouran, A., and Disi, A. M., 1995, The impact of development and population growth on ecological systems: Global and local issue, *Dirasat* **22**A(2):70–84.

Herrmann, S. M., and Hutchinson, C. F., 2005, The changing contexts of the desertification debate, *J. Arid Environ.* **63**(3):538–555.

HTS (Hunting Technical Services), 1956, *Report on the Range Classification Survey of the Hashemite Kingdom of Jordan,* Ministry of Agriculture, Amman-Jordan.

JMD (Jordan Meteorological Department), 2005, Jordan Climate, Meteorological Department, Amman-Jordan.

Kharin, N., Tateishi, R., and Harahsheh, H., 1999, *Degradation of the Drylands of Asia,* Center for Environmental Remote Sensing (CERS), Chiba University, Japan.

Koohafkan, A. P., 1996, *Desertification, Drought and their Consequences,* Environmental and Natural Resources Service (SDRN) FAO, Research, Extension and Training Division, EP: Analysis: Desertification, pp. 7.

Shakhatra, M., 1987, Desertification in the Arab World: Causes and Consequences, in: *Combating of Desertification in North African Countries. The Green Belt for North African Countries,* Arab League Educational, Cultural and Scientific Organization (ALECSO), Tunisia (in Arabic).

Tillawi, 1995, *Afforestation in Jordan,* 3rd scientific week, Higher Council for Science and Technology (HCST), Amman-Jordan, Vol. V. pp. 156–189.

UN, 1997, World Urbanization Prospects: The 1996 Revision, Annex Tables, United Nations Secretariat: Population Division, New York, p. 115.

UNCCD (United Nations Convention to Combat Desertification), 1992. http://www.fao.org/desertification/article_html/en/1.htm.

UNCCD (United Nations Convention to Combat Desertification), 2005, *What is desertification?*

UNEP, 1992, United Nations Conference on Environment & Development, Agenda 21, Chapter 12, Adopted at the Rio Conference June, UN, New York.

UNEP, 1997, *World Atlas of Desertification,* 2nd edition, United Nations Environment Programme, UN, Rome.

UNEP, 1999, GEO-2000. *Global Environmental Outlook,* United Nations Environment Programme, Nairobi, Kenya.

Walker, A.S., 1998, *Deserts: Geology and Resources,* United States Geologic Survey: Washington, DC.

APPENDIX. Matrix of Critical Issues Leading to Desertification in Jordan.

Problem type	Evidence	Sources	Impacts
Limited water and arable land resources	1. Low per capita available water and arable land and decrease with time	1. Aridity 2. Intensive agricultural schemes 3. Rapid population growth 4. Present agricultural and water policies	1. Slow down socioeconomic development plans 2. Decrease of food self-sufficiency
Water quality deterioration	1. Groundwater and surface water depletion 2. Salt water intrusion (sea and saline water) 3. Contamination of surface water and shallow water aquifers	1. Overexploitation of water 2. Irrational water demand 3. Discharge of wastewater 4. Lack of awareness 5. Lack of integrated agricultural and water policies	1. Increasing water scarcity 2. Salinization and loss of agricultural lands 3. Negative impact on human and environment
Progressive deterioration of irrigated lands	1. Increase of salinized cultivated areas and increase of soil salinity 2. Decrease of soil fertility 3. Decrease of agricultural productivity	1. Extensive traditional irrigation practices 2. Intensive agricultural schemes 3. Intensive use of agro-chemicals 4. Lack of awareness and participation of stakeholders 5. Lack of integrated agricultural development policies	1. Degradation of irrigated lands 2. Reduction of agricultural productivity 3. Degradation of water resources 4. Increase of poverty and rural urban migration
Cultivation of marginal lands (lands between arable and rangelands)	1. Low and unsustainable productivity of marginal lands 2. Increase of sand drifts and dust storms	1. Population pressure and increased food demand 2. Inappropriate agricultural practices 3. Lack of awarness	1. Loss of rangelands and food production 2. Increase of soil erosion 3. Escalation of sand dune encroachment

(continued)

APPENDIX. (continued).

Problem type	Evidence	Sources	Impacts
Deterioration of rangelands	1. Change of natural plant cover 2. Deterioration of land productivity 3. Increased soil erosion, sand drafts and dust storms	1. Mismanagement of rangelands (over-grazing, wood cutting, tourism) 2. Socio-economic factors 3. Increasing pressure for food security 4. Adopted policies and programs for development	1. Loss of Productivity of rangelands 2. Loss of Food resources 3. Loss of Biodiversity 4. Soil by wind and water erosion
Development policies and interventions	1. Appropriation of productive 2. lands for urban and industrial development 3. Development of large-scale private agricultural	1. Increasing pressure on land at local and regional levels 2. Sedentarization of nomads and pastoral communities 3. Local effects of structural adjustment program 4. Inappropriate interventions such as top-down planning and implementation	1. Reduction of agricultural productivity 2. Increasing poverty 3. Loss of rural income 4. Increasing rural-urban migration 5. Marginalization of rural areas
Weak institutions	1. Continuing degradation of water and land resources 2. Lack of integrated agricultural and water policies 3. Duplication of efforts 4. Lack of awareness	1. Lack of financial resources 2. Inadequate technical capabilities, training and research 3. Lack of coordinating effort and net-working for combating desertification 4. Lack of sustainable development plans for desertified areas	1. Poor water and land development and management 2. Wasteful uses of resources 3. Progressive expansion of desertified areas

GLOBAL ENVIRONMENTAL CHANGE
AND THE INTERNATIONAL EFFORTS CONCERNING
ENVIRONMENTAL CONSERVATION

LIVIU-DANIEL GALATCHI*

Ovidius University of Constanta

Abstract: Global environmental problems have surfaced in recent years in the midst of a general upgrading of economic standards in advanced countries. This has been paralleled by rapid growth of poverty, population, and urbanization in less developed countries, and resulted in broadening mutually dependent international relationships. The United Nations Conference on Environment and Development (UNCED/Earth Summit) was convened in June 1992; among other events, the summit marked the end of the East-West Cold War framework that had existed since the end of World War II, and now encourages international cooperation and action relative to global environmental issues. International efforts concerning environmental conservation relate to global warming, protection of the ozone layer, acid deposition, forests, wildlife, marine environment, transboundary movement of hazardous wastes, desertification, and pollution problems in developing countries.

Keywords: Environmental change; conservation; climate change; ozone layer; forests; wildlife; aquatic environment; wastes; desertification; pollution; United Nations

1. Introduction

The Earth Summit formulated such worldwide agreements for sustainable development as the Rio Declaration on Environment and Development and Agenda 21, and a number of efforts relating to agreements made at this Earth Summit were conducted. In May 1994, the second meeting of the Commission on Sustainable Development (CSD), which was established at the UN Economic and Social Council in order to follow up on the Earth

---

*Address correspondence to Dr. Liviu-Daniel Galatchi, Associate Professor, Ovidius University of Constanta, Department of Ecology and Environmental Protection, Mamaia Blvd 124, RO 900527 Constanta-3, Romania; e-mail: galatchi@univ-ovidius.ro

P.H. Liotta et al. (eds.), Environmental Change and Human Security, 103–115.
© 2008 *Springer Science + Business Media B.V.*

Summit, convened to direct studies on the problems of trade relative to the environment and of hazardous chemical substances. In late April and early May of that year, the Global Conference on the Sustainable Development of Small Island Developing States was held in support of the Earth Summit, and the Barbados Declaration, which resulted in the adoption of an action plan.

There have also been various efforts to pursue separate treaties in several environmental problem categories. The United Nations Convention to Combat Desertification, which was agreed upon at the Earth Summit as needing adoption, was ratified at the fifth substantive meeting of the inter-governmental negotiating conference in June 1994. The new International Tropical Timber Agreement (ITTA) was adopted in January 1994, and took steps in December of that year to apply the new agreement on a temporary basis. Also, the UN Convention on the Law of the Sea went into effect in November 1994. In addition, the sixth meeting of member states of the Montreal Protocol was held in October 1994, while the first meeting of member states of the Convention on Biological Diversity was held in November of that year, and the first conference of the parties to the UN Framework Convention on Climate Change was held in March and April of 1995.

2. Global Warming

The atmosphere includes carbon dioxide, methane, water vapor, and other greenhouse gases, and their activity creates a livable environment for humans, plants, and animals. In recent years, however, human activity has resulted in the release of large volumes of greenhouse gases such as carbon dioxide and methane into the atmosphere. As a result, the greenhouse effect has been bolstered and the threat of global warming has arisen.

The main breakdown of activities to be carried out are

- Creation of an international framework
- Monitoring, observations, surveys, and research
- Technology development and dissemination
- Official development assistance, etc., in the environmental field
- Environmental considerations
- Global environmental conservation-oriented socioeconomic activities, popularization, and public awareness
- Global warming countermeasures
- Countermeasures to depletion of ozone layer
- Acid deposit countermeasures
- Marine pollution countermeasures

- Countermeasures to transboundary movements of hazardous wastes
- Countermeasures to declines in wildlife species
- Countermeasures to desertification
- Countermeasures to pollution problems in developing countries
- General expenditure related to global environmental conservation
- Satellites and other research and development–related expenses
- Energy policy-related expenses

According to reports released by the Intergovernmental Panel on Climate Change (IPCC), if the current rate of increase of greenhouse gases continues, the average air temperature at ground level is forecast to rise by about 3°C by the end of the twenty-first century. This temperature rise is believed to be an extremely rapid change of a sort not seen in the last 10,000 years. Moreover, sea levels are expected to rise by about 65 cm (and a maximum of 1 m) by the end of the twenty-first century (IPCC Workshop on New Emission Scenarios, 2005).

The United Nations Framework Convention on Climate Change went into effect on 21 March 1994. This convention sets as its ultimate objective a standard to stabilize greenhouse gas densities in the atmosphere so as to prevent dangerous human interference in the climate system. It makes various requirements on member states, including preparation of a list of emission and absorption sources for greenhouse gases and drawing up of national programs for countermeasures to warming. In particular, the advanced member states must be making policies and actions with the awareness that reducing emission volumes of carbon dioxide and other greenhouse gases to levels at the beginning of the millennium would contribute to the goals of the convention.

As of December 1994, fifteen countries had sent information. In December of that year, the temporary secretariat of the convention prepared a combined report based on the reports from these advanced countries, with detailed examinations of these country reports to be completed for each country by the end of 1995.

The tenth meeting of the convention negotiating committee was held in August 1994 and the eleventh meeting in February 1995 to focus on implementation. The first Conference of the Parties was held in March and April in Berlin, where the participants agreed on policies and measures and also aimed for the establishment of numerical restraint and reduction goals within the specific periods of 2005, 2010, and 2020. They also set in motion a process for studying conclusions to be considered at the third Conference of the Parties held in 1997, and to consider efforts for the periods beyond 2000 that do not yet have clear stipulations under the convention. These efforts were to be strengthened by including them as promises in protocols and other documents to be ratified. Moreover, a number of contracting parties agreed to introduce the concept of "Activities Implemented Jointly" in efforts toward arresting global warming, and to establish a permanent secretariat. Furthermore, the advanced nations proposed the Climate Technology Initiative (CTI) as a first step toward serious implementation of the convention.

The Intergovernmental Panel for Climate Change (IPCC) is an international organization established jointly by the United Nations Environment Programme (UNEP) and the World Meteorological Organization (WMO), whose goal is to make forecasts on global warming, to gather the latest knowledge about its effects and countermeasures, and to provide a basis for scientific policies for arresting global warming. IPCC's first evaluation report was prepared in August 1990, while a supplementary report came out in February 1992. Later, it underwent reorganization, and engaged in preparing a second evaluation report by the end of 1995. A portion of the second evaluation report was prepared in October 1994 and was included in a special 1994 report that incorporated the latest knowledge on the warming effects of greenhouse gases and on possible emissions scenarios. This report was prepared at the request of the negotiating committee for the Framework Convention on Climate Change, and was presented at the first Conference of the Parties of the Framework Convention on Climate Change held in March and April 1995 (IPCC Workshop on New Emission Scenarios, 2005).

3. Protection of the Ozone Layer

Most of the Earth's ozone exists in the stratosphere, in the region called the ozone layer. The ozone layer absorbs most of the harmful ultraviolet rays in solar light, protecting life on the Earth. It has become apparent that this ozone is being depleted by artificial chemical substances such as chlorofluoro-carbons (CFCs) and halon. It is worried that the depletion of the ozone layer and the accompanying increase in ultraviolet rays will result in such health risks to people as skin cancers and cataracts, and will also hinder the growth of plants and plankton.

CFCs are substances made up of carbon, fluorine, and chlorine that are useful for such things as solvents, refrigerants, foaming agents, and propellants, while halon, which contains bromine, is mainly used as a fire extinguisher. Since these are chemically stable substances, when they are released into the atmosphere they fail to break down in the troposphere and so reach the strat-osphere where they are decomposed under the strong ultraviolet rays from the sun. This releases chlorine and bromine atoms which serve as catalysts to break down the ozone.

Once the depletion of the ozone layer by CFCs and other substances occurs, recovery of the layer requires a long period of time. Moreover, the damage inflicted is an environmental problem on a global scale.

To prevent depletion of the ozone layer, the Vienna Convention for the Protection of the Ozone Layer was adopted in March 1985, and the Montreal Protocol on Substances that Deplete the Ozone Layer was adopted in September 1987. Later, this protocol was amended in June 1990, and again amended in November 1992.

With the focus on the Ozone Layer Protection Law, the European Union has implemented the following policies actively to deal with global-scale air pollution problems:

- Controls on production of CFCs, etc.

- Emission control and rational use of CFCs, etc.

- Promotion of recovery, recycling, and destruction of CFCs, etc.

- Promotion of monitoring, observation, and research on ozone layer depletion

4. Acid Deposition

Acid deposition results primarily from the presence of sulfuric and/or nitric acids in the atmosphere. These acids are formed in the atmosphere from pollutants such as sulfur oxides (SO_x) and nitrogen oxides (NO_x), generated mainly from the combustion of fossil fuels. Acid deposition includes deposition in various forms of precipitation, such as rain, fog, mist, or snow (wet deposition), and deposition of acidic gases and aerosols without the intervention of precipitation (dry deposition). The principal problems induced by acid deposition occur when air pollutants such as SO_x and NO_x pass from the air into precipitation and are wet deposited on water, soil, or plant surfaces, or when such pollutants are directly dry deposited on such surfaces, and result in acidification of the associated aquatic or terrestrial ecosystems.

Acid deposition has far-reaching impacts. It affects fish and other aquatic life forms as a result of the progressive acidification of inland water bodies, including marshes, lakes, rivers, and streams. It threatens forests as a result of the acidification of the soil. It accelerates and aggravates the decline and decay of monuments and buildings of cultural importance. For these reasons the problem of acid deposition is of deep concern. The problem first came to worldwide attention in Northern Europe and North America. Here the problems of acidification of lakes and the decline or dieback of forests has been especially prominent.

One of the distinctive features of the acid deposition problem is the fact that deposition of acidifying substances can occur thousands of kilometers from the original sources of precursor emissions. Thus, acid deposition does not respect national boundaries, and has accordingly become an international problem. In Europe the transnational aspect of the problem led to the signing of the Long-Range Transboundary Air Pollution Convention (LRTAP) in 1979. This convention expressed a common commitment on the part of the signing parties to reduce acid precursor emissions and jointly monitor emissions and precipitation chemistry.

In the past, acid deposition was considered a problem specific to industrial nations. However, in recent years air pollutant emissions have increased dramatically in many developing countries as a result of their rapid rate of economic growth and industrialization. This will almost certainly have the effect of aggravating already existing local-scale air pollution problems, and creating regional-scale ones.

In 1992 at the Earth Summit in Rio de Janeiro the document known as Agenda 21 enjoined both advanced industrial and developing nations to make a stronger commitment to tackling large-scale environmental problems such as acid deposition.

5. Forests

Various types of forests are distributed around the world according to the climate characteristics of each region. Total forest area is about 3.88 billion hectares, covering about 29 percent of total land area (including inland waters) (FAO, *Production Yearbook, 2005*). The forests are natural resources that provide multiple environmental services, including wildlife habitat, soil conservation, water sources, absorption and fixing of carbon dioxide, and supply sources for lumber and charcoal, as well as supplying non-lumber products such as raw resources for pharmaceuticals.

In recent years, forest areas in the advanced regions have either leveled off or increased while forests in the tropical regions of developing countries have declined drastically. According to the final report of the Forest Resources Assessment Project, the latest report by the Food and Agriculture Organization of the United Nations (FAO), an estimated annual average of 15.4 million hectares of tropical forest was lost in the 10 years from 1991 to 2000. The tropical forests are said to support about half of the world's wildlife species, making them a storehouse of genetic resources. But the loss of large swathes of area threatens many wildlife species with the danger of extinction. Moreover, it has also been pointed out that the loss of forests releases large amounts of carbon dioxide, thus becoming a factor in the acceleration of global warming. Causes of tropical forest loss vary depending on the region, including nontraditional slash-and-burn agriculture, excessive collection of wood for fuel, inappropriate commercial logging, and overgrazing. Behind these causes lurk various socioeconomic factors, including population growth, poverty, and land control systems.

Countermeasures for forest decline agreed upon in the Statement of Forest Principles, the first global agreement on forests which was reached at the Earth Summit and in Agenda 21, provided some of the first debates on conservation and sustainable management of the world's forests, including tropical forests. The new International Tropical Timber Agreement (ITTA) concluded in 1994, which replaced the 1983 International Tropical Timber Agreement, was the first agreement adopted following the end of the Earth Summit. One objective of the ITTA was to provide support for producer countries to implement "The Year 2000 Objective," which is a strategy for achieving exports of tropical timber and timber products from sustainable managed sources by the year 2000. The International Tropical Timber Organization (ITTO) that was set up by the ITTA is actively seeking the cooperation of both producer and consumer countries for conservation, sustainable management, and utilization of tropical timber. In addition to the planning strategy for "The Year 2000 Objective," and drawing up of guidelines, the organization has implemented more than 200 projects.

Preparations of criteria and indicators to understand and study the sustainability of forest management, other than tropical forests, are being debated by European countries and non-European countries alike.

First, studies into criteria and indicators for forests within the European region have been proceeding in European countries, with temporary agreement reached on criteria and quantitative indicators. Studies have continued since that time into descriptive indicators.

In the non-European countries, including Japan, Canada, the United States, and others, work had been progressing on the preparation of criteria and indicators that target all forests other than tropical forests, and an agreement was reached at the Sixth International Working Group Meeting held in Chile in February 1995.

Again, in preparation for the third meeting of the UN Commission on Sustainable Development (CSD) held in April 1995, the Intergovernmental Working Group on Forests (IWGF) met twice at the behest of Canada and Malaysia, in April and October 1994. Many countries and international organizations participated in the IWGF to deal with such issues to be debated at the CSD as forest conservation, improvement of forest cover, and forest conservation and trade.

For surveys and research into tropical forests, national experimental research institutions are conducting research into tropical forest ecologies at field sites in the Malaysian rain forest, using the Global Environment Research Fund. The national experimental research institutions are also conducting monitoring and research into changes in the tropical forest and their effects at field sites in the Thai tropical forest, using the Ocean Development and Earth Science Research Fund.

The private sector, as well, is working to conserve the tropical forest by providing support for a sapling experiment to plant *Dipterocarpaceae* trees to rejuvenate the tropical forests in Sarawak State, Malaysia. They also plan to plant 15,000 ha of native tree varieties in Papua New Guinea, for the purpose of sustainable forest production.

6. Wildlife

At present, decline in wildlife species is proceeding at a speed never before seen in the history of the world because of habitat destruction through human activity and because of overharvesting. It is predicted that 5–15 percent of the world's total species could become extinct by 2020.

Wildlife is a basic structural component forming the global ecosystem, and its existence is essential as an important resource for humans and for making life more pleasant and relaxing. Prevention of species extinction is a worldwide emergency issue.

7. Marine Environment

The ocean covers three-fourths of the Earth's total surface, is an important location for life form production, and has a major affect on climate, through mutual interaction with the atmosphere, making it an essential element in maintaining all life on the Earth.

The characteristics and resources that the ocean possesses have been utilized and developed by humans since ancient times. But in recent times, with the increasing reliance on marine resources and increases in all types of pollution that occur with human activities, conservation of the marine environment has become an important issue. Since regional ocean surveys tend to be conducted in seas near developed countries, the overall picture of the state of global marine pollution is not necessarily clear. Nevertheless, in closed seas such as the North Sea, Baltic Sea, Black Sea, and Mediterranean Sea, the occurrence of red tide is increasing, along with pollution from hazardous substances such as heavy metals (Vadineanu, 2000). Moreover, because the threat of major marine pollution exists from supertanker navigation and the development of sea bottom oil fields, and because damage incurred from the occurrence of a single accident can spread over large areas for a long period of time, conservation of the marine environment has received global attention. In particular, a succession of major oil spills in recent years caused by supertanker accidents, and large-scale oil spills that occurred during the Gulf War at the end of the twentieth century, have had serious effects on the marine environment, again reminding international opinion of the importance of marine environment conservation.

Marine pollution problems include the flow of pollutants off the land, discharges of oil and other substances from ships at sea, and dumping of wastes at sea. Effective results in preventing this pollution can only be expected with cooperation from all countries in the world (Galatchi, 2006). International cooperation in the form of adoption of treaties for prevention measures against marine pollution, including the International Maritime Organization (IMO), are being actively promoted, and countries are developing their marine pollution countermeasures in line with these activities.

8. Transboundary Movements of Hazardous Wastes

Movement of hazardous wastes across national boundaries originally took place as an ordinary occurrence in places like Europe where many nations border each other and commercial travel is common (Galatchi and Tudor, 2006). Nevertheless, this movement could not be said to have been appropriately done in environmental terms. The EC and OECD commenced moves to create transboundary control systems as an outcome of a major incident in 1982 related to the disappearance of dioxin-polluted soil that had been left over after an accidental explosion at an Italian agricultural chemicals factory in 1976 (Seveso Polluted Soil Shipment Incident), and which was then discovered in France the following year.

Again, a number of incidents occurred in the latter 1980s where hazardous wastes were being exported from developed countries to developing countries, thus causing environmental pollution in those countries, including the Coco Incident (1988), where discarded transformers containing PCBs from Italy, Norway, and elsewhere were discovered to be dumped in Nigeria. These types of incidents apparently began to occur when it became difficult for hazardous waste owners to dispose of it in advanced countries, and they turned to exporting to less developed countries where the regulations were more relaxed and disposal costs lower. In this way, people became aware that the problem of transboundary movement of hazardous wastes was no longer limited to the advanced countries but became a problem that involved the developing countries and that required responses on a global scale.

In response to these problems, UNEP worked to conclude the Basel Convention on the Control of Transboundary Movements of Hazardous Wastes and Their Disposal in March 1989, which mandated a system of permits and prior notification for the export of hazardous wastes and required re-export where inappropriate exports and disposal of wastes had occurred; the convention went into effect on 5 May 1992.

Moreover, the problem of transboundary movement of hazardous wastes was taken up at the Earth Summit held in Brazil in 1992 as one important theme in global environmental problems, and its importance as a problem needing attention was also noted in Agenda 21.

9. Desertification

Desertification, as defined by Agenda 21 adopted at the June 1992 Earth Summit and the Convention on Prevention of Desertification is *"land degradation in arid, semi-arid, and dry sub-humid areas resulting from various factors, including climatic variations and human activities."* In this case, *"land"* is a concept that includes the soil, water resources, and vegetation, while *"degradation"* is the decline of a land's resource potential due to a single or multiple processes, such as erosion decline of the natural variation of vegetation over long period of time, and salinization of land. According to a survey on the current state of desertification conducted by UNEP in 1991, desertification is progressing and covers about one-quarter of the Earth's total land area, and about 70 percent of arid land. This is equal to 3.6 billion hectares and affects about one-sixth of the world's population.

Major causes of desertification include livestock overgrazing, overharvest of agricultural land, over-collection of wood for charcoal and firewood, and rising salinization of agricultural land caused by inappropriate irrigation. Lurking in the background are such socioeconomic factors as local poverty and rising population in less developed countries, which make the desertification problem more complex.

International efforts for the prevention of desertification centered around UNEP and the United Nations Conference on Desertification (UNCOD), held in 1977. Furthermore, in response to the Earth Summit's Agenda 21, the 47th UN General Assembly in 1992 established the Intergovernmental Negotiating Committee on Desertification for the adoption of the Convention on Prevention of Desertification. The Convention Negotiation Conference met five times through June 1994, and in the fifth meeting adopted two resolutions concerning the Convention on Prevention of Desertification (main document plus four regional implementation annexes), i.e., the Resolution concerning Temporary Measures until the Convention Takes Effect and the Resolution Concerning Emergency Action for Africa. In October 1994, the signing ceremony for the convention was held in Paris, and eighty-six countries (including the EU), signed the convention.

10. Pollution Problems in Developing Countries

In developing countries, poverty and population pressures are contributing factors to deforestation and desertification whereas urbanization and industrialization have contributed to problems related to air, water, and solid waste pollution. With the end of the Cold War structure, it has become apparent that the countries of Eastern Europe and the former Soviet Union are faced with serious pollution problems left behind by planned economies that placed little regard on environmental conservation policies. Pollution problems are arising all over the world, and from the global point of view, the pollution problems in developing countries can no longer be considered as locally confined.

Moreover, many of these developing countries do not possess the economic, technological, personnel, or institutional foundations for coping with pollution problems and they require the cooperation of the advanced nations.

11. Conclusions

Before the effects of global warming begin to appear and it is too late to do anything about them, it is important to immediately promote policies that are executable (Guvernul Romaniei, 2001).

Some of the most important measures to be taken are as follows:

1. To promote global warming prevention countermeasures in local areas, assistance must be provided to local public groups for adoption of the master plan (Plan for Local Promotion of Countermeasures to Global Warming).

2. To strengthen efforts by businesses to save energy, guidance must be given to businesses for saving energy at factories and other places with determination standards based on the laws for rationalization of energy utilization. Additionally, low-interest financing and tax incentives should be provided to specific business activities (introduction of energy-saving facilities at factories, etc.).

3. To reduce and recycle wastes, and to effectively utilize excess heat from garbage incineration and sewage, model projects for sewage channel heat utilization must be promoted, including power generation projects involving solid garbage fuel.

4. To form transport systems that can reduce and restrict carbon dioxide emissions, modal shifts (persuasion to switch to rail transport or to

coastal sea transport) in the main transport lines between distribution centers of medium and long distances should be promoted. This should include development of more efficient distribution system configurations and development of bypass roads. Technology evaluations should also be conducted and practical utilization surveys conducted on low-pollution vehicles for the purpose of achieving their broader use and the subsequent policies on how to increase their distribution and use through society over the medium and long term. Assistance must also be provided for the introduction of low-pollution vehicles by implementing aid for their use for municipal purposes, e.g., pollution patrol cars.

5. To form supply structures for energy that emits few greenhouse gases, the development and utilization of (1) nuclear power with the presumption of guaranteed safety; (2) utilization of hydropower and geothermal power; (3) introduction of combined cycle power generation and of solar power generation should be promoted. Moreover, the long-term energy supply and demand forecasts must be revised after considering changes related to the Framework Convention on Climate Change. Introduction and spread of energy sources that leave few burdens on the environment, such as solar power and other types of natural energy, and fuel cells, must be promoted to achieve the supply targets for energy alternatives to oil.

6. To reduce the uncertainties related to global warming, and to adopt appropriate policies based on scientific knowledge, research should be conducted into elucidating natural phenomena, forecasting the future, developing measures for evaluating effects, monitoring greenhouse gases, and developing monitoring technologies utilizing satellites, etc. Additionally, steps need to be taken to expand the global environment research program budget to promote surveys and research in these areas.

7. To find ways to restrict greenhouse gas emissions, research must be promoted into solutions such as the development of advanced new energy and energy saving technologies, and to creating revolutionary technologies such as fixing or effectively utilizing carbon dioxide.

8. To make people aware of the action program to arrest global warming and of the policies based on that program, pamphlets should be distributed and meetings organized with local governments.

9. To provide continuing support in the area of international cooperation, especially in regard to the developing countries of the world, support should be provided for the adoption of strategies that were undertaken by the developed countries in response to warming.

Aknowledgment

We thank Professor Dr. Stoica-Preda Godeanu for helpful suggestions during our research.

References

Agenda 21; http://www.un.org/esa/sustdev/documents/agenda21/english/.
FAO, *Production Yearbook, 2005;* http://www.fao.org/.
Galatchi, L. D., 2006, "The Romanian National Accidental and Intentional Polluted Water Management System," in: G. Dura et al., eds., *Management of Intentional and Accidental Water Pollution*, Springer, New York, pp. 181–184.
Galatchi, L. D., and Tudor, M., 2006, "Europe as a Source of Pollution—the Main Factor for the Eutrophication on the Danube Delta and Black Sea," in: L. Simeonov and E. Chirila, eds., *Chemicals as Intentional and Accidental Global Environmental Threats*, Springer, New York, pp. 57–63.
Guvernul Romaniei, 2001, "Strategia Nationala pentru Dezvoltare Durabila," Bucharest, pp. 1–120.
IPCC, 2005, Workshop on New Emission Scenarios.
Vadineanu, A., 2000, "*Dezvoltarea durabila*," University of Bucharest Press, Bucharest, pp. 1–224.

MANAGEMENT OF ENVIRONMENTAL CHALLENGES AND SUSTAINABILITY OF BULGARIAN AGRICULTURE

HRABRIN BACHEV*

Institute of Agricultural Economics, 125 Tzarigradsko shosse Blvd. Block 1, 1113 Sofia, Bulgaria

Abstract: This paper presents major environmental challenges in Bulgarian agriculture, identifies and assesses specific modes for environmental governance in the farming sector, and estimates prospects for changing environmental performance of farms under conditions of EU integration and Common Agricultural Policy implementation. We adapt the framework of the New Institutional and Transaction Costs Economics, and assess the efficiency of diverse market, private, and public modes for environmental governance. Our analysis of the post-communist transformation of agriculture shows that it has changed the state of the environment and brought some new challenges such as degradation and contamination of farmland, pollution of surface and ground water, loss of biodiversity, and significant greenhouse gas emissions. Badly defined and enforced environmental rights; a prolonged process of privatization of agrarian resources; use of farming practices that do not motivate long-term investment; high uncertainty assets with low frequency specificity inappropriate for environmentally related transactions—all of these factors are responsible for the failure of market and private modes of environmental management. The strong need for public intervention has not been met by effective government, community, or international assistance; consequently agrarian sustainability has been compromised. The assessment of possible impact of EU CAP implementation under Bulgarian conditions indicates that the main beneficiary of various new support measures will be the biggest operators. Income, technological and environmental discrepancy between different farms, sub-sectors, and regions will be further enhanced.

Keywords: Market, private, public and hybrid modes of agrarian and environmental reforms; impact of agriculture on environment; transformation of farming structures; impact of EU integration and CAP implementation; Bulgaria

*Address correspondence to Hrabrin Bachev, Institute of Agricultural Economics, 125 Tzarigradso Shosse Blvd. Block 1, 1113 Sofia, Bulgaria; e-mail: hbachev@yahoo.com

P.H. Liotta et al. (eds.), Environmental Change and Human Security, 117–142.

1. Introduction

The post-communist development in Eastern Europe has been associated with specific challenges for agrarian sustainability and its environmental dimension (Bentcheva and Georgiev, 1999; Gatzweiler and Hagedorn, 2003; Sumelius, 2000). That is a result of the fundamental transformation and the specific governing structures dominating the sector (Bachev, 2007a).

Bulgarian agriculture has been a good example of common challenges for most of transitional and new EU member states from the region. During the communist period farming was carried out in about 300 public farms averaging tens of thousands hectares. Over-intensification of production and huge overloading of animals were typical; pressure on the environment, with associated pollution and degradation of nature, was significant.

Unprecedented post-1989 reforms transferred the management of agrarian resources into the hands of more than two million new stakeholders. The transition to a modern market economy lasted twelve years and was associated with the evolution of a quite specific governing structure. It consists of a huge subsistence and part-time farming, enormous enhancement of land management, limited livestock operations in a few big farms, domination of primitive and "grey" structures, inefficient formal institutions, big reliance on personal relations, and massive corruption. (Bachev, 2005, 2006). Transformation of the sector has also been associated with significant changes in resource use, technologies, and production structure, as well as the emergence of new environmental problems.

This paper *first* presents the major environmental challenges in Bulgarian agriculture; *next,* identifies and assesses the specific modes for environmental management in the farming sector; and *finally,* gives prospects for changing environmental performance of farms under the conditions of EU integration and Common Agricultural Policy (CAP) implementation.

An interdisciplinary framework of the new institutional and transaction cost economics is incorporated (Furuboth and Richter, 1998; North, 1990; Williamson, 1996). It gives a new insight on the comparative efficiency of various market, private, public, and hybrid modes of management, and their potential to deal with various environmental challenges (Bachev, 2004, 2007a). Furthermore, it includes in its analysis the formal and informal institutions (structure of property rights and system of their enforcement), and the transaction costs associated with all modes of governance. The specific factors of governance choice—*institutional, behavioral* (bounded rationality, tendency for opportunism), *dimensional* (frequency, uncertainty, assets specificity, and appropriability of transactions), and *technological* are identified. Based on this, we assess possible prospects of modernization of environmental management in the specific economic, institutional, and natural environment of Bulgarian agriculture.

2. Major Environmental Challenges of Bulgarian Agriculture

The post-communist transformation of Bulgarian agriculture has changed significantly the state of the environmental situation and brought to light a number of new challenges.

The total amount of chemical fertilizers and pesticides used has declined considerably (Figure 1). Compared to 1989, the application of fertilizers and pesticides per hectare now represents merely 22 percent and 31 percent respectively.[1] That sharp reduction in chemical use has diminished drastically the risk of chemical contamination of soil, water, and farm produce.

However, a negative rate of fertilizer application of N, P, and K shows low phosphorus and potassium levels (Figure 2). Consequently, an average of 23,595.4t N, 61,033.3t P_2O_5 and 184,392t K_2O have been irreversibly removed annually from soils since 1990. Furthermore, an imbalance of nutrient components has been typical with application of 5.3 times less phosphorus and 6.7 times less potassium with the appropriate rate for nitrogen used during that period.[2] Moreover, a monoculture or simple rotation has been constantly practiced by most large farm managers concentrating on a few profitable crops (such as sunflower and wheat).[3] All of these practices further contributed to deterioration of soil quality and soil organic matter content in the country.

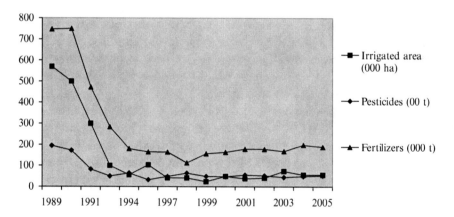

Figure 1. Use of Chemicals and Irrigation in Bulgarian Agriculture.
Source: National Statistical Institute, Annual Statistical Indicators.

---

[1] Intensity of usage of chemicals per hectare in Bulgaria ranks among the lowest in Europe with an average annual level for the past ten years of 30.6kg N, 3.6kg P_2O_5, 0.6kg K_2O, and 49.8kg pesticides (including 25.6kg herbicides) (NSI).

[2] The actual ratio between N:P:K has been 1:0.15:0.06 while the optimal one is 1:0.8:0.4.

[3] Since 1991 the share of oil crops have been twice that of cereals (MAF).

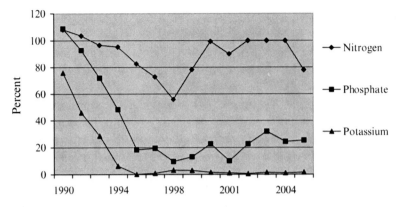

Figure 2. Rate of Fertilizer Application of Basic Elements in Bulgarian Agriculture.
Source: Ministry of Agriculture and Forestry (MAF), Annual Agrarian Reports.

There has been a considerable increase in agricultural land affected by acidification in the country (Figure 3). Soil acidification occurs as a result of a long-term application of specified nitrate fertilizers[4] and imbalanced fertilizer application without the adequate input of phosphorus and potassium. Currently almost 25 percent of soils are acidified as the percentage of degraded farmland acidified soils reach 4.5 percent of total lands. After 1994 the percentage of acidified soil began to decrease; however, in recent years there is a reverse tendency along with the gradual augmentation of the use of nitrates. The percentage of salt affected land has doubled after 1989 but it is still an insignificant (mostly abandoned or intensively irrigated) part of the total farmland. During the entire period no effective measures have been taken to normalize soil acidity and salinity.[5]

Erosion has been a major factor for land degradation in the country (Figure 3). Around one-third of the arable lands are subjected to wind erosion and 70 percent to water erosion.[6] Soil erosion intensity and scope depend on topographic attributes, soil type, and climate factors (intensity of rainfall, frequency of dust storms). Since 1990 water and wind erosion affect

---

[4] Consisting mostly of ammonium nitrate (70–80 percent) and carbamide (20–30 percent) (EEA).

[5] For instance, during transition now the limed acidified lands comprises far below 2 percent of the areas limed until 1990. And no chemical amelioration or drainage of salt affected land has been effectively implemented (MAF).

[6] All terrains with a slope over 20° are affected by water erosion. There is a risk of erosion of more than 5 t/ha/year for over a quarter of arable lands (including 65 percent of orchards and vineyards) (EEA, 2006).

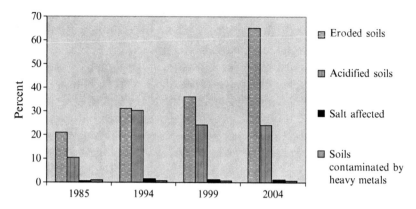

Figure 3. Categories of Degraded Agricultural Lands in Bulgaria.
Source: Executive Environment Agency (EEA), Environmental Report 2006.

between 25 and 65 percent of farmland; total losses vary from 0.2 to 40 t/ha in different years (EEA). Annual losses of earth mass from water erosion are estimated at 136 Mt while wind erodes between 30 and 60 Mt.[7] The progressing level of erosion is a result of the extreme weather but it has been also adversely affected by dominant agro-techniques, deficiency of anti-erosion measures, and uncontrolled deforestation.

The impact of irrigation on erosion and salinization has been significantly diminished since 1990. There has been a sharp reduction of irrigated farmland (Figure 1) as merely 2–4.7 percent of existing irrigation network has been practicably used. Moreover, a considerable physical distortion of irrigation facilities has taken place affecting 80 percent of the internal canals (MAF). The harmful effect of the latter has been combined with a decrease in rainfall in recent years and shortage of irrigation water in reservoirs.[8] The decline in irrigation has had a direct negative effect on crop yields and the structure of the crop rotation. In addition, irrigation has not been effectively used to counterbalance the effect of global warming on farming (extension of farm season and increased water requirements)[9] and further degradation of agricultural land. Bulgarian farming has not been adapted to cope with

---

[7] Two-thirds of the water erosion losses and almost all of the wind erosion losses come from the arable land (EEA).

[8] The year 2007 is an example in this respect. According to the authority, the reservoirs for irrigation are half full in the middle of May (when they should be full) before the start of the irrigation season.

[9] During the last twenty years the farming season in Europe has been extended by ten days, and 1990, 1992, and 2000 were among the ten driest years of the last century (EEA).

new environmental challenges caused by the alterations of extreme drought and rainstorms, increased risk of dry winds and reduction of rainfall, and unfavorable modifications in soils quality.

The areas of agricultural land industrially polluted by heavy metals generated by industrial wastes have fallen since 1990 (Figure 3); they are not significant, and only about 30 percent of the affected soil needs special monitoring.[10] However, pollution of soil and water from industrial activities, waste management, and improper farming activities present risks for the environment and human health. Data shows that in 6.6 percent of the tested soil, the concentration of pollutants is higher than critical contamination limits (EEA, 2006).

Around a quarter of the riverlength does not meet the normal standards for good water quality (MAF, 2006). The nitrates content in groundwater has been decreasing, and now only 0.7 percent of samples exceed the ecological limit value of 10 mg/l (EEA, 2006). Nevertheless, monitoring of water for irrigation shows that in 45 percent of water samples, the nitrates concentration exceeds the contamination limit value by 2- to 20-fold (MAF, 2006). Nitrates are also the most common polluter of underground water for the last five years with a slight excess over the ecological limit (EEA, 2006). Recently defined Nitrate Vulnerable Zones (2004) cover 60 percent of the territory of the country and less than 7 percent of agricultural land use. The lack of effective manure storage capacities and sewer systems in the majority of farms contributes significantly to the persistence of the problem. According to the last census only 0.1 percent of livestock farms possess safe manure-pile sites; around 81 percent of them use primitive dunghills, and as many as 116,000 holdings have no facilities at all (MAF, 2003). A major part of the post-communist livestock practices are carried out by a great number of small and primitive holdings often located within village and town borders. This contributes significantly to the pollution of air, water, and soil and to population discomfort (unpleasant noise and odor, dirty roads).[11]

A serious environmental challenge has been caused by the deficiency in storage and disposal of the out-of-date or prohibited pesticides of the ancient public farms. Currently those chemicals account for 11,079 t, and a large proportion of them are not stored in safe places. As much as 82 percent of all polluted localities in the country are associated with these dangerous chemicals, and only a tiny portion of them have gone through the entire cycle of examination (EEA, 2006). There are 477 abandoned storehouses

---

[10] Pollution is localized in ecological problematic regions (EEA).

[11] Conflicts between bigger livestock operators and rural populations are common in recent years.

registered for such pesticides, situated in 460 locations around the country, and just 38 percent of them are guarded. In addition, general levels of pollutants exceeding the ecological limit for triasine pesticides in underground water have been reported through the past five years, which is a consequence of the increased use of these chemicals. Additionally, the illegal garbage yards in rural areas have noticeably increased.[12] Farms contribute extensively to waste "production" of both organic and industrial materials, leading not only to negative changes in the beauty of scenery but also bringing about air, soil, and water pollution.

There has been significant degrading impacts of agriculture on biodiversity in the country. The policy toward intensification and introduction of foreign varieties and breeds during communist period, and the lack of any policy toward protection of biodiversity afterwards have led to degradation of the rich diversity of local plants and animal breeds. According to the official data all 37 typical animal breeds in Bulgaria have been endangered during the last several decades as 6 among them are irreversibly extinct, 12 are almost extinct, 16 are endangered and 3 are potentially endangered (MEW, 2006).

Moreover, since 1990 a considerable portion of agricultural lands have been left uncultivated for a long period of time or entirely abandoned.[13] The latter has caused uncontrolled "development" of species, allowing development of some and suppression of others. Also, some of the most valuable ecosystems (such as permanent natural and semi-natural grassland)[14] have been severely damaged. Part of the meadowland has been left under-grazed or under-mowed, leading to the intrusion of shrubs and trees. Some of the fertile semi-natural grasslands have been converted to cultivation of crops, vineyards, or orchards. This has resulted in the irreversible disappearance of plant species diversity. Meanwhile, certain public (municipal, state) pastures have been degraded by unsustainable use (over-grazing) by private and domestic animals. In addition, a reckless collection of some valuable wild plants (berries, herbs, flowers) and animals (snail, snakes, fish) have led to the destruction of natural habitats. Furthermore, some genetically modified crops have been introduced without an

---

[12] The official figure for major illegal garbage locations is 4,000 (EEA).

[13] Currently, the "fallow land" accounts for 13.4 percent of the arable land and the "uncultivated" (for more than three years) farmland comprises 7.9 percent of all agricultural land (MAF, 2006). In some post-communist years the abandoned land reached up to 35 percent of the total agricultural land.

[14] The permanent grasslands comprise more than a third of Utilized Agricultural Land in the country (most of them are semi-natural) (MAF, 2006). Approximately 20 percent of the agricultural lands of Bulgaria are lands of High Nature Value according to the definition of the European Environment Agency (MAF).

independent assessment of possible hazards for traditional and organic production (genetic pollution) and human health, or providing appropriate safeguards or proper public and consumer information.

There has been a significant reduction of overall greenhouse gas (GHG) emissions from agriculture since 1988 (Figure 4). Moreover, the decline in the sector's contribution has been higher than the national. Since 2000 there has been a reverse trend toward a general increase in the total national emissions of GHG, and GHG from agriculture as well. In 2001–2004 agriculture released 1.8 percent of Nox, 9 percent of NMVOC, 16.9 percent of CH_4, 67.6 percent of NH_3, and 0.2 percent of CO in the country (EEA, 2005).

The N_2O emissions comprise 57.4 percent of the total emissions from agriculture (in CO_2 equivalent)[15] and there is a slight increase of the share in last five years. Agriculture has been a major ammonia source, accounting for two-thirds of the national emission. After 2000, the majority of NO_2 emissions comes from agricultural soils (86.7 percent)[16] and manure management (13.2 percent).

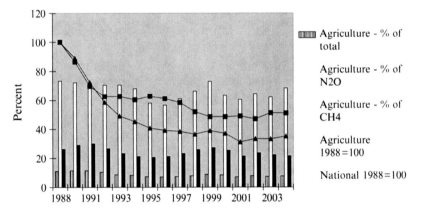

Figure 4. Trends and Components of Greenhouse Gas Emissions from Bulgarian Agriculture. *Source*: Vassilev et al. (2006).

---

[15] GHG emissions from the "agriculture sector" result from activities during the production and processing of agricultural products, soil fertilization, and animal manure processing and preservation. All emissions from combustion processes for energy production and from agricultural machines are reported in the "energy sector," but they represent insignificant amounts (EEA).

[16] N_2O emissions from "agricultural soils" include direct emissions (from soil fertilization with synthetic nitrogenous fertilizers; nitrogen input from manure applied to soils; decomposition of waste from N-fixing crops; decomposition of vegetable waste from other cultures' animals; cultivation of histosols; emissions from pasture animal)s, and indirect emissions (ammonia and nitrous oxides release in the ambient air after nitrogen fertilization; emissions from drawing of water) (EEA).

Methane emission from agriculture represents about 25 percent of the national. After 2000 the biggest portion of CH_4 comes from fermentation from domestic livestock (72.5 percent) and manure management (24.2 percent). There has been a fourfold increase of CH_4 from rice cultivation since 1999 as a result of the development of this farming sector. The share of GHGs from field burning of residues and crops (which also emits GHGs-precursors) is not significant but it has doubled since the period before 1990.

All of these figures show that Bulgarian agriculture is still not well adapted to environmental changes and the requirements of sustainable development.

3. Modes of Environmental Management in Bulgarian Agriculture

Throughout the fundamental transformation of Bulgarian agriculture all major modes (market, private, and public) have failed to provide an effective environmental management.

First, during most of the transition diverse environmental rights (on clean and aesthetic nature, preservation of natural resources, and biodiversity) were undefined or poorly defined and enforced.[17] Outdated systems of public regulations and control dominated until recently, which contributed little to contemporary needs of environmental management. In addition, there was no modern system for monitoring the state of soil, water, and air quality, and credible information on the extent of environmental pollution and degradation was unavailable. Furthermore, there existed neither social awareness of the "concept" of sustainable development nor any "need" to be included in public policy or private and community agenda. The lack of culture and knowledge of sustainability has also impeded evolution of voluntary measures and private and collective actions (institutions) for effective environmental management.

In the last few years before EU accession, the country's laws and standards were harmonized with the immense EU legislation.[18] *Acquis communautaire* has introduced a modern framework for the environmental governance including new rights (restrictions) on protection and improvement of the environment, preservation of traditional varieties and breeds, biodiversity, and animal welfare. However, a good part of these new "rules of the game"

---

[17] Lack enforcement of laws and absolute and contracted rights is universal in Bulgaria. The requirements for fighting against corruption and reforming administration and juridical system are underlined by the Monitoring Report for the Preparation for EU Membership (EU, 2006) and closely scrutinized by the EC after accession.
[18] Acquis communautaire contains 26,000 pieces of legislation accounting for 80,000 pages (EU, 2005).

are not well known or clearly understood by the various public authorities, private organizations, and individuals. Generally, there is not enough readiness for effective implementation of the new public order because of the lack of experience in agents, adequate administrative capacity, or practical possibility for enforcement of novel norms (lack of comprehension, deficient court system, and widespread corruption).[19] In many instances, the enforcement of environmental standards is difficult (practically impossible) since the costs of detection and penalizing offenders are very high, or there are no direct links between the performance and the environmental impact. For example, although the burning of (stubble) fields has been banned for many years since it is harmful for the environment, this practice is still widespread in the country.

The harmonization with the EU legislation and the emergence of environmental organizations also generate new conflicts between private, collective, and public interests. However, the results of public choices are not always to the advantage of effective environmental management. For instance, the strong lobbying efforts and profit-making interests of particular individuals (and groups) have led to a 20 percent reduction in numbers and 50 percent reduction in area of initially identified sites for the pan-European network for preservation of wild flora, fauna, and avifauna, NATURA 2000.[20]

Second, privatization of agricultural land and other land assets of ancient public farms took almost ten years to complete (Bachev, 2007b). During a substantial part of that period, the governance of critical agrarian resources was in ineffective and "temporary" structures (such as Privatization Boards, Liquidation Councils, Land Commissions). Sales and long-term lease markets for farmland did not emerge until 2000, and leasing on an annual basis was a major method for the extension of farm size until recently. That was combined with a high economic uncertainty and a big interdependency of agrarian assets (Bachev, 2006). A huge amount of part-time subsistence

---

[19] In Bulgaria in 2004 only 409 penalties for environmental pollution (amounting to €381,500) were imposed, including 201 for water pollution, 197 for air pollution, and 4 for soil pollution. In addition, for breaking environmental laws 993 fines were given for €276,500, including 111 in the water sector, 143—air, 330—waste, and others—409 (EEA).

[20] NATURA eco-network initially covered 225 zones with area corresponding to 36 percent of the country's surface. To date only 80 percent of the zones and 18 percent of the identified territory has been approved. The identified agricultural lands comprise 25.9 percent of the identified NATURA 2000 territory, and its share in total agricultural lands is 18 percent (MAF).

farming,[21] production cooperation on a large scale, and low sustainability of bigger farms (based on provisional lease-in contracts), all come as a result (Table 1). Specialized livestock farms comprise a tiny portion of all farms (mostly in poultry and swine) while 97 percent of the livestock holdings are miniature "unprofessional farms" breeding 96 percent of the goats, 86 percent of the sheep, 78 percent of the cattle, and 60 percent of the pigs in the country (MAF, 2005). Furthermore, market adjustment and intensifying competition has been associated with decrease in unregistered farms by 32 percent since 2000, the cooperatives by 51 percent, and the livestock holdings by 20 percent.

Dominating modes for carrying out farming activities have had little incentive for long-term investment to enhance productivity and environmental performance (Bachev, 2007b). The cooperative's large membership makes individual and collective control on management very difficult (costly). That focuses managerial efforts on current indicators, and gives a great possibility for using coops in the best private interests. Besides, there are differences in investment preferences of diverse members due to the non-tradable nature of the cooperative shares ("horizon problem"). Given the fact that most members are small shareholders, older in age, and nonpermanent employees, the incentives for long-term investment for land improvement and renovation of material and biological assets have been very low. Last but not least important, the "member-oriented" (not-for-profit) nature of coops prevents them from adapting to the diversified needs of members and markets (demand, competition).

TABLE 1. Number, Size, and Importance of Different Farm Types in Bulgaria.

Type of farm	Number		Percent of total number		Percentt of utilized farmland		Average size (ha)	
	2000	2005	2000	2005	2000	2005	2000	2005
Unregistered	755,300	515,300	99.3	99.0	19.7	33.5	0.9	1.8
Cooperatives	3,125	1,525	0.4	0.3	61.6	32.6	709.9	584.1
Firms	2,275	3,704	0.3	0.7	18.7	33.8	296.7	249.4
Total	760,700	520,529	100	100	100	100	4.7	5.2

Source: Ministry of Agriculture and Forestry.

---

[21] Currently, subsistence and semi-subsistence farms comprise the largest number of farms, and almost one million Bulgarians are involved in farming, mostly on a part-time basis and as a "supplementary" income source (MAF, 2006).

On the other hand, small-scale and subsistence farms possess insignificant internal capacity for investment, and small potential to explore economy of scale and scope (big fragmentation and inadequate scale). Besides, they have little incentive for nonproductive (e.g., eco-, animal welfare) investment. Moreover, there have been no administrative capacity and political will to enforce quality and eco-standards in that vast informal sector of the economy.

Likewise, the larger business farms operate mainly on leased land and concentrate on high pay-off investment with a short pay-back period. That has been coupled with ineffective outside pressure (by authority, community) for respecting official standards for ecology, land use (e.g., crop rotation, nutrition compensation), and biodiversity. In general, survivor tactics and behavior rather than a long-term strategy toward farm sustainability has been common among the commercial farms.

Furthermore, during the entire transition phase the agrarian long-term credit market was practically blocked due to the big institutional and market uncertainty, and the high specificity of much of the farm investments (Bachev, 2007b). In addition, newly evolving Bulgarian farming has been left as one of the least supported in Europe. Until 2000 the public aid was mainly in the form of preferential short-term credit for the grain producers and insignificant support to capital investments.[22] That policy additionally contributed to the destructive impact for unbalanced unilateral N fertilization by the biggest producers having access to the programs. Despite the considerable progress in public support since 2000 (most of which comes from EU Special Assistance Program for Agriculture and Rural Development) the overall support for agriculture is considered very low (Bachev, 2005). In addition, only a small proportion of the farms benefit from some form of public assistance, most of these farms being large enterprises from regions with fewer socioeconomic and environmental problems.[23] Basically, a publicly supported farm must meet the requirements for good environmental performance. However, the minor number of actually supported farms, the deficiency of clear criteria for eco-performance, and the lack of effective controls have led to a negligible contribution to overall improvement of

---

[22] Estimates demonstrate that the Aggregate Level of Support to Agriculture before 2000 was very low, close to zero or even negative (OECD, 2000).

[23] Under the SAPARD Measure "Investment in agricultural holdings" only 7.7 percent of all agro-firms, 2.3 percent of the cooperatives, and insignificant number of the unregistered farms got funding from the program as few projects are from the less-developed regions like South-West, North-West, and mountainous parts of the country (Interim Assessment of SAPARD Program in Bulgaria, MAF, 2004).

environmental situation in the country. What is more, the envisioned Special Accession Program for Agriculture and Rural Development measure "Agro-ecology" was not approved until the middle of 2006 and no project has been funded under that scheme.

Hence, since 1990 all "environmental management" has been left on farmers' "goodwill" and "market signals." Market governance (competition, marginal rule) has led to a significant decline in most of crop and livestock production (Figures 5 and 6). This is due to the several interrelations between

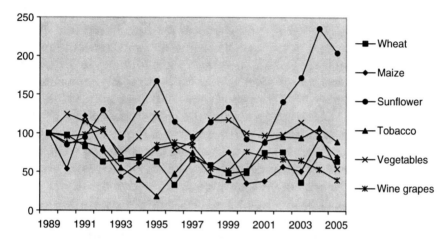

Figure 5. Dynamics of Crop Production in Bulgaria (1989 = 100).
Source: National Statistical Institute, Annual Statistical Indicators.

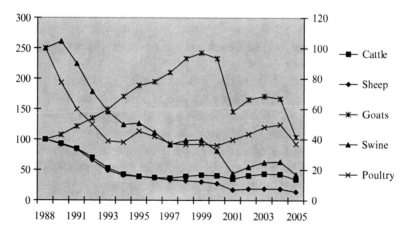

Figure 6. Dynamics of Number of Livestock in Bulgaria (1988 = 100).
Source: National Statistical Institute, Annual Statistical Indicators.

the rise in agrarian inputs (chemicals, water, machinery) and farm produce prices, a drop in internal and export demands, and the failure of a large number of farms to adapt to the new market environment. The smaller size and owner operating nature of the majority of farms avoided certain problems of large public enterprises from the past (e.g., lost natural landscape, biodiversity, nitrate and pesticide contamination, huge manure concentration, and uncontrolled erosion). Subsistence and small-scale farming has also revived some traditional (and more sustainable) technologies, varieties, and products. A by-product from that "market and private governance" was a considerable de-intensification of agriculture and an ease of general environmental pressure and generated pollution compared to the pre-reform level. Consequently, a good part of the farm production received an unintended organic character obtaining a good (internal and international) reputation for products of high quality and safety. In addition, the private mode has introduced incentives and possibilities for integral environmental management (including revival of eco- and cultural heritage, anti-pollution, aesthetic, and comfort measures) profiting from interdependent activities such as farming, fishing, agro-tourism and recreation, processing, and trade.

However, improved environmental stewardship to owned resources did not extend to nature in general (low appropriability of rights). Private management has often been associated with less concern for manure and garbage management, over-exploitation of leased and common resources, and contamination of air and groundwater. Basically, there is a considerable time and space gap for much of the environment-related costs and benefits. Little appropriability of such transactions (associated with positive or negative externalities) can be hardly governed though a market or private mode, with constant under-investment in positive externalities and over-production of negative externalities (Bachev, 2004). Subsequently, the adverse trend to further deterioration of soil quality (exhaustion, erosion, and acidification), distortion of eco-balance, and development of other environmental problems have taken place.

Market-driven organic farming has emerged in recent years as well (Figure 7). It is a fast growing approach but it is restricted to 92 farms and covers merely 0.23 percent of the Utilized Agricultural Area (MAF, 2006). In addition, 27,881 ha have been approved for gathering wild organic fruits and herbs. There are also five organic livestock farms with 722 animals (cattle, sheep, and goats) and 269 bio-apiaries with 23,883 bee families. The organic farm has been introduced by business entrepreneurs who managed to organize and fund this new venture, arranging needed independent certification[24]

---

[24] To date bio-certification has been done exclusively by certifying bogies from other countries. There are also local certifying companies approved by the authority.

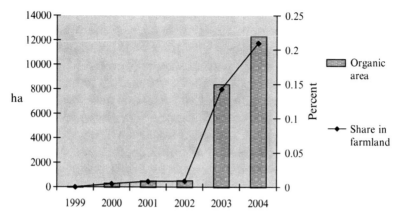

Figure 7. Development of Organic Farming in Bulgaria.
Source: Ministry of Agriculture and Forestry, Agrarian Report (2006).

and finding potential buyers for the highly specific output. Produced bio fruits, vegetables, essential oil plants, herbs, spices, and honey are entirely for export since no internal market exists in the country. The lack of organic market is not only because of the higher prices of organic products but also because of the limited consumer confidence in the authentic character of products and certification. In addition, eco-labeling of processed farm products (relying on self-regulation) has appeared, which has been more a part of the marketing strategy of certain companies than a genuine action for environmental improvement.

Since 2001 the assets of public-owned irrigation companies were transferred to the newly evolving Water Users Associations. However, the expected "boom" in efficiency (quantity, productivity) from a collective management of irrigation activities has not materialized. That is because of the semi-monopoly situation of regional state water suppliers (monopoly terms and pricing), little water-users' incentive to innovate facilities and expand irrigation, and still uncompleted privatization of state irrigation assets.

Generally, initiation, development, and maintenance of an organization of large groups is very costly, and such a coalition is not sustainable for a long time ("free rider" problem). In Bulgaria, the evolution of farmers and environmental associations has been additionally hampered by the large number of rural agents and their diversified interests (size of ownership and operation, type of farming, individual preferences, or different age and horizon).

Third, market and the private sector have failed to govern effectively the environment-related activities in Bulgarian agriculture. There has been a need for a third-party public intervention (assistance, regulation, support, partnership, in-house organization). However, the government and local authority involvement has not been significant, comprehensive, sustainable, or even related to the matter.[25] The total budget of the Ministry of Water and Environment accounts for just 1.5 percent of the national budget, and a sizable piece of it (more than 26 percent) comes from the EU (MEW, 2006). Besides, the agricultural sector gets a tiny portion of all public eco-spending.

In the past several years a number of programs have been developed to deal with the specific environmental challenges.[26] In addition, national monitoring systems of environment and biodiversity have been set up. Likewise, a mandatory ecological assessment of public programs was introduced in 2004. Nevertheless, the actual eco-policies rest fragmented and largely reactive to urgent environmental problems (natural disasters such floods, storms, drought) rather than based on a long-term strategy for sustainable development.[27] Moreover, there is no efficient coordination between different programs and management levels.[28] The programs and action plans are usually developed and executed in a highly centralized manner (by bureaucrats, foreign experts, and profit-making companies) without the involvement of independent local experts, stakeholders, or the public at large.[29] In addition, there is considerable deficiency in administrative capacity at local level in terms of staff, qualification, and material and financial means. As a result of all of this, inefficiency in priority setting and management (incompetence, corruption), and an inadequate impact of public programs prevails.

---

[25] For instance, a number of the programs of the Ministry of Labor and Social Policy aim to "create employment in environment activity" lasting one or two years.

[26] For example: National Strategy for Preservation of Biodiversity (1999); National Strategy for Environment (2000); National Plan for Agrarian and Rural Development (2000); National Programme for Limitation of Total Emissions of Sulphur Dioxide, VOC and Ammonia (2002); National Program for Waste Management Activities (2002); Environmental Strategy for the Instruments of ISPA (2003); National Strategy for Management and Development of Water Sector (2004); Strategy for Developing Organic Agriculture (2005).

[27] The most recent example is the decision of the Minister for Agriculture and Forestry to grant a free of charge irrigation for all crops but rice because of the forecast extremely dry summer and expected decrease in yield. At the same time no measures have been taken to prevent shortages of irrigation water in reservoirs, improve irrigation facilities, and induce incentives for farmers to increase effective irrigation.

[28] Even the specially established National Commission for Sustainable Development with representatives of different government agencies and NGOs does not meet regularly.

[29] A rare representative survey shows that 56 percent of the general public is "interested in problems of environment." However, almost 90 percent are "not familiar with the content of any of the major environmental programs," less than 2 percent have "taken part in public debates" while almost 80 percent do "not know they could participate in such discussion." And merely 3.4 percent have ever "participated in environmental protection activities" (National Survey on Population Attitude on Questions Related to Environment and Updating Action Plan 2005–2014, Forum Foundation, Sofia, 2004).

Moreover, a multifunctional role of farming has not been effectively recognized, proper system for its assessment (data, indicators) introduced, or provision of a public service "environmental preservation and improvement" funded by the society. Nor have the essential public institutions and infrastructure crucial for the sustainable farming development been built: the public system for enforcement of laws, regulations, and contracts does not work well; essential property rights (on environmental resources and biodiversity, special and organic products, GM products and intellectual agrarian property) are neither well defined nor properly enforced; public support programs are rarely governed effectively and in the best interest of the legitimate beneficiaries; agricultural research is under-funded and can hardly perform its function for innovation and independent expertise; the newly established agricultural advisory system does not serve the majority of farms and include environmental issues; an urgently needed public system for agrarian insurance has not been introduced; a crucial agrarian and rural infrastructure (wholesale markets, irrigation, roads, communications) has not been modernized; public support for initiating and developing farming associations has not been given (Bachev, 2007b).

A great number of international assistance projects (funded by the UN agencies, EU, Foreign Governments, or NGOs) have been carried out to "fill the gap" of the national government failures. They either focus on a specific issue (sustainable agriculture, desertification) or mobilize local actors for sustainable development. These programs introduce western experiences in governance and try to make a difference. However, they are limited in scale and unsustainable over time; in some cases overtaken by the local groups and funding improperly used; and above all with no significant impact. The endurance of environmental and other challenges demonstrates that an effective system of governance has not been put in place. Subsequently, the modernization of Bulgarian farms according to the EU standards (quality, safety, environmental, animal welfare) has been delayed; growth in farm productivity, competitiveness, and sustainability severely restricted; and technological, income, and eco-disparity between farms of different type, sub-sectors, and regions broadened (Bachev, 2007b).

4. Prospects of Environmental Performance of Bulgarian Farms

The EU integration and CAP implementation will provide new opportunities for Bulgarian farms (Bachev 2005, 2006). The EU funding alone, which agriculture will receive from 2007 on, will be 5.1 times higher than the overall level of support to farming before acceding. Besides, the EU accession introduces and enforces a "new order"—strict regulations and control; tough quality, food safety, and environmental standards; financial support and protection against market instability.[30] The external monitoring, pressure, and likely sanctions by the EU will lead to better enforcement of laws and standards in the country.[31] For instance, in March 2007 the EC started a procedure for sanctions against Bulgaria for not reducing emissions of greenhouse gases according to the EU Program for Environment and Combating Adverse Climate Changes.

Furthermore, huge EU markets will be opened that will enhance competition and let Bulgarian farms explore their comparative advantages (low costs; high quality, specificity, and purity of produce). The novel conditions of market competition and institutional restrictions will give strong incentives (pressure) for new investments for increasing productivity and conforming to higher product, technology, and environmental standards. The larger and business farms are most sensitive to the new market demand and institutional regulations since they largely benefit (or lose) from timely adaptation to new environmental regulations. In addition, they have a higher capacity to generate resources and find outside (credit, equity, public) funding to increase competitiveness and meet new institutional requirements. The process of adaptation will be associated with appropriate land management and the intensification of production. The latter could revive or deepen some of the environmental problems (erosion, acidification, pollution) unless preventative pro-environmental governance (public order, regulation) is put in place. On the other hand, small-scale producers and most livestock farms will have a hard time adapting to new competition pressure, investment needs, and new food safety, environmental, and animal-welfare standards (Bachev, 2006).

A significant number of the farms will be qualified to receive direct payments from the EU.[32] In view of the current (low) level of support, the direct payments will augment farm sustainability and give means for adaptation to the new standards. However, the EU support will benefit

---

[30] For 2007–2009 the EU funds allocated for agrarian and rural development, for direct payments, and for market support accounts for €733 million, €722 million, and €388 million accordingly. Besides, Bulgarian agriculture will receive resources from the EU Structural Funds and the national budget (MAF).

[31] The Monitoring Report for the Preparation for EU Membership has underlined the specific problems associated with CAP implementation in Bulgaria: operability of the Integrated Administration and Control System for CAP, nonexistence of Land Parcel Identification System, insufficient financial control capacity for Structural Funds implementation, needs for effective measures for collecting dead animals and animal by-products, and control on food safety (EU, 2006).

[32] Depending on the number of eligible applicants the level of subsidies will be between €69–74.2/ha in 2007, €82.8–89.1/ha in 2008, and €96.8–104.1/ha in 2009. Besides, certain farms may get top-ups from the national budget (Bachev, 2007b).

unevenly different types of farms, as no more than 3 percent of the farms (large farms, cooperatives, and agri-firms) will obtain more than 85 percent of the subsidies. Many effective small-scale operators (vegetables, tobacco, greenhouse) and subsistence farms will receive at most a tiny fraction of the direct payments.[33] Besides, specialized livestock farms will not be eligible to receive any payments under the "area based scheme." Above and beyond this, the bulk of subsidies will go to the more developed regions where the biggest farms and utilized farmland are located.

There will also be significant EU funds for rural development exceeding 4.7 times the relevant pre-accession level. This will let more and relatively smaller farms to access public support schemes and invest in modernization of enterprises. Furthermore, new essential activities will be effectively funded, such as commercialization and diversification of farming; introduction of organic farming; maintaining productivity; biodiversity; agri-environment protection; animal welfare; support for less-favored areas and regions with environmental restrictions.[34] All of these will bring additional employment and income for farmers, and increase economic and environmental sustainability of farms.

Mostly bigger farms would be able to participate in public support programs because they have superior managerial and entrepreneurial experience, available resources, possibilities for adaptation to new requirements for quality and other standards, and potential for preparing and winning projects. Besides, the actual system of governance of public programs would be less likely to change overnight. Therefore, agrarian and rural development funds will continue to benefit exclusively the largest structures and the richest regions of the country, and more abuses will likely take place; and CAP support will not contribute to decreasing economic and ecological discrepancy between farms, sectors, and regions.

The CAP implementation will improve the environmental performance of farms. There is a mandatory requirement for farms to "keep the farmland in a good agricultural and environmental status" in order to receive direct payments and participate in public programs. Moreover, direct payments would likely induce farming on currently abandoned lands, and improve environmental situation and biodiversity. Furthermore, there will be a huge budget allocated for special environmental measures (going beyond "good farming practices").[35] A large number of farms will have economic incentives to take part in various agri-environmental and animal welfare programs.

---

[33] A minimum farm size eligible for a direct payment is set on 1 and 0.5 ha for orchards and vineyards. A minimum of 0.1 ha for each parcel also applies.

[34] National Strategy Plan for Rural Development for 2007–2013 (NSPRD).

[35] NSPRD allocates a budget for "preservation of national resources and improvement of countryside" amounting €623.3 million (27.1 percent of the total funding) (MAF).

The CAP measures will positively affect the environmental performance of large business farms in that these enterprises (and potential big polluters) will be under constant administrative control and severe punishment (fines, loss of licenses, and closure) for disobeying new environmental and animal welfare standards. Therefore, they will be strongly interested in transforming their activities according to the new eco-norms, making necessary eco-investments, and changing production structures. Moreover, larger producers will be motivated to participate in special agro-environmental and biodiversity programs, since they have lower costs (potential for exploring economies of scale and scope) and higher benefits from such long-term public contracts.

Nevertheless, some of the terms of the specific contracts for environment and biodiversity preservation, animal welfare, and keeping tradition are very difficult (expensive) to enforce and dispute.[36] In Bulgaria the rate of compliance with these standards will be even lower because of the lack of readiness and awareness, insufficient control, an ineffective court system, and domination of "personal" relations and bribes. Correspondingly, more farms than would otherwise enroll will participate in such schemes (including the biggest polluters and offenders). Subsequently, the outcome of implementation of these sorts of instruments would be less than the desirable "European" level.

More to the point, direct costs and lost income for conforming to the requirements of the special programs in different farms will vary considerably, and they will have unequal incentives to participate. Having in mind the voluntary character of the most CAP support instruments, we should expect that the biggest producers of negative impacts (large polluters and noncompliant with modern quality, agronomic, biodiversity, and animal welfare standards) will stay outside of these schemes since they would have the highest environment enhancement costs. On the other hand, small contributors would like to join since they would not command great efforts (and additional costs) compared to the supplementary net benefit. Moreover, the government is less likely to set up high performance standards because of the perceived "insignificant" environmental challenges, the strong internal political pressure from farmers, and the possible external problems with the EU control (and sanctions) on cross-compliance. Therefore, CAP implementation will probably have a modest positive impact on the environmental performance of Bulgarian farms.

Public support and new public demand will give a push to further development of market modes such as organic farming, industry driven eco-initiatives (eco-labeling, standards, professional codes of behavior), protected high

---

[36] The low rate of farmers' compliance with environmental contracts is a serious problem even in some of the old EU member states (Dupraz et al., 2004).

quality products (protected designation of origin, protected geographical indication, and traditional specialty guaranteed), a system of fair trade, and production of alternative (wind, manure) energy at farms. For instance, the significant EU market and lower local costs will create strong incentives for investment in organic and specific productions by the large enterprises—farms, partnerships, and joint ventures (including with non-agrarian and foreign participants). The small farms would have less capacity to find necessary capital and expertise for initiating, developing, certifying, and marketing. In addition, the coalition (development, management, and exit) costs between small-scale producers are extremely high to reach an effective operation level. Therefore, the latter will either stay out of these new businesses or have to integrate into larger or non-farm ventures. However, assuring the effective traceability of the origin and quality for small farms is very costly and they are not preferable partners for integrators (processors, retailers, and exporters).[37] On the other hand, the internal market for organic and specialized farm products would be unlikely to develop fast,[38] having in mind the low income of the population and the lack of confidence in public and private systems of control.[39] Besides, when the appropriability of absolute and contracted rights is low (easy to copy a product, difficult to detect and punish offenders) private governance fails to organize activities in a socially optimal scale (underinvestment and under-supply of products).

Many semiprofessional and professional livestock farms will be less sustainable in the middle term because of low productivity and noncompliance with the EU quality, hygiene, animal welfare, and eco-standards. Few of these farms will be able to adapt through specialized investment for enlargement and conforming to the new institutional restrictions. Meanwhile, the EU pressure for enforcement of standards in the commercial sector will increase and lead to closure or takeover of a large number of livestock farms. The improvement of manure management and reduction of animals will be associated with a drop of the environmental burden by the formal sector (less overgrazing, manure production, and mismanagement).

---

[37] The high transaction costs for complying with new EU safety requirements is the reason for a gradual replacement of local milk suppliers by milk imported from EU.

[38] The National Action Plan for Developing Organic Agriculture envisions a very optimistic figure of 8 percent of the utilized agricultural land to be managed under organic production methods by 2013 (MAF, 2005).

[39] Detected cases of control failures are reported daily by the media. Recent European crises with mad cow disease and avian fever is also indicative. Despite the assurance of Bulgarian authorities, the consumption of beef and chicken decreased 40–50 percent in the country.

On the other hand, only a few subsistence farms will likely undertake market orientation and extend their present scale because of the lack of entrepreneurship and resources, the advanced age of farmers, and insufficient demand for farm products (Bachev, 2007b). It will be practically impossible for authorities (costly or politically undesirable) to enforce the official standards in that huge informal sector of the economy. Therefore, massive (semi)subsistence farming with primitive technologies, poor food safety, and low environmental and animal welfare standards will continue to exist in years to come.

Economies of scale and high interdependency of assets, nonetheless, change both size and governance of individual farms, evolution of group organization, cooperation, and joint ventures. For instance, interdependency of activities would require concerted actions for achieving certain eco-effect; a high asset dependency between livestock manure (over) supplier and nearby (manure demanding) organic crop farms would necessitate coordination. A special governing size or mode will be also imposed by some of the institutional requirements. For example, a mandatory minimum scale of activities is set for taking part in certain public programs (e.g., marketing, agri-ecology, biodiversity, organic farming, tradition and cultural heritage); signing a five-year public environmental contract would dictate a long-term lease or purchase of managed land.

Some production cooperatives would also profit from their comparative advantages (interdependency and complementarity to individual farms, potential for exploring economy of scale and scope on institutionally determined investment, adapting to formal requirements for support, using expertise, financing and executing projects, not-for-profit character), and extend their activities into eco-projects, environmental services, and eco-mediation between members.

Principally, the hybrid modes (public-private partnership) are much more efficient than the pure public forms given the coordination, incentives, and control advantages (Bachev, 2007a). In the majority of cases, the involvement of farmers, farmers' organizations, and other beneficiaries increases efficiency, decreases asymmetry of information, restricts opportunism, increases incentives for cost-sharing, and reduces management costs. For instance, a hybrid mode is more appropriate for supplying non-food services by farmers such as preservation and improvement of biodiversity, landscape, and historical and cultural heritages. That is determined by the farmers' information superiority, the strong interlinks of activity with the traditional food production (economy of scope), the high assets specificity to the farm (farmers' competence, a high cite-specificity of investments to the farm and land), and the spatial interdependency (the need for cooperation of farmers at a regional or wider scale), and the farm's origin of negative externalities. Furthermore, enforcement of most labor, animal welfare, and biodiversity standards is often very difficult or impossible. Therefore, public support to voluntary environmental initiatives of farmers and rural organizations (informing, training, assisting, funding) would be much more effective than mandatory public modes in terms of incentive, coordination, enforcement, and disputing costs. Furthermore, involvement of farmers, farmers' organizations, and interest groups in priority setting and management of public programs at different levels is to be institutionalized in order to decrease information asymmetry and the possibility for opportunism, diminish costs for coordination, implementation, and control, and increase overall efficiency and impact.

An immediate result of new market and public opportunities for getting additional income (profiting) from environmental products and services will be an amelioration of the economic performance and overall sustainability of many farms and rural households.

There are still a number of "blank points" in the adaptation of EU regulations in Bulgarian agriculture. For instance, "the whole farm" is a subject of support in agri-environmental measures (e.g., organic farming), but its borders are not defined at all in the national legislation. This will create serious difficulties since land and other resources of the majority of farms are considerably fragmented and geographically dispersed.

Furthermore, surveys show that many of the specific new regulations are not well known by the implementing authorities and the majority of farmers. The lack of readiness and experience would require some lag time until the "full" implementation of the CAP. The latter will depend on the pace of building an effective public and private capacity, and training (by experience) of bureaucrats, farmers, and other agrarian agents. Besides, most of the farm managers have no adequate training and managerial capability and are older[40] with a low learning and adaptation potential. Therefore, there will be significant inequalities in application (enforcement) of new laws and standards in diverse sectors of agriculture, farms of different type and size, and various regions of the country.

Last but not least, there will be enhanced competition for environmental resources between different industries and interests. That will further push overtaking natural resources away from the farm governance into non-agricultural (urban, tourism, transport, industry) use. The need to compete for and share resources would also deepen conflicts between various interests and social groups, regions, and even neighboring states. All this will require a special governance (cooperation, public order, hybrid form) at local,

---

[40] Farm managers older than 45 and 65 are 85 percent and 40 percent respectively (MAF, 2004).

TABLE 2. Experts Assessment on Likely Short-Term Impact of EU CAP Implementation
on Bulgarian Commercial Farms[a].

Type farms	Farm income	Amount of investment	Product quality	Access to public programs	Environmental performance	Animal welfare
USL	0	0 0	0	0	0	0 0 0
USC	0	0 0	0 0	0	0	0 0 0
UBL	+ +	+ + +	+ +	+ + +	+	+ + +
UBC	+ + +	+ + +	+	+ + +	+	0 0
CoSL	−	0 0	+	0	+	0 0 0
CoSC	+	0 0	−	0	+	0 0 0
CoBL	+ +	+ + +	+ + +	+ + +	+ +	+ + +
CoBC	+ + +	+ + +	+ +	+ + +	+ +	0 0 0
FSL	+	0	+	+	−	+
FSC	+	+ +	0	+	−	0 0 0
FBL	+ +	+ + +	+ + +	+ + +	+ + +	+ + +
FBC	+ + +	+ + +	+ +	+ + +	+ +	0 0 0

[a](one digit): 43–60 percent consensus, (two digits): 61–80 percent consensus, (three digits): 81–100 percent consensus, (+)—positive effect, (0)—neutral effect, (−)—negative effect, (U)—unregistered farms, (Co)—cooperatives, (F)—firms, (S)—relatively small, (B)—relatively big, (L)—livestock farms, (C)—crop farms.
Source: Interviews with leading Bulgarian academic experts in farm structures.

national, and transnational levels to reconcile conflicts for the benefit of effective environmental management.

Our analysis on likely short-term (2007–2009) impact of CAP implementation on sustainability of farms is also supported by the leading experts on farm structures in Bulgaria.[41] Most experts believe that the introduction of CAP measures will affect positively the income of relatively large farms while the impact on income of smaller holdings and most livestock farms is expected to be neutral or even negative (Table 2). Furthermore, it is considered that CAP will have a positive effect on investment activities of the bigger farms, while no impact is envisaged for others.

Experts estimate that all farms will be forced to adapt to superior EU quality standards and improve product quality. However, CAP implementation will boost the access to public support programs of relatively big farms while no significant changes in distribution of public funds to smaller farms is anticipated. Expectation is that CAP will improve the environmental activities of relatively big firms, unregistered farms, and the largest part of

[41] Our survey was done in autumn 2006, just on the eve of the country's accession to the EU.

the cooperatives. Most experts imply that new rules will positively affect animal welfare in the "professional" operators—bigger livestock farms of all types, and smaller livestock firms.

5. Conclusions

The analysis of environmental management in Bulgarian agriculture has let us identify the major environmental problems and risks, specify the driving factors for their emergence and persistence, and make a more realistic forecast about future eco-development.

We have proved that the contemporary development of Bulgarian agriculture is associated with specific environmental challenges quite different from old EU member states. Some of the environmental problems and risks associated with agriculture reach the point of limited or nonexistent management—e.g., degradation of soil quality, erosion, and water pollution. This is the result of the specific institutional and governing structure evolving in the agrarian sector during the past twenty years. Furthermore, the comparative institutional and transaction costs analysis shows that implementation of the common EU policies will give unequal results in "Bulgarian" conditions. In the short and medium terms it most likely will enlarge income and technological and environmental discrepancy between different farms, sub-sectors, and regions. In the longer term, environmental hazard(s) caused by agricultural development will increase unless effective public and private measures are taken to mitigate the existing and emerging environmental problems (risks).

The identification of efficiency, complementarities, and sustainability of different (market, private, public, and hybrid) modes of governance for the specific environmental problems and risks in each country is a big challenge of modern development. In particular Bulgarian conditions it has a substantial importance for the modernization of public policies and property rights structure, and individual and business plans of improvement of eco-management. Firstly, it helps anticipate possible cases of market, private sector, and public (community, government, and international assistance) "failures," and designs appropriate modes for new public intervention (assistance, regulation, in-house organization, partnership with private sector, and fundamental institutional and property rights modernization). In particular, it formulates specific policies and institutional frameworks to overcome the existing environmental problems and safeguard against the possible eco-risks, and avoid the severe environmental challenges in developed countries. Next, it could assist various agrarian and rural agents' organizations, and individual and collective actions allowing a successful adaptation to changing economic, institutional, and natural environmental conditions.

References

Bachev, H., 2004, Efficiency of agrarian organizations, in: *Farm Management and Rural Planning 5*, Kyushu University, Fukuoka, pp. 135–150.

Bachev, H., 2005, National policies related to farming structures and sustainability in Bulgaria, in: *Implementing the CAP Reform in the New Member States: Impact on the Sustainability of Farming System*, EU JRC—IPTS, Seville.

Bachev, H., 2006, Governing of Bulgarian farms—Modes, efficiency, impact of EU accession, in: *Agriculture in the Face of Changing Markets, Institutions and Policies: Challenges and Strategies*, J. Curtiss, A. Balmann, K. Dautzenberg and K. Happe, eds., IAMO, Halle (Saale), pp. 133–149.

Bachev, H., 2007a, Governing of agrarian sustainability, *ICFAI Journal of Environmental Law*, Vol. VI, 2, Hyderabad, pp. 7–25.

Bachev, H., 2007b, Post-communist transition in Bulgaria—Implications for development of agricultural specialization and farming structures, in: *Agricultural Specialization and Rural Patterns of Development*, University of Rennes, Rennes.

Bentcheva, N., and Georgiev, S., 1999, Country report on the present environmental situation in agriculture—Bulgaria, in: *Central and Eastern European Sustainable Agriculture Network*, proceedings FAO workshop, Gödöllo.

Dupraz, P., Latouch, K., and Bonnieux, F., 2004, Economic implications of scale and threshold effects in agri-environmental processes, paper presented at the 90 EAAE Seminar, October 27–29, Rennes.

EEA, 2005, *Annual State of the Environment Report 2005*, Executive Environment Agency, Sofia.

EEA, 2006, *Annual State of the Environment Report 2006*, Executive Environment Agency, Sofia.

EU, 2005, *Europe in 12 Lessons*, EC, Brussels.

Furuboth, E., and Richter, R., 1998, *Institutions and Economic Theory: The Contribution of the New Institutional Economics*, University of Michigan Press, Ann Arbor, MI.

Gatzweiler, F., and Hagedorn, K., eds., 2003, *Institutional Change in Central and Eastern European Agriculture and Environment*, CEESA/FAO Series, FAO and Humboldt University of Berlin, Berlin.

MAF, 2003, *Agricultural Census*, Ministry of Agriculture and Forestry, Sofia.

MAF, 2005, *Agrarian Report for 2005*, Ministry of Agriculture and Forestry, Sofia.

MAF, 2006, *Agrarian Report for 2006*, Ministry of Agriculture and Forestry, Sofia.

MEW, 2006, *Annual Report for 2006*, Ministry of Environment and Water, Sofia.

North, D., 1990, *Institutions, Institutional Change and Economic Performance*, Cambridge University Press, Cambridge.

OECD, 2000, *Review of Agricultural Policies: Bulgaria*, OECD, Paris and Sofia.

Sumelius, J., 2000, *A Review of State of Sustainability of Farming Systems in Selected Central and Eastern European Countries*, FAO, Rome.

Vassilev, H., Christov, C., Hristova, V., and Neshev, B., 2006, *Greenhouse Gas Emissions in Republic of Bulgaria 1988, 1990–2004, National Inventory Report 2004*, Ministry of Environment and Water (MEW), Sofia.

Williamson, O., 1996, *The Mechanisms of Governance*, Oxford University Press, New York.

REGIONAL ASSESSMENT OF LANDSCAPE AND LAND USE CHANGE IN THE MEDITERRANEAN REGION

Morocco Case Study (1981–2003)

MALIHA S. NASH[1], DEBORAH J. CHALOUD[2],
WILLIAM G. KEPNER[3] AND SAMUEL SARRI[4]

[1-3] *U.S. Environmental Protection Agency**
[4] *College of Southern Nevada***

Abstract: The ability to analyze and report changes in our environment and relate them to causative factors provides an important strategic capability to environmental decision and policy makers throughout the world. In this study, we linked changes in land cover with changes in human demographics and natural phenomena, including rainfall. The methodology presented here allows users to locate and map changes in vegetation cover over large areas quickly and inexpensively. Thus it provides policy makers with the capability to assess areas undergoing environmental change and improve their ability to positively respond or adapt to change. Morocco was used as an example and changes in vegetation cover were assessed over a twenty-three-year period (1981–2003) using 8-km Normalized Difference Vegetation Index (NDVI) data derived from the Advanced Very High Resolution Radiometer (AVHRR). A regression model of NDVI over time was developed to identify long-term trends in vegetation cover for each pixel in the study area. Patches of changes in vegetation cover were identified using ArcView for visualization of specific areas. A decreasing trend in vegetation cover is an indicator of some type of stress, either natural (e.g., drought, fire) or anthropogenic (e.g., excessive grazing, urban growth), that affects the life-support function of the environment for humans. Although Morocco was the only country used in this case study, the described approach has broad application

*Address correspondence to Maliha S. Nash, Deborah J. Chaloud, and William G. Kepner, U.S. Environmental Protection Agency, Office of Research and Development, 944 East Harmon Avenue, Las Vegas, Nevada, 89119, USA
**Address correspondence to Samuel Sarri, College of Southern Nevada, Department of Philosophical and Regional Studies, 333 Pavilion Center Drive, Las Vegas, Nevada, 89144, USA

P.H. Liotta et al. (eds.), Environmental Change and Human Security, 143–165.
© 2008 *Springer Science + Business Media B.V.*

throughout the world and offers an opportunity for combating changing ecological conditions that affect populations.

Keywords: Morocco; Southern Mediterranean; anthropogenic; natural and socio-economic factors; AVHRR; changes in NDVI; autoregression; ecological services; desertification

1. Introduction

Unplanned and uncontrolled use of land may exacerbate environmental degradation and have profound consequences on human well-being. For local communities to thrive in an area, the natural systems that support them with good soil, water, and vegetation need to be maintained and monitored. When the natural resource support system degrades because of climate or anthropogenic causes, local communities begin to search for alternatives to support their needs. While some communities migrate to nearby or distant locations, others are attached by their heritage so they remain in their native land and continue using local resources.

The Argane Forest, located in southwest Morocco, is an example, where the Argane tree species (*Argania spinosa*) has an essential role in the ecology, economy, and social relations of the local communities (Belyazid, 2000). Felling (timbering) of trees as a source of income, drought episodes, and the slow regeneration of the Argane trees have degraded most of this habitat. The Argane Forest provides an important example of the necessity of monitoring that can be used to assess current conditions and past trends. Similar relationships between the ecological state and the socioeconomic prosperity of communities that depend on keystone species, such as the Argane tree, can be found at other sites in Morocco.

Changes in vegetation cover over time can be used to assess the ecological condition of a given area. Reforestation and restoration of grazed areas, for instance, will increase vegetation cover, while increasing impervious surface area due to urban growth will decrease vegetation cover (Nash et al., 2006). Mapping changes in vegetation cover using the Advanced Very High Resolution Radiometer (AVHRR) is an inexpensive technique to monitor and assess changes in environmental conditions over large areas.

Changes in vegetation were studied to monitor spatial-temporal dynamics of vegetation using the Normalized Difference Vegetation Index (NDVI) (Minor et al., 1999; Lanfredi et al., 2003; Gurgel and Ferreira, 2003; Nash et al., 2006). Long-term monitoring using a simple and inexpensive method may help provide environmental decision makers with early warning signals of areas where land sustainability is being degraded, or alternatively, to confirm improvement in land cover. Changes in vegetation cover and environmental conditions can be detected and quantified using current communication technology combined with remote sensing data, historical data, published research, and expert knowledge (Schmidtlein, 2005; Nash et al., 2006). NDVI has been used in desert and semidesert areas in many African countries and over broad spatial scales. For the African Sahel, data from 1981–2003 were used to examine the synchronization between drought and the decrease in greenness for specific time and spatial scales (Anyamba and Tucker, 2005). Our study focused on Morocco as a case study that could be used as an approach to examine the entire region (Figure 1). Morocco was

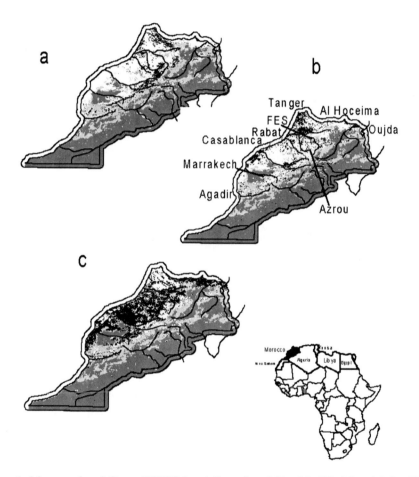

Figure 1. Morocco Land Cover (USGS Land Cover/Land Use Modified Level 2 for 1993), Major Cities, and Geographical Location in Africa. Desert and Semidesert Are Represented by Dark- and Light-Gray Shading, Respectively. Black Represents (a) Evergreen Needle Leaf Trees, Deciduous Broadleaf Trees, Evergreen Broadleaf Trees, and Interrupted Forest, (b) Crops, Short and Tall Grass, and (c) Evergreen Shrubs. Lines Represent Rivers. A Buffer of 25 km Surrounds Morocco.

chosen for a number of reasons, including: (1) it is moderate in size, (2) it has variation in topography and land cover types, (3) it experienced significant landscape changes over the last several decades, (4) it has a history of increasing socioeconomic pressures in parts of the country, and (5) primary and ancillary data for the country are available. Time-series analyses were used to quantify changes per pixel. The changes in vegetation cover were determined using the slope value of the regression model for NDVI (n = 810/pixel). Statistical analyses were performed for each pixel over a twenty-three-year period, and results were mapped using ArcView 3.3 for visualization and assessment at local and regional scales. The objective of the work presented in this paper was to use an approach that could have broad applications relative to identifying areas of environmental instability and trends over multiple spatial and time scales. Further, the research presented in this paper was intended to explore the links between natural and social sciences as they relate to solving contemporary environmental problems, especially those that are trans-boundary in nature. Specifically, this research was used to locate areas in Morocco that experienced significant changes in vegetation cover, as indicated by significant changes in NDVI values, and to determine, where possible, the likely cause of the change. Ancillary data such as human population distribution and change were mapped to be synchronous with that of NDVI changes. However, the overall intent was to develop a relatively inexpensive and simple process that could be employed to examine environmental sustainability at multiple scales, including catchments, national scale, or regions such as the Mediterranean. Collectively, the combination of socioeconomic data with NDVI change data will assist in promoting trans-boundary cooperation and may result in providing a coordinated response to environmental challenges.

2. Data and Methods

2.1. STUDY AREA DESCRIPTION

Morocco covers an area of 44,630,000 ha with elevation ranging from 4,165 m at the Jebel Toubkal (south of Marrakech) to 55 m below sea level at the Sabkha Tah (close to the Western Sahara and 22 km south of the city of Tarfaya) (Google Map & Google Earth).

Morocco is within the Mediterranean bioclimatic classification (Berkat and Tazi, 2004). The combined influence of the sea and ocean and the presence of the Atlas Mountains that transect the land from north (Rif Mountains) to south in the Sahara Desert create a diverse climate range for Morocco. Within this topographically diverse region, normal annual precipitation varies from less than 100 mm/year in the deserts and coastal plains to 1,200 mm/year in the Rif and the Middle Atlas mountains and exhibits a bimodal precipitation pattern. Most precipitation occurs as rainfall during October–December and March–April.

Water comprises only 0.06 percent of the total Moroccan land cover. Many of the rivers are intermittent and are mainly used for irrigation and generating electricity. The Moulouya River drains to the Mediterranean Sea while most of the other major rivers (Sebou, Bou Regre, Oum Rabia, Tansift, and Souss) drain to the Atlantic Ocean. The largest river in Morocco is the Drâa River, which originates in the Atlas Mountains and flows southeast through the cities of Ouarzazate and Zagora before crossing the border into Algeria and eventually winding west toward the Atlantic. It drains into the ocean only when river flows are sustained by prolonged periods of precipitation.

Morocco experienced periods of drought in 1979–1984 and throughout most of the 1990s. Socioeconomic conditions are directly related to precipitation in that most of the arable lands are located in semiarid areas (precipitation 250–500 mm/year). Approximately 22 percent of the land use in Morocco is arable and contains permanent crops (MONGABAY. com, 2000). While crops are primarily located along rivers in the coastal plain and semidesert Saharan areas, much of the low-productivity Saharan steppes are utilized as rangeland, accounting for approximately 75 percent of the land use. Forests account for less than 10 percent of the land cover (Figure 1).

2.2. DATA

The primary data source for this study was NDVI composite images generated by the National Oceanic and Atmospheric Administration (NOAA). The NDVI is calculated as:

$$NDVI = (IR - R) / (IR + R)$$

where IR is the Near Infrared Layer and R is the Visible Red Layer of the electromagnetic spectrum (Rouse et al., 1973; Tucker, 1979). The near infrared band of the spectrum emphasizes the contrast between vegetation and water. In the Visible Red Layer, vegetation appears darker than man-made structures. The calculated NDVI ranges from -1 to 1. For our study we used AVHRR data available from the United States Geological Survey (USGS) (ftp://edcftp.cr.usgs.gov). The images (*.bil) for the continent of Africa were converted to a grid, defined to a common projection, and clipped to a buffered (25 km) boundary of Morocco. NDVI were scaled to range from zero to 255, where 255 is water and 253 is a designated "invalid" or missing value. Therefore, NDVI values between 0 and 252 were used in the analyses. The NDVI data set consisted of ten-day composite NDVI data over a twenty-three-year period (1981–2003). This provided a maximum of 810 observations per pixel (8 × 8 km), which resulted in a total of 6,289,650 observations.

Another important data set used for this study was the Gridded Population of the World, version 3 (GPWv3, 2005). Population data used in the analyses were adjusted to match the UN total population figures. Change in populations between 1990 and 2000 were mapped and used as overlays to areas of significant change in vegetation cover.

2.3. ANALYSES

Time series regression (autoregression) was used to estimate significant trends in NDVI at the single-pixel level. This type of analysis was selected because errors in temporal data may be dependent (i.e., autocorrelated). If such dependency exists, then the standard error of the estimate (e.g., slope) will be inflated, and the significance level of the slope will not be correct. The Statistical Analysis Software (SAS) was used for all analyses (proc autoreg; *SAS/ETS*, 1999). In addition to the time series analysis that used all available data points over the full period, maps of averaged NDVI were created for four-year intervals (1981–1984, 1985–1988, 1989–1992, 1993–1996, 1997–2000, 2001–2003), roughly corresponding to known periods of drought and above-average rainfall (Figure 2). These maps facilitate observations of vegetation patterns over the study area and changes in the level of greenness or photosynthetic activity over time. Average NDVI values are divided into five groups of increasing greenness, starting with "no vegetation" (NDVI ≤ 25) and ending with "dense vegetation" (NDVI ≥ 126; Table 1).

The complete time series includes 810 observations for each 8 × 8 km grid cell and a total of 7,765 grid cells were utilized in the analyses. The slope value obtained from the regression corresponds to changes in vegetation cover. A negative slope indicates loss of vegetative cover while positive slope indicates an increase in vegetation cover. We use a value of 0.05 as a significance level for the probability of the slope. The significant slopes for the NDVI are mapped for the study area. From the map, patches of both positive and negative significant NDVI change over time are identified and marked. Ancillary data sources are then consulted to assist in identifying the probable causes of the significant changes. These ancillary sources include literature, maps, regional experts, satellite aerial photography from Google_Earth and Google Maps, Interagency Vegetation Mapping Project (IVMP) utility land cover datasets, and datasets from the World Database on Protected Areas (WDPA).

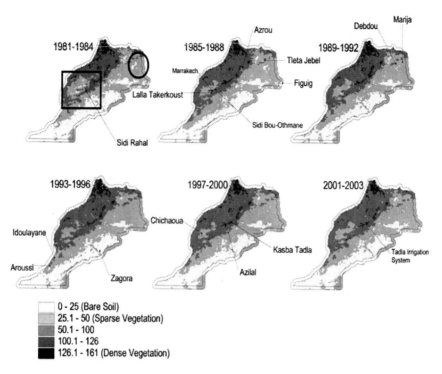

Figure 2. Average NDVI Across Morocco for Each Four-Year Temporal Group. The Marked Circle and Rectangle Designate Specific Areas Discussed in Text and to Follow Changes Over Time. Cities, Rivers, and Irrigation Systems Were Posted on the Maps for Geographical Positing and Referencing. There Were Several Drought Periods Since 1970; Drought Was Recorded in Year Groups (e.g., 1980–1985, 1990–1995) and a Few Wet Years Were Also Observed During the Mid-1980s and 1990s, Especially Along the Atlantic Coast and South of the Atlas Mountains.

TABLE 1. Changes in the Relative Distribution of Greenness with Time in Morocco. NDVI Group 0–25 Is Characterized by Desert (Bare) Area. NDVI Values ≥25.1 Are an Indication of Vegetation Presence. NDVI of 126 and More Indicate Dense Vegetation. Values Are Percent of Land Surface.

	0–25.0	25.1–50	50.1–100	100.1–126	126.1–157
Years			(%)		
1981–1984	34.9	41.0	17.7	3.8	2.7
1985–1988	31.6	42.1	18.0	4.4	4.0
1989–1992	29.9	42.5	19.5	4.7	3.4
1993–1996	29.4	43.5	19.7	4.0	3.4
1997–2000	28.7	43.2	20.6	4.6	3.0
2001–2003	29.3	44.1	19.3	4.5	2.8

3. Results

Figure 2 provides a spatial representation of the distribution of average NDVI for each of the four-year intervals roughly corresponding to
periods of drought and adequate rainfall. It is evident in this figure that
the majority of land in Morocco falls into the sparsely vegetated (NDVI
values between 25 and 50) and bare (NDVI < 25) classifications. The
highest NDVI values (>126), representing dense vegetation, are located in
the mountains and in urban areas on the Mediterranean coast; for example,
where the cities of Elbiutz, Faham, Mraheddebane, and Beneelouidane
are located. South of the mountains, NDVI values decrease, indicating
the sparser vegetation cover of the desert lands. Within the desert areas
close to the Algerian border, the Drâa River valley supports oases of date
palms around the cities of Zagora, Benizouli, and Asrir n'llemchance
(Figure 2). In addition to date palms, agriculture in the area includes
vegetable farms, primarily located around the town of Ouarzazate near
the Algerian border.

The relative distribution of vegetation groups for each of the four-year
intervals is quantified and summarized in Table 1. Bare (NDVI < 25) and
sparsely vegetated (NDVI between 25 and 50) together account for more
than 70 percent of the land area in each interval. The bare classification
has decreased slightly over time, from 35 percent in the first interval to
29 percent in the 1997–2000 interval, while the sparsely vegetated category has increased, from 41 percent in the 1981–1984 interval to 44
percent in the 2001–2003 interval. This shift in greenness is likely due to
increased irrigation and drought recovery. Dense vegetation, primarily
forested land in the mountainous areas where rainfall is generally higher,
increased from 2.7 percent to 4 percent between the first two intervals,
but has been decreasing in each subsequent interval. The initial increase
is likely due to drought recovery in the latter half of the 1980s, while the
decreases since that decade are likely due to a combination of drought
and tree cutting for firewood.

Of particular interest over the four-year intervals are the areas denoted
with a rectangle and a circle in Figure 2. Within the rectangle, a large patch
of sparsely vegetated area (NDVI 25–50) is evident in the first four-year
interval (1981–1984). In the next interval (1985–1988) it appears to have
shrunk considerably, replaced on the margins with the next-higher average
NDVI group. This corresponds very well to the conditions of that time, that
is, 1974–1984 was a period of drought throughout the inland desert areas
of Morocco and the African Sahel while 1985 and 1986 had higher-than-
average rainfall. In the subsequent two time intervals, this patch remains

roughly the same size and then shrinks to its smallest size in the 1997–2000 interval. In the most recent time interval (2001–2003), the patch appears to have enlarged. Similarly, the circled area shows a patch with the lowest average NDVI in the early time intervals, which has been gradually replaced over time with the next-higher average NDVI group. This patch is located in a desert area (Figure 1) within the high plains eco-region and is affected by the Mediterranean coastal climate. It is possible that this area received enough rainfall in the years following the drought of 1974–1984 to nourish vegetation. This increase in greenness is consistent up through the 1993–1996 interval, after which these areas shrank. In the last available time interval (2001–2003), the areas of increased vegetation are still larger than those of the drought years of 1981–1984. Landscape classes in this patch range from desert to evergreen land cover types (Figure 1).

In the time series regression using all available data for each pixel, the slope of NDVI over time represents the direction of change in vegetation cover (Figure 3). In this analysis, vegetation increases are indicated for 79 percent of the study area, although only 20 percent of the area has experienced a significant increase. In a complementary fashion, 21 percent of the area has experienced vegetation decreases, and 3 percent has significantly decreased. Mapping pixels with significant slopes identifies locations where significant changes have occurred (Figure 3). Most of the significant increases in vegetation cover are in the Rif Mountains, adjacent to the Atlas Mountains, and in the Drâa River valley. Additionally, clusters of pixels with significant increase in greenness are found along the northeast border with Algeria. Areas exhibiting significant decreases in vegetation cover are primarily in the northwestern part of Morocco where a large part of the Moroccan population resides; the dominant land use in this area are urban and agriculture, including animal grazing. Particular areas of significant vegetation increases are shown in circles in Figure 3 and labeled A1 through A6 while the five areas of significant vegetation decrease are depicted within rectangles in Figure 3 and labeled B1 through B5.

Population distribution for 1990 and 2000 (Figure 4a and b) was most dense in areas close to the coast in the northern and western parts of the country and much less dense in the desert areas, especially in southern Morocco north of Western Sahara. Population density increased and the boundaries of the major cities of Agadir, Casablanca, Mohammedia, Rabat, Tanger, and Oujden expanded between the two censuses. The spatial pattern and variability of population change over time (Figure 4c and d) show the nonuniform distribution of population around major cities (Agadir, Marrakech, Casablanca, Mohammedia, Rabat, Tanger, Meknes, Fes, and Oujda) and expansion into the mountain and valley areas. In Figure 3, significant decreases in NDVI (gray pixels) coincide with the expansion of the urban areas.

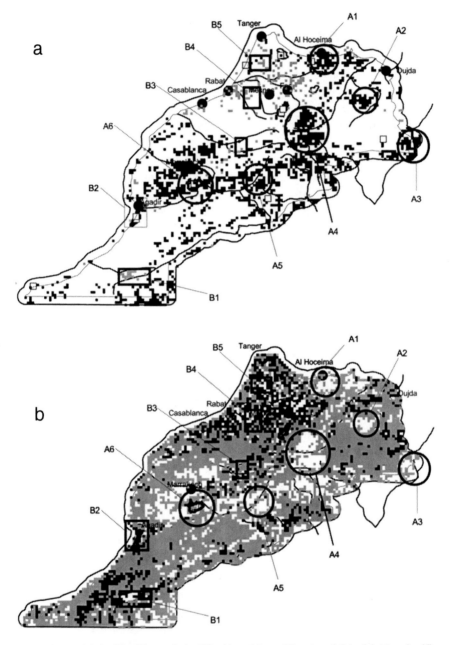

Figure 3. Pixels with (a) Significant Gain (Black) and Loss (Gray) and (b) with Nonsignificant Gain (Gray) and Loss (Black) in Greenness for Morocco, Based on Time Series Analysis for the Years 1981–2003. Marked Areas A1–A6 and B1–B5 Designate Specific Areas of Significant Vegetation Increase and Decrease, Respectively, Evaluated Using Ancillary Data.

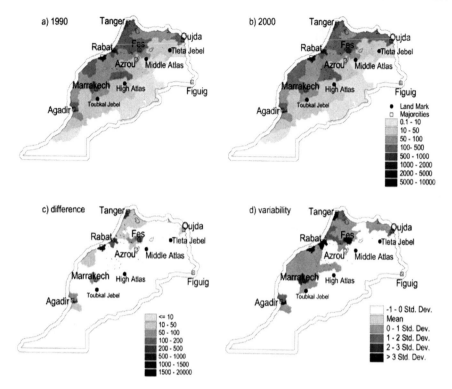

Figure 4. Population Density for (a) 1990 and (b) 2000, and (c) Difference and (d) Spatial Variability of the Population Density (person/km^2) Distribution Over Time. Small Polygons Represent National Reserves and Parks. "Std. Dev" Is Standard Deviation.

4. Discussion of Changes in Greenness

The results of the two different analyses (i.e., mapping/averaging NDVI by time interval and time series analysis of NDVI) each reveal distinct features of changes in the landscape over time. The maps of NDVI average per four-year interval highlight temporal contrast between areas with high and low vegetation cover. Slope maps, on the other hand, show the rate of change in vegetation cover over the entire twenty-three years contained in the data set and whether these changes are significant or not.

4.1. MAPS OF NDVI FOUR-YEAR INTERVALS

Morocco, as with other African countries, experienced a period of drought in the 1970s followed by a period of drought recovery in the years between

1982 and 1985, when rainfall was close to the long-term precipitation mean (Nicholson, 2005). However, in complex topography, precipitation can vary widely across an area the size of Morocco even in periods of widespread drought. By using rainfall data from several rainfall stations in Oum Er Rbia Basin, Chaponniere and Smakhtin (2006) demonstrate the spatial and temporal variability of rainfall in the basin area represented in the rectangle in Figure 2. Near Chichacouca, the average annual rainfall over thirty-four years (1965–2001) was 175 mm. Annual mean rainfalls 1981–1987 and 1999–2000 was lower than normal but exceeded the normal in 1988 through 1998. Sidi Rahal, located 40 km east of Marrakech, had an annual mean rainfall approximately double that of Chichacouca, 344 mm/ year for 1970–2001.

Sidi Rahal is located within areas with NDVI 50–100 (Figure 2) reflecting sparse vegetation cover. It is also 6 km northeast of an agricultural area. Higher rainfall and presence of annual crops in Sidi Rahal may be the reasons behind the greenness over the years in Figure 2. The size of the patches of low NDVI within the rectangle area increased and decreased with time reflecting the response to the rainfall for these years. From a rainfall station (Sidi Bou-Othmane) located in the Jiblet Mountains, 25 km north of Marrakech, Znari et al. (2002) reported the extent in annual rainfall variation from 1993 to 1998. While the mean and standard deviation were 290 and 126 mm, respectively, annual rainfall varied from 460 mm in 1995 to 103 mm in 1998. Although the trend in rainfall over time was not significant, the increasing trend in annual rainfall between 1993 through 1996 (slope = 67 mm/year, R^2 = 0.42, p = 0.35) may be the driving force behind the shrinkage of the patches with low NDVI values within the rectangle in Figure 2.

While rainfall is generally the primary factor in vegetation changes affecting NDVI greenness, irrigation may also result in vegetation increases. Reservoirs are the principal source of water for irrigation systems in Morocco. In the area bounded by the rectangle in Figure 2, the Azib Douirani dam was built on the Douirane River near Chichacouca in 1987 to provide stable water supplies and flood control. The Ait Quarda and Bin el Ouidane dams were established in 1953 to provide irrigation and hydroelectricity; they feed the Tadla irrigated area that is located north of Azilal and southeast of Kasba-Tadla. In 2001, the Dchar El Oued dam was built to feed the northern part of the Tadla agricultural area. The irrigated area can be seen in Figure 2 in patches of pixels with NDVI values 100–126. The irrigated agricultural land produces citrus, olives, and sugar beets.

Patches of greenness in the circle (Figure 2) increased over time. The circle spans an area near or encompassing the cities of Marija, Elrahebat,

Fouchel, Bordj-Doglat-Sedra, and Sidi-Ali-Doglat. This area contains desert, semidesert, evergreen shrub, and evergreen needleleaf tree land cover types. The dam establishment database (FAO, 2007a) indicates no recent dam construction; hence gains in greenness within the circle are not linked to irrigation. While the villages of Esraf and Sidi-Bou-Djemila are located within cultivated land and noncultivated rigid terrain, respectively, the village of Debdou is an oasis within the middle Atlas Mountains surrounded by lush orchards, oaks, and thuja forest. Debdou is in the subhumid climate region, so this village and surrounding area receives higher rainfall. Gains in greenness to the west of Debdou over time (Figure 2) may also be linked to growth of the cities Mahirija and Ain-et-Guettara where agricultural land is a dominant land cover.

4.2. TIME SERIES ANALYSES AND AREAS OF SIGNIFICANT CHANGE

The result of the time series analysis is a slope map (Figure 3) showing the direction of change over the study period. Significant positive increases in NDVI suggest a pronounced recovery from deforestation, restoration of degraded land, introduction of irrigation, or preservation of land natural resources. Significant negative slopes indicate a loss of vegetation; common causes include fire, overgrazing, urbanization, and forest cutting. Positive nonsignificant trends in NDVI appear to indicate that greenness has stabilized over time. Negative nonsignificant trends appear to indicate that degradation is occurring. The nonsignificant changes, both positive and negative, may signal a cautionary call for monitoring to further investigate the conditions underlying gradual vegetation changes that may threaten long-term sustainability.

Changes in vegetation cover of Morocco over the years have been driven by many factors. Drought is a major stress factor that Morocco and other surrounding countries have experienced. These countries need to enhance the areas of irrigated lands. Dams were constructed to aid irrigation by diverting water from regions with surplus to regions with water deficits. A total of 105 dams were constructed in Morocco between 1929 and 2003 (FAO, 2007a); 68 percent of these dams were built between 1980 and 2000. Irrigation has permitted expansion of agriculture into sparsely vegetated semidesert and desert regions. Urban areas have expanded to accommodate population growth. Depletion of the natural resources (e.g., trees) for human use in the Atlas Mountains helped influence migration of some of the nomadic people to urban centers or areas outside the country.

Eleven areas were identified for further evaluation, six with decreased vegetation cover and five with increased vegetation cover (marked areas;

Figure 3). Ancillary data were consulted to ascertain a specific cause of change. Area A1 is located near the town of Al Hoceima, a port on the Mediterranean Sea and one of the major towns within the Rif. The Rif is part of the Mediterranean conifer and mixed forest eco-region that extends from north Morocco to northwestern Tunisia. Tourism is the main revenue for Al Hoceima. This town is within Al Hoceima National Park, which was established on 470 km^2 by the World Bank Funds in 1991. The vegetation in this park is Mediterranean forest (evergreen sclerophyllous), woodland, and scrub. The park encompasses many marine and coastal ecosystems with diverse species. The local population has utilized the land for grazing, firewood, and fishing for years. These resources began to decline and became insufficient for the local inhabitants, which led to migration to neighboring countries and Europe in search of better economic opportunities (UNDP-GEF SGP, 2000). Preservation of habitat for humans, animals, and vegetation was recommended in 1995 via ASASHA (L'Association Solidarité pour l'Action Sociale et Humanitaire d'Al Hoceima), a nongovernment organization. Consequently, a Small Grants Program (SGP) through the United Nations Development Program (UNDP) was established in 2000 to educate local populations regarding natural resources and preserving vegetation to provide income from ecotourism and the sale of medicinal plants (UNDP-GEF SGP, 2000). A decline in forest cover in areas within the Rif Mountains in 1997–1999 was found to be minimal (Louakfaoui and Casanova, 2000). Preservation of the natural resources in this area may help enhancing greenness, causing the significant NDVI increase. Both Figures 2 and 3 show an increase over time in the number and size of green patches in this area.

The areas marked A2 through A5 in Figure 3 are all located near rivers in mountainous areas and the observed vegetation increases are likely due to increased rainfall and irrigation diversions from the river. Area A2 is located north of the Oued Charef River in a mountainous region. Area A3 is bounded from the south by Jebel Melah and Ras el Yhoudi and encompasses a number of mountains such as Jebel es Seffah and Rokna el Kahla. The Oued Moulouya River flows through area A4 in a mountainous location. The mountains north of the river are: Jebel Ouchilas, Jebel Hariga, and El Hajj. Mountains south of the Oued Moulouya River are: Jebel Bourr and Bou Tazert. The Assif Iriri Dades River passes through area A5 just south of the mountains of Jebel Talat n' Mensour, Jebel Arg, and Jebel Tilohah, and just north of the mountains of Jebel Bou Tabrha, Amlal, and Iskin n' Abid.

Area A6 includes the mountain peak Jebel Toubkal (4,200 m) at the end of the High Atlas Mountains. The High Atlas Mountains consist of a series of longitudinal crests with a SW–NE orientation. The Middle

Atlas Mountains are located at the center of the state, south of the Rif Mountains between the central and high plateaus north of the High Atlas Mountains with a SW–NE orientation. Thuriferous juniper, evergreen oaks (*Quercus ilex*), and Cedar (*Cedrus atlantica*) grow on these mountains. The thuriferous juniper is known for its tolerance to the extreme climatic conditions in these mountains and it plays an essential role for the surviving nomadic Berber population. The height and circumference of this tree can reach 19 m and 16 m, respectively. The foliage and wood provide a critical source of income for the Berber population. This area is densely populated compared to the Moroccan average. The population in the villages of Azzaden, N'Fis, Ourika, and Ait Bou Gamez increased 30 percent between 1971 and 1981. The thuriferous juniper cover has been reduced 90 percent due to grazing and human use since 1938 (Fromard and Gauquelin, 1993). This degradation has endangered the livelihood of mountain dwellers, leading to their emigration to other parts of the country. In an attempt to conserve the remaining forest, Toubkal National Park (polygon inside area A6, Figure 3) was established in 1942 encompassing an area of 380 km² (15 × 25 km). It is located 110 km west of Ouarzazate and 70 km south of Marrakech. Ifni Lake is within this park located southeast of Jebel Toubkal (PAWHP, 1988). In 1997, studies were initiated in an attempt to protect Toubkal National Park and help to revegetate the land with thuriferous juniper seedlings. Many field studies were conducted in the High Atlas of Azzaden Valley and Marrakech (Gauquelin, 1988; Badri, 1990).

The area marked B1 in Figure 3 is in a desert landscape situated between the intermittent streams of the Cap Drâa (drain to the Atlantic Ocean), Drâa, and Oued Tigzerte rivers. Mountains such as Lahouid and Taskalouine Jebel are to the north of B1 patch. This area is within the region just south of the Atlas where yearly rainfall variability is twice that of the area north and adjacent to the coast (Knippertz et al., 2003). The decrease in vegetation cover may be related to a decrease in rainfall in the mountains necessary to support stream flow at this downstream location. Rainfall data are needed to verify this causal effect.

Although not significant, the decrease in NDVI within the marked area B2 is notable. The marked area B2 contains the Arganeraie Biosphere Reserve, situated on 25,687 km² within the 828,000 ha Argan Forest in southwest Morocco around the city of Agadir. This forested area encompasses many reservations and national parks. This area is bordered to the north and east by the High Atlas and Anti Atlas mountains, to the south by the Sahara Desert, and to the west by the Atlantic Ocean. It covers several different land cover types: forest, cultivated, steppe, dunes, cliffs, sandy beaches, and wetlands. The natural cover in this ecosystem is rich in plant species with important ecological and socioeconomic roles. The local population utilizes this vegetation for pasture, wood, medicine, oil, and cosmetics.

Faced with desertification from the south and overutilization of the land for agriculture, in 1998 Morocco initiated the preservation of the Argan tree and designated the Arganeraie Biosphere (MAB-UNESCO, 1998) within the previously established (1991) Souss-Massa National Park. The Argan trees are tolerant to drought and other environmental stress factors; hence conservation, preservation, and perhaps enhancing their establishment especially to the south as a wind barrier will have a positive impact on the environment.

The areas marked B3, B4, and B5 in Figure 3 all show significant decreases in vegetation change, apparently due to urbanization. In a study assessing urbanization impact on agricultural lands, urban and rural growth were monitored in three cities, Béni Mellal (B3), Khémisset (B4) and Ksar El Kébir (B5) over two decades (Belaid, 2003). An area of 30 × 30 km² was centered on each city and growth was monitored for three decades. The most significant expansion in urban and rural areas was in Ksar El Kébir city where urban and rural classifications increased by 344 percent and 160 percent, respectively. Urban and rural areas were the result of converting rain-fed fruit orchards and rangelands. In 1979, Oued El Mkhazine Dam was constructed near this city to permit irrigation of land that was previously rain-fed agricultural land. While Béni Mellal experienced 180 percent expansion in urban and 200 percent in rural areas, less expansion was noted in Khémisset (70 percent in urban and 60 percent in rural). Northwest of Rabat along the Atlantic coast, the Maâmoura National Forest, which covers 150,000 ha, experienced a decline in cork oak tree cover from 1955 to 2000 at a rate of 1,600 ha/year (Assali and Falca, 2006). Declines in forest cover in Maâmoura National Forest were evident in 1997 and 1999 using NDVI (Louakfaoui and Casanova, 2000). Felling, grazing, and drought are among stress factors that reduced greenness in this national forest. Monitoring changes in forest cover in this national forest using remote sensing from 1989 to 1991, changes can be linked to felling or other stress factors (FAO, 2008). To remediate degradation, a program was established by Morocco through an international partnership to plant 20,000 ha with cork oak trees in Maâmoura Forest (Maghreb Arab Presse, 2007). The severity of changes is less with proximity of B3 where Béni Mellal city is located (B3, Figure 3). Béni Mellal city is within the Tadla irrigation system in the Oum Er Rbia watershed. As mentioned earlier, Ait Ouarda Dam was built in 1953 and another nearby dam (Ait Messaoud) was built on 2003 to accommodate the expanding irrigated fields. Bouknadel Dam was established near the town of Khémisset in 1998 on the Serou River.

5. General Discussion

5.1. LANDSCAPE STRESS FACTORS

More than 85 percent of Morocco is classified as rangeland, which includes forests, steppes, high meadows, and Saharan lands. These rangelands provide food for livestock. As a result of intensive grazing without proper management, the state of Morocco realized that millions of hectares of rangeland are being degraded, especially in the northeastern zone, the *Argania spinosa* ecosystem, and the sub-Saharan and Saharan zones. Grazing on steep slopes in the Rif Mountains (northeastern zone) accelerated the removal of vegetation and exposed soil to erosion. One hundred kilometers south of Marrakech, the Quneine watershed spans 20 villages where 10,000 Berbers live, utilizing the land for agriculture and grazing. The watershed area is 200 km² located in the High Atlas Mountains, with an elevation that varies between 832 m and 2,746 m. Natural vegetation includes trees (thujas, junipers, chestnut, oaks), and aromatic and medical herbs. The vegetation cover pattern is patchy, exposing much of the soil surface to erosion. Local residents utilize a small proportion of the watershed land for agriculture using spring water; the remainder of the 200 km² watershed is used as a rangeland. In sparsely vegetated land, grazing and other domestic uses of trees and vegetation without proper management exacerbates land degradation. Rainfall can influence land degradation over time, resulting in gullies and rills reflecting the severity of the erosion process. From three years' study using the Revised Universal Soil Loss Equation (RUSLE) model, Klik et al. (2002) found that average soil loss was 33.7 t/ha/year in this watershed. Erosion was spatially variable within the watershed and associated with grazing and timbering. Several land management strategies were recommended to reduce the effect of erosion and to preserve sustainable land production for the local communities.

Another concern is deforestation. A study of the Tleta watershed (180 km²), located between the cities of Tangeris and Tetouan upstream of the Ibn Battouta Dam, was established to collect monitoring data for assessing the sustainability of the western part of the Rif Mountain forests and rangeland. The land use/land cover of 1977 indicated that cereal crops comprised 50 percent of the watershed, forests comprised 40 percent of the watershed, and the remaining was scrubland used for grazing. The cultivated land increased to 65 percent and to 66 percent in 1991 and 1996, respectively, with a proportional decrease in forested area. The land was transformed by 1 percent of the total area per year from natural forest/scrubland to cropland between 1963 and 1987 in a nearby watershed

(Merzouk et al., 2003). Cultivation, grazing, and logging were cited as the main reasons for the decline in forested areas within 1977–1988.

5.2. POPULATION STRESSORS

The 2006 population estimate for Morocco is 33.2 million (LC-FRD). While the total population of Morocco at the turn of the last century (1900) was estimated to be five million, the predicted total population by the year 2044 is estimated at 46 million (Bennis and Sadeq, 1998). The estimated rate of population growth was the highest between 1952 and 1960 (2.8 percent) but has since declined, reaching a rate of 2.06 percent in 1994. The decline in population growth rate may be attributed to environmental stress, an increase in emigration, and many social factors. More than five million registered Moroccans were recorded in Europe between 1982 and 2004 (Haas, 2005).

While the growth rate has decreased, the distribution of the population within the country is changing and becoming more concentrated within the proximity of major cities. During the 1980s and periods of drought, the rural population migrated to urban areas seeking jobs. The population growth rate in urban areas was five times that of rural areas throughout the 1980s. Additionally, immigration and population displacement from neighboring Sub-Saharan countries (due to drought, famine, poverty, survival), as well as illegal transition toward Europe, are all posing a tremendous cost burden, on both the Moroccan environment and economy. Specifically, sporadic and somewhat orderly economically driven urbanization ensues. Take, for instance, the recent precipitous growth of impoverished areas in Marrakech and vicinity; the build-up of a whole city, Tamsluht, at 250,000, just a few miles north of Marrakech; and the construction in such large cities as Casablanca, Tangiers, Rabat, Agadir, Kenitra, Oujda, El Jadida, Fes, and Meknes.

To support the increasing urban population with food and other daily life commodities, it is necessary to increase or create new sources of these goods. This led to increases in irrigated agriculture, deforestation, and overuse of rangeland for livestock, thereby jeopardizing the long-term sustainability of the environment. Belyazid (2000) reported that as a result of overgrazing, deforestation, and conversion to agriculture to meet the needs of the populations of Oujida and Rabat, the sustainability of the Argan will not last more than two decades if proper management steps are not taken.

5.3. SOCIOECONOMIC FACTORS AND ENVIRONMENT

While welfare is often measured by economic factors such as growth rate of per capita Gross Domestic Product (GDP), and population growth, social issues are essential factors too. Education, health, and unemployment are some of the social indicators that are recently being considered in many economic studies (Sekkat, 2004). Environmental variables have also become important factors and were incorporated in growth accounting regression models. Sekkat quantified the effect of lagged income, investment rate, population growth, secondary school, inflation rate, and drought on GDP per capita growth rate from 1960 to 1997. Population growth rate and drought had a negative impact on economic growth due to the higher population density in rural areas where agriculture is the main commodity. Agriculture is an activity that is less conducive to economic growth because of its dependence on rainfall. The establishment of dams over the years did not meet the water demand for available agricultural land and the majority of crops are produced on nonirrigated land. In comparison, only 13 percent of Moroccan agricultural produce comes from irrigated land versus other countries that may reach as high as 40 percent. Hence, drought plays a major role in affecting economic development as indicated by GDP. To overcome losses from the agricultural sector and to preserve reasonable GDP, alternatives to agriculture such as industry or services have been considered for introduction.

6. Conclusions

We presented a simple method to locate and map changes in vegetation cover, which can be used to identify areas under stress. The method only requires free or inexpensive NDVI data (which can be derived from many sources) and basic statistical and mapping software. AVHRR data are useful for evaluating large areas, but finer-scale studies can be performed using higher resolution imagery. The use of remotely sensed data is far more cost effective than field studies and can be performed more quickly. Use of data over long periods permits analyses of historical change and identification of long-term trends.

Although acquisition of all ancillary data for the entirety of Morocco was not possible, we incorporated the available ancillary data, including some precipitation data, to identify rainfall trends. Drought appeared to be the most common natural cause of decreased vegetation cover. Areas with decreasing vegetation cover and unchanged or increasing rainfall are likely under stress from a source that can be managed, such as excessive timber harvesting, overgrazing, or urban growth. These areas may represent optimal locations for decision makers to take protective or remedial actions. In arid and semiarid ecosystems, degradation represented by decreasing vegetation cover may lead to irreversible desertification unless action is taken. Our analyses offer results that can be mapped at different scales (locally and regionally) and can be used to assess trends over both time and space; plus the approach can easily be applied to other locations.

New challenges are emerging in the relationship between the environment and the issue of security and are being given important consideration throughout the world. The question of the relationship between the environment and security is now a primary interest among both the scientific and policy communities. Landscapes change continually as an outcome of both natural and human-induced stress (i.e., land use is the primary result of the interaction of humans and natural ecosystems). The decrease in both quantity and quality of ecological resources, population growth, and equal access to resources are important factors related to environmental security risks. Renewable resources such as water and vegetation in arid environments are crucial factors in security issues, especially with respect to instability and migration. Thus both scarcity and degradation of resources leading to the risk of losing ecosystem goods and support services can contribute to instability, migration, and social conflict in both the national and international context.

Environmental degradation has various impacts on perceptions, behavior, and human social interactions. The NDVI change analysis approach provides an excellent process for statistically measuring and visualizing environmental change (i.e., both degradation and improvement) over broad geographical areas. However, it will undoubtedly take the combined thinking from both the environmental and social sciences to fully understand human-environment interactions, linkages, and their true meaning to security and world peace.

Acknowledgments

We are very grateful for the inputs of Dr. Bradford and Dr. Pitchford (U.S. Environmental Protection Agency), Rick VanRemortel and Tim Ehli (Lockheed Martin). We also thank Hy Dao (Metadata & Socio-Economics Unit, UNEP/GRID-Geneva, Switzerland) who directed us to the Africa Data Dissemination Service. The U.S. Environmental Protection Agency, through its Office of Research and Development, funded the research described herein. Although this work was reviewed by EPA and approved for publication, it may not necessarily reflect official agency policy. Mention of trade names or commercial products does not constitute endorsement or recommendation for use.

References

Anyamba, A., and Tucker, C. J., 2005, Analysis of Sahelian vegetation dynamics using NOAA-AVHRR NDVI data from 1982–2003, *J. Arid Environ.* **63**:596–614.

Assali, F., and Falca, K., 2006, *The Validity of Cork and Holm Oak Stands and Forests,* Report on the Évora Conference Meeting, 25–27 October, Lisbon, Portugal; http://www .aifm.org/page/doc/Evora06_rapport_gb.pdf, pp. 13–15, accessed 22 August 2007.

Badri, W., 1990, Cycle hydrologique et biogéochimique et influence du couvert sur la strate herbacée dans un peuplement à genévrier thurifère du Haut Atlas de Marrakech (Maroc), thesis, Marrakech, Université Cadi Ayyad.

Belaid, M. A., 2003, Urban-rural use detection and analysis using GIS & RS technologies, 2nd FIG Regional Conference, Marrakech, Morocco, December 2–5.

Belyazid, S., 2000, Achieving sustainability in the Argan Forest, Morocco, master thesis, Lund University Master's Program in Environmental Sciences (LUMES), Sweden.

Bennis, A., and Sadeq, H. T., 1998, *Case Study: Morocco, Population and Irrigation Water Management: General Data and Case Studies;* http://www.aaas.org/international/ehn/waterpop/Morroc.htm, accessed 15 August 2007.

Berkat, O., and Tazi, M., 2004, *Country Pasture/ Forage Resources Profiles: Morocco;* www.fao.org/ag/agp/agpc/doc/counprof/morocco/Morocco.htm, accessed August 14 2007.

Chaponniere, A., and Smakhtin, V., 2006, A review of climate change scenarios and preliminary rainfall trend analysis in the Oum Er Rbia Basin, Morocco, Working paper 110 (Drought Series: paper 8) International Water Management Institute, Colombo, Sri Lanka.

FAO, Food and Agriculture Organization of the United Nations, Land and Water development Division, 2007a, Georeferenced Database on Africa Dams; http://www.fao .org/ag/agl/aglw/aquastat/damsafrica/african_dams060908.xls, accessed 15 August 2007.

FAO, Food and Agriculture Organization of the United Nations, 2008; http://www.fao.org/docrep/W7825E/W7825E00.htm#Contents, accessed 21 May 2008.

Fromard, F., and Gauquelin, T., 1993, *Thuriferous Juniper Stands in Morocco: Research and Conservation for an Endangered Environment and Species;* http://www.fao.org/docrep/u8520e/u8520e0a.htm, accessed 15 August 2007.

Gauquelin, T., 1988, Dynamique de la végétation et des formations superficielles dans les montagnes du bassin occidental de la Méditerranée: étude des formations d genévrier thurifère et à xérophytes épineuses en coussinet des Atlas marocains. Thesis. Toulouse, Université Paul Sabatier.

GPWv3, *Gridded Population of the World Version 3,* 19 December 2005; http://sedac.ciesin .columbia.edu/gpw, accessed 13 August 2007.

Gurgel, H. C., and Ferreira, N. J., 2003, Annual and interannual variability of NDVI in Brazil and its connections with climate, *Int. J. Remote Sens.* **24**(18): 3595–3609.

Haas, H., 2005, *Morocco: From Emigration Country to Africa's Migration Passage to Europe;* http://www.migrationinformation.org/Profiles/display.cfm?ID=339, accessed 13 August 2007.

Klik, A., Kaitna, R., and Badraoui, M., 2002, *Desertification Hazard in a Mountainous Ecosystem in the High Atlas Region, Morocco,* 12th International Soil Conservation Organization Conference Proceedings, volume IV, pp. 636–644, Beijing, China, May 26–31.

Knippertz, P., Christoph, M., and Speth, P., 2003, Long-term precipitation variability in Morocco and the link to the large-scale circulation in recent and future climates, *Meteorol. Atmos. Phys.* **83**:67–88.

Lanfredi, M., Lasaponara, R., Simoniello, T., Cuomo, V., and Macchiato, M., 2003, Multiresolution spatial characterization of land degradation phenomena in southern Italy from 1985 to 1999 using NOAA-AVHRR NDVI data—art. no. 1069, *Geophys. Res. Lett.* **30**(2):1069.

LC-FRD, Library of Congress—Federal Research Division, 2006, Country Profile: Morocco, May 2006; http://lcweb2.loc.gov/frd/cs/profiles/Morocco.pdf, accessed 13 August 2007.

Louakfaoui, E. M., and Casanova, J. L., eds., 2000, Analysis of the state of forest in Morocco using the change-vector and NOAA-AVHRR imagery, 1st Workshop EARSeL Special Interest Group on Remote Sensing for Developing Countries, University of Gent, Belgium 13–15 September; http://www.earsel. org/pdf/DevCountries-Ghent2000/Proceedings/ Vegetation-Land%20cover.pdf, pp. 173–182, accessed 23 August 2007.

Maghreb Arab Presse, 2007, Morocco to Plant 20,000 ha of Cork Oak Trees Yearly Over a Decade, Lisbon, Portugal June 11 2007; http://www.map.ma/eng/sections/imp_economy/morocco_to_plant_209714/view, accessed 24 August 2007.

MAB-UNESCO, Man and Biosphere—United Nations Educational Scientific and Cultural Organization, 1998, comme réserve de biosphère en décembre 1998; http://www.arabmab.net/ countrydetails.cfm?cid=8.

Merzouk, A., Alami, M. M., Berkat, O., and Sabir, O. M., 2003, Effects of Rangeland Changes on Water Balances and Water Quality in Morocco: A Rif Mountains Case Study; http:// mediasfrance.org/Reseau/Lettre/13bis/en/03.MAROC.RIF.PDF, accessed 15 August 2007.

Minor, T. B., Lancaster, J., Wade, T. G., Wickham, J. D., Whitford, W., and Jones, K. B., 1999, Evaluating change in rangeland condition using multitemporal AVHRR data and geographic information system analysis, *Environ. Monit. Assess.* **59**:211–223.

MONGABAY.com, 2000, *Morocco Deforestation Rates and Related Figures*; http:// rainforests.mongabay.com/deforestation/2000/Morocco.htm, accessed 13 August 2007.

Nash, M. S., Wade, T., Heggem, D. T., and Wickham, J., 2006, Does anthropogenic activities or nature dominate the shaping of the landscape in the Oregon pilot study area for 1990–1999? in: *Desertification in the Mediterranean Region: A security Issue.* Series C: *Environmental Security,* Vol. 3., W. G. Kepner, J. L. Rubio, D. A. Mouat and F. Pedrazzini, eds., Springer, The Netherlands, pp. 305–323.

Nicholson, S., 2005, On the question of the "recovery" of the rains in the West African Sahel, *J. Arid Environ.* **63**:615–641.

PAWHP, Protected Areas and World Heritage Programme, 1988, *Protected Areas Programme: Morocco;* http://sea.unep-wcmc.org/sites/pa/1120v.htm, accessed 15 August 2007.

Rouse, J. W., Hass, R. H., Schell, J. A., and Deering, D. W., 1973, Monitoring vegetation systems in the great plains with ERTS, in: *Third ERTS Symposium,* NASA-351, Vol. 1, NASA, Washington, DC, pp. 309–317.

SAS/ETS, 1999, *User's Guide,* SAS Institute, Cary, NC.

Schmidtlein, S., 2005, Imaging spectroscopy as a tool for mapping Ellenberg indicator values, *J. Appl. Ecol.* **42**:966–974.

Sekkat, K., 2004, *The Sources of Growth in Morocco: An Empirical Analysis in a Regional Perspective,* Taylor & Francis Journals, vol. 2(1), pp. 1–17.

Tucker, C. J., 1979, Red and photographic infrared linear combinations for monitoring vegetation, *Remote Sens. Environ.* **8**:127–150.

UNDP-GEF, SGP, United Nations Development Programme, Global Environment Facility, Small Grants Programme, 2000, Diversity: Bringing Stakeholders together to Protect Natural Resources: Integrated Efforts to Preserve Biodiversity in Morocco's: Al Hoceima National Park; http://sgp.undp.org/downloads/Biodiversity%20-%20morrocco %20Ea. pdf, accessed 22 August 2007.

WDPA, World Database on Protected Areas; http://www.unep-wcmc.org/wdpa, accessed 13 August 2007; choose "search for site"; select Morocco for country.

Znari, M., El Mouden, E., and Francillon-Vieillot, H., 2002, Long term variation in reproductive traits of Bibron's Agama, Agama impalearis, in western Morocco. *Afr. J. Herpetol.* **51**:57–68.

SECTION III

HUMAN CHALLENGES: CASE STUDIES

HUMAN SECURITY FOR AN URBAN CENTURY

Local Challenges, Global Perspectives

ROBERT J. LAWSON[1]*, MACIEK HAWRYLAK[2]
AND SARAH HOUGHTON[3]

[1-3] *Foreign Affairs and International Trade Canada, 125 Sussex Dr., Ottawa, Ontario K1A0G2*

Abstract: Just as traditional environmental concerns—water, desertification, balanced development—can pose challenges to human security, so can the built environment. For the first time in history, the majority of people now live in cities, and rapid urbanization is already shaping trends in global peace and security. Armed violence is increasingly taking place in sprawling hillside slums, involving adolescent boys with automatic weapons, corrupt police officers determined to "clean up" city streets, or vigilante groups who take justice into their own hands. The violence feeds on the toxic mix of transnational criminal organizations and failed public security. But the built environment can also be a starting point for conflict resilience. Secure cities—cities with effective public security; inclusive, participatory governance; and positive social capital—will be critical to the prevention of armed violence and the protection of civilian populations from such violence when prevention fails.

Keywords: Human security; cities; public security; slums; urban violence; open armed conflict; endemic community violence; organized crime; gangs; privatization of security; small arms; street children; children in gangs; displaced persons; migrants; conflict resilience; social capital; democratization; built environment

---

* Address correspondence to Bob Lawson, Senior Policy Advisor, Human Security Policy Division (GHS), Foreign Affairs and International Trade Canada, 125 Sussex Dr., Ottawa, Ontario K1A0G2

P.H. Liotta et al. (eds.), Environmental Change and Human Security, 169–202.

1. Human Security at the Dawn of an Urban Century

For the first time in human history as many people now live in cities
as in rural areas. By 2030, it is estimated that 60 percent of the world's
population will live in cities.[1] We have entered what some are calling
the "Urban Century."[2] Conceptions of local and global governance are
changing in an age when the mayor of a megacity such as Mexico City
now governs more people than the leaders of 75 percent of the world's
states. With a third of all urban dwellers living in slums, rapid urbanization is also reshaping the security and development challenges facing the
global community.

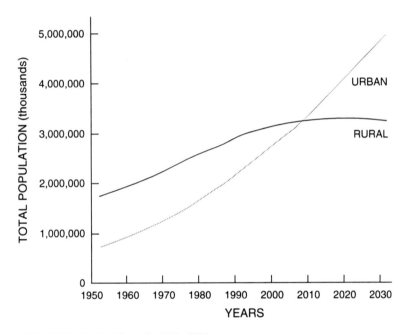

Figure 1. World Population Growth 1950–2020.

---

[1] Debate surrounds the exact timing of when the proportion of the world's population living in cities reaches 50 percent. Depending on the source, the date when the urban population surpasses the rural population varies from 2005 to 2007.

[2] The term "Urban Century" is not attributed to any single source, but has been used
by UN-HABITAT, Stephen Graham, Jane Jacobs, the World Bank, and others. See, for
example, Graham, 2002.

The speed of this demographic shift is without precedent in human history (see Figure 1).[3] The population of metropolitan Dhaka, Bangladesh, for example, exploded from 400,000 in 1950 to almost ten million in 2006.[4]

Today's urbanization is also occurring primarily in the cities of the developing world, which now account for over 90 percent of global urban growth (UNHSP, 2006: 4). This figure will represent almost all population growth on the planet in the next quarter-century, as rural populations are expected to decline after 2015 (UNHSP, 2006: 4). However, much of this urban expansion is occurring in the context of rapid but highly inequitable economic growth. With many municipal governments already lacking the capacity to provide basic security to all urban dwellers, such rapid urbanization means that each year more and more people are living in impoverished, informal slums in and around urban areas.

1.1. HUMAN SECURITY IN URBAN SPACES

The term "human security" has now been in widespread use for about a decade (see, for example, UNDP, 1994). It emerged as a critique of approaches to the promotion of international peace and security that focused almost exclusively on the security of states and their governments. The essential idea of human security is that people rather than nation-states are the principal point of reference and that the security of states is a means to an end rather than an end in itself. In an increasingly interdependent world, the security of people in one part of the world depends on the security of people elsewhere. International peace and security is ultimately constructed on the foundation of people who are secure (McRae, 2001).

The term itself has been associated with efforts to reduce people's vulnerability to a broad array of risks ranging from attacks on civilian populations in civil wars to people's social-psychological well-being. Whatever the breadth of the definition, one thing is clear: any conception of human security must address the question of safety from physical violence for people and their communities.

One fundamental objective in the pursuit of human security is reducing the human costs of war. This is achieved by creating and strengthening international humanitarian standards, enforcing the rule of law, promoting the peaceful resolution of conflicts where they exist, and preventing their

---

[3] The urbanization that accompanied the Industrial Revolution in Europe and North America recorded slightly smaller figures. See LeGates and Stout, 2003: 31.

[4] Based on data compiled by the National Geographic Society and the United Nations Population Division.

reemergence. Since the end of the Cold War, human security has been shaped less by wars between states and more by armed conflict within states. With 90 percent of conflicts now taking place within states, people are now much more likely to be killed or injured as a result of the failure of a state to maintain the rule of law within its own territory than its inability to defend its borders from attacks by other states.

A closer look at the violent threats faced by people living in major cities and slums suggests a need to focus on reducing the risk of physical violence in situations outside of formal armed conflicts as well. Extraordinarily high levels of violence are also affecting cities—prominent hubs of power that can become flashpoints of large-scale violence between groups.[5] As a result, human security is increasingly at risk in urban environments.

1.2. UNDERSTANDING VIOLENCE AND CONFLICT IN URBAN AREAS

The loss of territorial control by the state is a defining feature of civil wars, often due to the existence of rebel armies, insurgents, or paramilitaries which physically exert control over part of the country, and are engaged in open armed conflict with state forces. Such violence may be fueled by groups with competing political, ideological, or economic ambitions, in societies deeply divided by ethnic or religious differences, or with high levels of social inequality. As Figure 2 illustrates, **open armed conflict** in urban areas, with its profoundly negative human security impacts, is generally carried out by highly organized groups and characterized by a high level of intensity (as measured by human casualties).

The horrific acts of violence wrought by these actors—atrocities, war crimes, genocide, use of landmines and improvised explosives, recruitment of child soldiers, gender-based violence, and small arms deaths—are felt in places such as the Democratic Republic of the Congo (DRC), where 3.3 million people are estimated to have died since 1998, or the Darfur region of Sudan, where 146,000 have perished in battle-related deaths since 2003.[6] But the effects of open armed conflict are also felt in urban areas. The pillaging

---

[5] Conflict is defined here as "the instrumental use of armed violence by one group against another in order to achieve political, economic or social objectives." This definition is adapted from the definition of collective violence developed by the World Health Organization. See Small Arms Survey, 2005. On crime, violence, and cities, see Brennan-Galvin, 2002: 123–145.

[6] For the Democratic Republic of Congo, see Human Security Centre, 2005: 125. The Darfur figure is a conservative estimate by the United States Department of State. See U.S. Department of State, 2005.

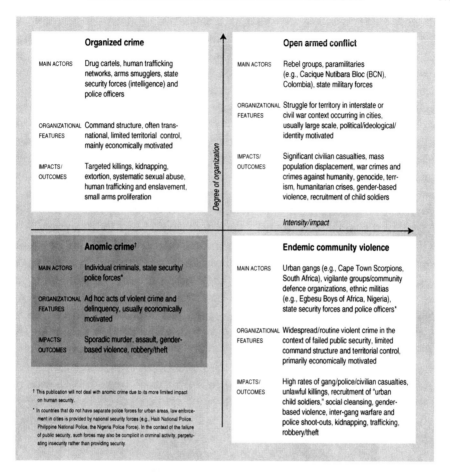

Figure 2. Typology of Urban Violence.

of Kindu in eastern DRC in 2001 and the Srebrenica massacre in Bosnia and Herzegovina in 1995 are two examples.

The failure of the state to control urban spaces within its territory can lead to **endemic community violence**, with devastating impacts on civilians. When the state cannot provide for the needs of its citizens, the security void is increasingly filled by private actors—vigilante groups, gangs, and militia groups that seek to exert control over defined urban spaces. Areas of cities in Afghanistan, Colombia, the DRC, Jamaica, Pakistan, Somalia, and South Africa have, at some point, fallen under the control of gangs with cohesive organization and demarcated territory (Rapley, 2006). Unlawful killings, exploitation, the use of children in armed gangs, and rape are just some of the consequences of failed public security in fragile cities.

Failed public security can produce levels of violence comparable to a civil war. High rates of gang, police, and civilian casualties; recruitment of "urban child soldiers"; social cleansing (systematic violence against "undesirable" social groups perpetrated by criminal groups or security forces); and gender-based violence are just some of the symptoms of endemic community violence, which can result in fatality rates comparable to those in situations of open armed conflict. A 2002 case study found that between 1978 and 2000, more people, particularly children, died in armed violence in the slums of Rio de Janeiro, Brazil (49,913), than in all of Colombia (39,000), a country that is actually experiencing civil conflict (Dowdney, 2004: 12). Endemic violent crime in El Salvador resulted in more violent deaths in the years following its civil war than during the war itself (Rodgers, 2003: 2).

Organized crime, such as illicit activities carried out by drug cartels and human trafficking networks, flourishes in the context of failed public security. Transnational criminal networks threaten people's safety and lives by carrying out targeted executions, trafficking and enslaving humans, and smuggling small arms across borders. Although they often feature a higher degree of organization and permanence than urban gangs or militias, this type of crime may have less severe human security impacts (as measured by civilian casualties) than open armed conflict and endemic community violence in urban areas.

A final type of urban violence worthy of mention is anomic crime—crime committed by individual actors unaffiliated with an organized group on a random, ad hoc basis. Although this type of crime can be found in virtually all cities, its relatively small overall impact and unorganized nature do not render it a significant threat to human security.

2. Armed Conflict and Failed Public Security in Cities

Today one billion people live in slums.[7] This is expected to increase to two billion by 2030. Slums are largely poor, densely populated, unplanned, and informal communities in urban or peri-urban areas.[8] In some cities, most of the population lives in slums: 60 percent of Nairobi's population lives in

---

[7] The term "slum," while not universally agreed upon, broadly describes informal or illegal settlements in and around cities, also known as "squatter communities" or "shantytowns." It is important not to perceive slum dwellers as criminals or helpless victims. However, while it should be recognized that many slums are peaceful places with dynamic, adaptive, and lively communities, they also house some of the poorest members of society in unsanitary areas that are under-serviced by infrastructure and under-protected by state security forces, which can be a catalyst for conflict.

[8] For a more detailed definition of slum, see UNHSP, 2006: 19.

slums on only 5 percent of the city's land (vom Hove, 2006). In other cases, slums have gradually merged into each other, spanning hundreds of square kilometers and housing millions. UN-HABITAT estimates that in 2005, 57.4 percent of South Asia's urban population and 71.8 percent of sub-Saharan Africa's urban population were living in slums (UNHSP, 2006: 16).

The growth of slums can breed violence and insecurity largely for two reasons. First, their sheer size and population can stretch state capacity to the point at which the state is unable to provide these areas with basic public security. Second, state security forces may be unwilling to provide security in slums, due to, for example, a lack of incentive to risk their lives entering dangerous areas. In the absence of effective public security, slum residents and urban elites alike may seek ways to protect themselves, resulting in the privatization of security. This can, in turn, contribute to a process illustrated in Figure 3, in

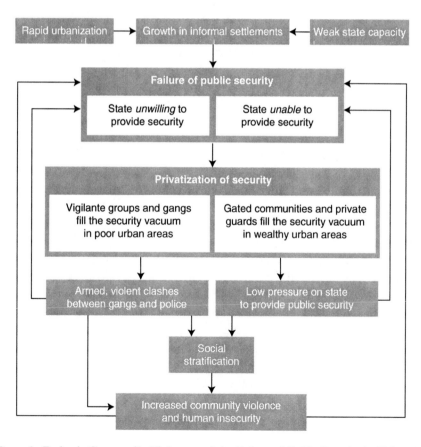

Figure 3. Endemic Community Violence and the Failure of Public Security in Urban Spaces.

which the failure of public security and the rapid growth of urban slums feed into a cycle in which community security is continually undermined.

2.1. THE FAILURE OF PUBLIC SECURITY

Many local governments lack the capacity to provide security for rapidly growing urban populations. In some cities, security forces include teenage boys who have had only a few days of training, and lack basic equipment such as handcuffs, flashlights, and helmets (see, for example, Weeks, 2006). Many security forces are also unable to recruit enough officers to keep up with the needs of growing cities, producing alarmingly low police-to-citizen ratios. The population of Cité Soleil, a 2-km^2 slum in the Haitian capital of Port-au-Prince, grew from 1,000 in the 1960s to an estimated 350,000 in 2003.[9] This growth strained the central government's ability to meet the needs of its citizens in a context of already weak state capacity.

The failure of public security also occurs in some slums because security forces are unwilling to provide it. Some urban areas are considered simply too dangerous to enter. In 2001, almost half the cities in Latin America and the Caribbean had areas considered inaccessible or dangerous to the police due to organized violence.[10] In 1995, Mexico City was reportedly divided among 1,500 competing gangs (UNDPI, 1995). Officers hired by the state to provide security may lack incentives to take the risks necessary to maintain public security in these areas because they are often paid meager salaries and enjoy little job security. Police in Kabul, Afghanistan, for example, earned as little as US $16–18 a month in 2004 (Transitional Islamic Government of Afghanistan, 2004).

Contributing to the unwillingness to provide security is a prevailing culture of impunity often found in the security sector. In many cities, police are known to use unnecessary force, including torture and unlawful killings, without legal ramifications. The failure of public security and public distrust of security forces can be mutually reinforcing phenomena. Widespread distrust of security forces operating in urban areas can stem from three main sources. The first is their known or suspected collusion with gangs in criminal activity. Security officers have long been known to participate in the illicit drug trade, and to provide arms or information—or turn a blind eye—to criminal activity in many cities. The second is an excessive use of force against people on the streets and in police custody. In extreme cases, this includes torture, rape, and unlawful killings.

---

[9] The International Committee of the Red Cross estimated Cité Soleil's population to be 200,000 in 2004. See ICRC, 2005.
[10] The figure is 48 percent according to UN-HABITAT, 2001: 9.

The third is the practice of targeting minorities, young people, and marginalized groups for security crackdowns. For example, in 2002 Kenyan state forces arbitrarily arrested hundreds of refugees from Ethiopia, Somalia, Sudan, and the Democratic Republic of Congo who were living in Nairobi's approximately 130 slums, in a massive military-style operation (Human Rights Watch, 2002). Children and youth are also frequently targets of excessive force used by security forces. Although these campaigns are often pursued under the guise of enhancing public security by ridding the streets of "objectionable" individuals, they are perhaps one of the most flagrant examples of the failure of public security resulting from an unwillingness to protect vulnerable groups. Such operations not only exacerbate insecurity and mistrust in communities, but also undermine the legitimacy of state security forces. Because an excessive use of force is sometimes used by those who patrol slum communities, it is not always clear if these security forces are enhancing or undermining human security. The use of torture, unlawful killings, and routinely high levels of corruption among security forces have been reported in cities throughout the world, in both developed and developing countries.

2.2. THE PRIVATIZATION OF SECURITY

When security forces are unable or unwilling to protect cities, residents are left to provide their own security. Thus, in many cities, security has become a private commodity among wealthy elites. In South Africa, for example, the number of private security guards has increased by 150 percent since 1997, compared to a 2.2 percent decrease in the number of police officials in the same period (SAPA, 2006). Even state police forces there have turned to private security companies to protect some police stations.

At the same time, gated residential enclaves—heavily guarded urban fortresses with sophisticated alarm systems, electric fences, surveillance cameras, and private security guards—are increasingly common in societies that are highly divided, whether along racial lines (such as Cape Town) or along income lines (such as Managua). São Paolo boasts 240 helipads—compared to ten in New York City—which shuttle the rich from the city to walled compounds such as Alphaville, an exclusive suburb patrolled by a private army of 1,100 guards (Faiola, 2002: A01). Ever-growing demand for elite security has fueled the growth of a lucrative, but often unregulated, private security sector in many countries.

Privatization can undermine public security in two contrasting ways. In some cases, private security guards are better armed than public forces, but less bound by standards of conduct, and disgruntled from inadequate pay or from having been discharged from military service, leading to abusive and unlawful behavior. In other cases, the allure of jobs in the private sector may attract the most skilled or ambitious individuals—including those who have already received training in the public forces—reducing the effectiveness of publicly provided security.

On the other end of the spectrum, those who are unable to afford private security services may develop adaptive strategies to fill the security vacuum. Community organizations are frequently formed to provide protection for residents. These groups are often peaceful and inclusive, such as neighborhood watch groups, but in other cases they may employ armed violence as a means of community defense—for example, citizens' militias and protection rackets. In Nicaragua, youth gangs claiming to protect residents from inter-gang warfare have emerged, in some cases becoming an institutionalized presence in poor communities (Moser and Rodgers, 2005). In Haiti's capital of Port-au-Prince, armed insurgents have forcibly occupied police stations and assumed law enforcement responsibilities. Filling the security void led one gang member from August Town, Jamaica to comment, "We have our own justice, the state does not provide justice."[11]

Slum dwellers living in security voids are particularly vulnerable to extortion and corruption by gangs and corrupt public authorities, as well as being caught in the midst of violent disputes between groups competing for power. When poor people cannot afford to pay protection fees to police or other, informal authorities, they may face violent reprisals, such as having their houses set ablaze. During a 2001 clash between landlords and gangs of tenants in Nairobi's largest slum, Kibera, 12 people were killed, about 100 women and girls were raped, hundreds more were injured, and thousands were displaced as houses were burned to the ground (McKinley, 2001).

When security is provided privately—by individuals or other groups in lieu of the state—it can lead to greater insecurity for the urban poor. In communities home to thousands of bored, poor, young people (particularly men), local gangs or public officials have been known to recruit from their ranks to establish vigilante groups that conduct violent counterattacks on gangs. The distinction between these well-armed vigilante groups and community gangs is not always easy to make. In slums outside Cape Town, for example, a vigilante group known as People Against Gangsterism and Drugs (Pagad) was formed to rid the community of murdering gang leaders. This led gangs to seek vengeance on Pagad, exacerbating inter-gang warfare and creating a vicious circle of violence (Botha, 2001).

---

[11] Quote from a female gang leader in August Town, Jamaica, cited in Dowdney, 2004: 237.

Security privatization can exacerbate the gap between the rich and the poor both physically, through the erection of elite gated communities, and socially, by aggravating a sense of grievance among people living in violent urban environments. Heightened hostilities between security forces, gangs, and vigilantes can manifest in armed, violent clashes, deepening social stratification and worsening community violence. At the same time, the growth of the private security sector in many countries—symptomatic of the state's inability to protect its population—reduces pressure on the state to provide these services publicly, which can contribute to state security forces' unwillingness to protect poor populations.[12]

2.3. SLUM INSECURITY: GANGS AND GUNS

The impact of failed public security in slums is not limited to actors involved in local law enforcement. Some armed criminal groups have expanded into organized criminal empires with thousands of members, complex internal organization, and aggressive recruitment strategies, which compete in open armed combat with gangs or authorities for territorial control. Politicians reportedly enter Rio de Janeiro's dangerous favelas only with permission from gang leaders; the groups control territory while operating a kind of parallel government that interacts with the state only occasionally. The gangs can, in effect, "negotiate the terms of [their] sovereignty" (Rapley, 2006: 95–103), undermining wider state authority and legitimacy.

The proliferation and ease of availability of small arms in urban areas has compounded the challenge that organized criminal violence presents to human security. The widespread use of guns in encounters between gangs, police, and vigilante groups has been on the rise in recent years, increasing the lethality of urban violence (International Action Network on Small Arms, 2006). At the same time, other armed groups have moved into some urban areas, taking advantage of the impunity afforded to them in anonymous, overcrowded, and under-policed slums. This phenomenon has been referred to as the "urbanization of conflict," or the "urbanization of insurgency" (Morrison Taw and Hoffmann, 1994), seriously threatening the security of urban dwellers, particularly youth. In Medellín, for example, the lines between gangs and other armed groups have blurred as paramilitary groups such as the United Self-Defence Forces of Colombia's Bloque

---

[12] "The persistence of crime and the decline of trust in cities have serious implications for governance, in that local governments are increasingly regarded as losing control, thus eroding confidence in leadership and city governments" McCarney, 2006: 4.

Metro have trained and hired several of the approximately 300 armed gangs, or "combos," for political murders and to battle guerrilla militias over turf (Hodgson, 2001; see also U.S. Citizenship and Immigration Services, 2001).

Gun violence linked to urban gangs is particularly severe in cities in Latin America. In Brazil, more than 100 people are killed by firearms every day, and the gun death rate in Rio de Janeiro is more than double the national average (Viva Rio, 2005). Youth are often particularly affected by guns: the Pan American Health Organization estimates that one in five schoolboys in the Caribbean has brought a gun to school and admitted to past or present involvement in a gang (Pan American Health Organization, 2000: 17). In Colombia, firearm-related deaths among youth under eighteen increased by 284.7 percent between 1979 and 2001 (Dowdney, 2003: 130).

The availability of small arms in many cities means that ordinary civilians, gang members, and private security guards are often better armed than state security forces. Guns in many cities are illegally purchased at low prices, sometimes from current or former state security personnel. Not only has the number of illegally held firearms been on the rise in many cities, but more lethal weapons such as assault rifles, machine guns, and submachine guns are becoming increasingly common on city streets (Dowdney, 2004: 31).

Gangs and organized crime cartels are often heavily involved in the global illicit drug trade, perpetuating urban gang violence by providing the finances needed to purchase firearms, bribe authorities, and pay gang members' salaries. Increases in addictive drug use are correlated with spikes in violence as rival factions fight for turf and control over lucrative trade connections (Dowdney, 2004: 30).

3. The Human Face of Urban Insecurity

The failure of public security in cities has significant negative impacts on the lives of people living in cities. Organized armed groups, such as gangs and paramilitaries, derive income from illicit drug and arms trades, the sex industry, human trafficking, and ransomed kidnapping, which flourish in many large, urban areas. In some cities, violence is so widespread that it has become a normal part of everyday life, with certain groups of the population—children, women, the poor, refugees, and internally displaced persons (IDPs)—particularly vulnerable to pervasive insecurity (Moser, 2004: 6).

Children growing up in cities are particularly vulnerable to threats of armed violence. Many children living in slums are recruited into armed urban gangs. Boys and girls who live and work on the streets may be targeted for violence, discrimination, and abuse, in some cases by the very people who are supposed to protect them. Poor men and women in urban areas are susceptible to gender-based violence, and women in particular face threats of sexual exploitation in cities. IDPs, refugees, and migrants who move to cities seeking shelter from conflict can face violent threats from state security forces, urban gangs, and hostile incumbents. Understanding the challenges faced by people found in the most vulnerable situations in cities is vital to improving human security in urban spaces.

3.1. CHILDREN AND YOUTH IN URBAN GANGS

Children growing up and living in impoverished slum settlements face a host of violent threats. They are often at heightened risk of being recruited into armed criminal groups, targeted by social cleansing campaigns, trafficked, and forced into the sex trade or domestic servitude. The "youth bulge" problem—that a disproportionate number of underemployed youth in a population has been historically linked with social upheaval—is now becoming more identified with urban areas in an age of rapid urbanization. Many cities in the developing world are made up of significant youth populations—about half of the urban population in Africa, for example, is under nineteen (UNHSP, 2004).

In poor urban slums, disaffected kids growing up in crowded households (a common feature of slums) tend to spend more time on the streets. With few jobs available and poor prospects for the future, organized criminal gangs can be perceived as appealing opportunities to provide them with an income, social network, and sense of security.

Despite the very real dangers of armed urban gangs, the average age of recruitment for child gang members is between eleven and fourteen, and has been falling in recent years.[13] In some gangs, children make up a significant minority or even a majority of the membership. In Medellín, for instance, an estimated 60–70 percent of gang members are children (Dowdney, 2004: 182). These child gang members are often considered expendable; they are commonly relegated to the lowest organizational tasks in the gang, and are often the first to be sent into armed confrontation with rival gangs or state security forces.

Targeted assaults and executions carried out by state security forces against children known or alleged to associate with armed gangs can occur in the culture of impunity that prevails in some major urban areas. Reports from nongovernmental organizations (NGOs) working in Honduras, for example, have indicated

---

[13] Dowdney finds that "A common theme in many of the groups investigated in this study is the decreasing age of child and youth members" Dowdney, 2004: 71.

that more than 1,200 children and youth died between 1998 and 2002 due to urban violence and social cleansing campaigns (Amnesty International, 2003).

The intensity of the violence faced by children in organized, armed urban gangs is in many ways comparable to that faced by child soldiers operating in rural areas. For example, both child soldiers and children in armed gangs face violent threats associated with armed combat against rival groups; they are often recruited against their will, are at heightened risk of sexual abuse, suffer psychological trauma, and find defecting difficult, if not impossible.

Like child soldiers, children in gangs are often viewed as armed combatants and targeted by state security forces because they pose a violent threat.[14] The Children and Youth in Organised Armed Violence project of Brazilian NGO Viva Rio warns that classifying children in gangs as urban "child soldiers" could compound this problem (Dowdney, 2004: 12), placing boys and girls in greater danger. However, appreciating the similarities that children face in these very different contexts may help to inform future child protection strategies.

In some cases, there are strong criminal and organizational linkages between rural-based insurgent groups and urban-based gangs. Youth gang members in Medellín, for example, are not considered "combatants," yet it is believed that most of the city's youth gangs have been co-opted by paramilitary and guerrilla armies (Ceballos Melguizo, 2001: 110–131; Gutiérrez Sanín and Jaramillo, 2004: 17–30). In Port-au-Prince, gangs engage in violent clashes over urban turf, politics, and criminal enterprises with rival factions, UN peacekeepers, and state security forces, which have claimed hundreds of lives (International Crisis Group, 2005). In Sierra Leone, following the country's civil war, slums materialized in the capital, Freetown, that were made up almost entirely of youth ex-combatants. This created an environment in which violent crime in and around the city flourished as disenchanted and psychologically traumatized young people struggled to survive (Save the Children Sweden, 2005: 16). Children who lack alternate livelihood options and are already armed, trained combatants are at heightened risk of re-recruitment into armed groups.

3.2. STREET CHILDREN: LIVING AND WORKING IN THE CITY

Whether or not they are experiencing armed conflict, cities can be sites of violence and insecurity for children growing up on the streets. An estimated 100 million children live and work on the streets in cities around the world,

---

[14] According to the Geneva Conventions, in situations of conflict child soldiers are armed combatants and their classification as such makes them legitimate targets for lethal force.

and about 40 percent of these children are homeless (Canadian International Development Agency, 2006). In Egypt, it is estimated that there are between 200,000 and one million homeless children, most of them in the cities of Cairo and Alexandria (UNICEF, n.d.). Street children range in age from three to eighteen, and are mainly boys (CIDA, 2006). They resort to life on the streets for many reasons, often having arrived in cities from rural areas in search of employment and security. Some have fled situations of domestic abuse or intractable poverty, while others are orphans who have lost their parents to violent conflict or HIV/AIDS or other illnesses.

Children living in the streets face grave threats to their security. They often face the possibility of recruitment into armed gangs, or are unlawfully detained and harassed by state security officers. Others are shipped away to rural areas. Because of the perception that they are criminals or associated with gangs, street children may be targets for social cleansing campaigns carried out by both state and non-state actors. For example, reports have indicated that in 2005, 431 street children and youths in Honduras were murdered,[15] while in Guatemala, 124 street children were killed in only three months (Casa Alianza UK, 2005). In Lagos, security forces and the Bakassi Boys, a well-known vigilante group, have engaged in violent campaigns aimed at cleansing the city of street orphans (Human Rights Watch and the Centre for Law Enforcement Education, 2002). In Harare, some 10,000 street children and vagrants were detained by state forces in 2005, as part of what the government called a "crime-fighting measure" (Blair, 2006). In November 2006, street children labeled as "wanderers" were reportedly arrested and subjected to physical abuse in crowded detention centers as part of a "round-up campaign" in Vietnam's capital city, Hanoi (Human Rights Watch, 2006b).

Boys and girls living on the streets are also extremely vulnerable to physical and sexual abuse and exploitation. Many suffer from addiction to drugs and chemical substances, live in unsanitary conditions, and face human rights violations and abuse on a daily basis. In 2000, 86 percent of Egyptian street children surveyed identified violence as a major problem in their lives, while 50 percent stated that they had experienced rape (UNICEF, n.d.).

The situation of women and girls can sometimes be overlooked in discussions of gangs and urban violence. After all, men are far more likely than women to be both victims and perpetrators of organized armed violence (Small Arms Survey, 2006). In El Salvador, for example, the ratio of males to females who died of gunshot wounds in 2000 was 35:1 (Dowdney, 2004: 117).

---

[15] According to Casa Alianza, an international NGO that works with street children, "In 2005, Casa Alianza Honduras recorded the violent death or arbitrary execution of 431 children and young people" Casa Alianza UK, 2006.

In Brazil, an estimated 94 percent of gun deaths were among men in 2000 (Eberwine, 2003).

However, these statistics do not justify neglect of the impact of urban armed violence on women and girls, who are affected by male-dominated organized violence in different ways. In some cities, organized violence against women and girls is systemic and widespread. Threats and intimidation, psychological abuse, and community insecurity disproportionately affect females. Further, rape is sometimes used by gangs as a systematic tool of intimidation and subjugation, particularly against young girls. This is also true for girls who are gang members themselves (Miller, 1998: 445). A March 2006 USAID gender-based study identified violence against women as one of the main means of terrorization used by urban gangs in Port-au-Prince, while a November 2005 United Nations Office for the Coordination of Humanitarian Affairs report on Haiti noted that "a worrying percentage of rapes [were] gang rapes."[16]

While females are less likely to join an organized armed gang than males, women and girls are more likely to fall prey to human trafficking circles or to be caught up in the sex industry. Cities such as Bangkok, Lagos, and Medellín have become focal points for organized trafficking in persons and the sex trade. According to Nigeria's National Agency for Prohibition and Trafficking in Persons and Other Related Matters, an estimated 15 million Nigerian children are being transported from rural to urban areas for child labor, slavery, or to work as prostitutes (Child Rights Information Network, 2005). In India, Pakistan, and Middle Eastern countries, an estimated 10,000–20,000 women and girls are trafficked and subjected to bond servitude and forced prostitution annually (U.S. Department of State, 2004). Many trafficking victims are sent to large cities in Asia, the Middle East, Western Europe, and North America, where they are sold and exploited (Orhant, 2001). Human trafficking not only perpetuates physical and gender-based violence, but is also closely tied to the spread of HIV/AIDS.

3.3. DISPLACED PERSONS AND VOLUNTARY MIGRANTS

Every year, millions of men, women, and children migrate to the world's burgeoning cities in search of economic opportunity, shelter, and improved security. In regions of the world that are experiencing conflict, a growing

---

[16] USAID, 2006; UNOCHA, 2005. While some rapes were likely acts of terror, others may have been acts of reprisal committed by gang members against women associated by blood or acquaintance to men in rival gangs. Women are rarely, if ever, full-fledged members of gangs in Haiti, though they may play auxiliary roles as wives, girlfriends, relatives, and friends of gang members.

percentage of the world's displaced—some 8.4 million refugees and 23.7 million IDPs—are moving to cities for the same reasons (United Nations High Commissioner for Refugees, 2005; Internal Displacement Monitoring Centre, 2006; see also Sommers, 1999: 22–24; Juma and Kagwanja, 2003; and Jacobsen et al., 2003). In some countries, IDPs travel to urban areas because it is there that humanitarian assistance is accessible (United Nations Sub-Committee on Nutrition, 2001: 36), although in general there are fewer formal assistance programs for refugees living in urban areas (United Nations High Commissioner for Refugees, 2006c). While the majority of refugees settle in temporary camps located in rural areas along international borders, IDPs are more likely to relocate to urban or peri-urban settlements. It is estimated, for instance, that 40 percent of the 1.5 million registered IDPs in Colombia live in ten major cities (United Nations High Commissioner for Refugees, 2005).

At the same time, many people move to cities voluntarily in search of employment, but find that once there, they remain unable to rise out of poverty. Each year, many thousands of migrant workers find themselves forced by economic necessity to squat in crowded, impoverished, and insecure slums.

Rapid population influxes, whether of voluntary migrants or displaced people, can overwhelm a city's infrastructure and service provision, including policing. This strains a city's ability to protect its residents, and may contribute to the failure of public security. Violent threats to urban migrants are compounded by discrimination, stigmatization, and poverty as new migrants adapt to unfamiliar urban environments. Women and girls are particularly vulnerable to human security threats perpetrated by human traffickers, organized crime syndicates, and even state security forces. The dangers posed to female migrant workers are amplified by widespread, traditional patriarchal norms that do not acknowledge and respect women's rights.

4. Conflict-Resilient Cities

In the twenty-first century, efforts to prevent and mitigate violent conflict will need to adapt to the increasingly urban nature of armed violence. Fortunately, several unique characteristics of cities have the potential to make them conflict-resilient: city dwellers enjoy varied tiers of government representation and close proximity to authorities, frequent interaction and ease of association with community groups, and visible sites of peaceful protest. Effective, inclusive, and responsive local-level political institutions can play a central role in preventing tensions between groups from escalating into organized violence. Population density can also potentially promote positive social capital in cities, which in turn can allow cities to act as buffers against national- or regional-level conflict.

Interventions that seek to build urban conflict resilience should focus on strengthening the capacity of cities to provide basic services. Effective and equitable service provision builds state legitimacy at the level of governance closest to the people. Cities are also important sites for rebuilding state capacity and preventing security failures in post-conflict environments. A specific focus on peacebuilding at the urban level—or "city-building"—to reestablish confidence in the state is potentially both a manageable and scalable approach to reconstruction in the aftermath of conflict.

4.1. GOVERNANCE AND DEMOCRACY AT THE LOCAL LEVEL

Effective, inclusive, and responsive governance at the local level is a key ingredient in building conflict-resilient cities. Flexibility and proximity to communities places local leaders in a unique position to be sensitive to the interests of their constituents, enabling them to respond to the needs of the community in an inclusive, participatory, and transparent fashion. Involving civil society actors in decision making through consultations and dialogue increases transparency and accountability in governance and fosters stakeholder responsibility and civic empowerment (Bush, 2004: 5). This may help to resolve disputes before they erupt into conflict by engaging groups in dialogue and ensuring that minority views are represented. In the city of Durban, for instance, involving traditional chiefs in policy planning helped to defuse conflict in the wake of South Africa's move from apartheid to democratic rule (Beall, 2005).

Community engagement in decision making offers a way to connect people to their city and to each other. The Canada-Bosnia and Herzegovina Local and Cantonal Government Cooperation Program, for example, is a project that seeks to strengthen effectiveness and transparency of local government institutions in the Municipality of Tuzla in a post-conflict environment characterized by high levels of poverty, displacement, and ethnic heterogeneity.[17] This approach can help to reduce tensions between urban populations and authorities by bridging gaps between those who govern and the governed (Goldfrank, 2002).

Inclusive institutions allow marginalized communities to be represented in talks with local officials, planners, and donors. Shack/Slum Dwellers International is one civil society group that facilitates such partnerships. Participatory budgeting schemes, which bring together local politicians and

---

[17] The project is managed by the Canadian Urban Institute with funding from the Canadian International Development Agency. See Bush, 2004: 16–25.

community members in budget formulation, provide slum communities with access to public resources and decision making. These schemes have received international acclaim, and have been adopted in more than 300 cities throughout the world, allowing marginalized populations to improve living conditions and security in their communities. Other community-driven approaches have helped to empower people in decentralized decision making, such as the Community Organizations Development Institute in Thailand, which has undertaken sustainable, participatory slum upgrading and low-cost housing construction activities in recent years (UNHSP, 2006: 169; see also Asian Coalition for Housing Rights, 2003).

Democratic governance can play an important role in building strong cities. By empowering groups through representation and opening lines of communication between groups based on principles of tolerance, inclusion, and respect for minority rights, democratic governance can help to reduce the likelihood of inter-group violence in urban spaces (Bush, 2004: 4).

However, effective local governance and democracy are frustrated by deeply entrenched corruption among authorities in many cities, which often has a disproportionately harmful effect on the poor. High levels of corruption are particularly common in slum settlements. According to UN-HABITAT, cities with a larger proportion of slum dwellers tend to score higher on indices of corruption and lower on government effectiveness. Combating corruption, therefore, is an important step in improving urban livelihoods and building conflict-resilient cities.

4.2. STRENGTHENING SOCIAL CAPITAL FOR CONFLICT RESILIENCE

Although slums can be dangerous places to live, slum dwellers can also form valuable social networks that enable them to cope with economic hardships and security challenges. With cities home to diverse arrays of people, positive social capital can build trust between groups of people of different backgrounds and power levels. The unique geophysical characteristics of cities, combined with population density, offer opportunities for the creation of positive social capital.

The term social capital, while not universally agreed upon, generally refers to the relationships or networks between people in communities, and the resources that permit cooperation between them (Judge, 2003). Three types of social capital—bonding, bridging and linking—together capture the elements of building trust within and across groups (see Figure 4) (Snoxell et al., 2006: 68).

Engendering positive social capital requires that all three types of social capital are fostered together. Activities that aim to build only bonding social

	Description	Example *(Positive)*	Example *(Negative)*
Bonding	Relationships among people who see themselves as sharing a common background	Slum dwellers working together on community upgrading projects	Ethnicity-based armed street gangs
Bridging	Relationships among people without a common background	Inter-ethnic peace negotiations	International, multi-city human trafficking networks
Linking	Relationships among people of different power levels	Community policing alliances between state authorities and slum residents	Children employed as drug-runners by organized criminal cartels

Figure 4. Three Kinds of Social Capital.

capital between members of the same age group, for example, may only serve to strengthen the cohesion of urban gangs. Likewise, social isolation can result if bridging and linking capital between groups are not built simultaneously, frustrating the potentially positive contribution to conflict resilience (Snoxell et al., 2006: 77). Such approaches can produce negative social capital—namely, when bonds are formed between groups to produce outcomes that can exacerbate human insecurity.

Building positive social capital between specific groups can be particularly effective in creative problem-solving to foster peace and cohesion at the local level. GROOTS International is one example of a grassroots women's organization that develops partnerships for development and problem solving, including post-disaster rebuilding.[18] In post-conflict environments, people turn to social networks first to provide security and basic services, such as neighborhood watch groups and garbage collection initiatives (Interpeace, 2006: 9). In Somalia, the Somali Youth Development Network, based in Mogadishu, allows young people to work together in peacebuilding dialogue and activities (Ali, 2006: 16).

Targeted social capital interventions have demonstrated that concerted attempts to build relationships between discrete groups in cities can help reduce violence. The Gender, Peace and Development Project in the conflict-affected region of Muslim Mindanao in the Philippines, for example, builds positive social capital by bringing together Christians, Muslims, and indigenous Lumads in community peacebuilding activities (Bush, 2004: 41–50). An intervention designed to build linking social capital between youth groups and municipal authorities in Cali, Colombia, led one official to comment,

---

[18] For more information, see http://www.groots.org.

"We can now say that the young population is very close to us and we can count on them to undertake awareness activities" (Snoxell et al., 2006: 75).

Cooperation between governments and slum dwellers, including community associations, builds trust between people with different levels of power, thus forging linking capital and enhancing conflict resilience. Squatters in Nairobi's Huruma Ghetto, who had previously been forced to pay exorbitant bribes or face violent evictions, joined together in 2002 to collectively engage city authorities in negotiations for housing construction (O'Meara Sheehan, 2002: 32). By bringing slum residents into local urban democratic processes, such as participatory budgeting in Porto Alegre, Brazil, bridging capital can also be enhanced, as people from different parts of the city work together to improve human security and quality of living (Mitlin, 2001: 162).

Urban social capital developed between potentially antagonistic communities can help to buffer cities against internal conflict before it emerges or escalates, and provide protection against outside identity-based violence at the state or global level. For example, Lucknow and Surat, two Indian cities with vibrant and mixed civil societies, were able to avoid the violence that plagued more segregated Hindu-Muslim cities such as Aligarh and Ahmedabad from the 1960s to the 1990s (Varshney, 2002: 228).

Social capital can also contribute to conflict resilience in times of political upheaval and transformation, as illustrated by peaceful democratic protests in the Ukraine in 2004. By mobilizing civil society networks, half a million protesters converged in the capital, Kiev, and successfully pressured the government into reversing election results that were largely seen as fraudulent (Kuzio, 2005: 128; see also Karatnycky, 2005: 43). This example shows that the cohesion that can result from positive social capital can help to engender dramatic change in cities without compromising human security.

4.3. PROVIDING THE BASICS: SERVICES AND SECURITY

A third component of urban conflict resilience is the provision of basic services. The failure of local authorities to provide basic services to marginalized populations can exacerbate inequalities and fuel tensions between groups. On the other hand, municipal authorities are well placed to respond to the immediate needs of their constituents in emergency situations or in early stages of conflict, and in so doing may help to prevent tensions from mounting. In conflict situations, effective municipal governments can mitigate the impact of violence and insecurity by providing effective law enforcement and access to justice, and by protecting vulnerable populations through emergency shelters and assistance (Bush, 2004). Local-level governments also typically make decisions about access to land, which can often ignite or defuse conflict. Decision making about land-use planning that is participatory and transparent can be key to helping prevent conflict (Bush, 2004: 11).

The longer that slum residents remain in the same place, the more time they have to form networks that build social capital among neighbors and acquaintances. Policies such as extending land tenure in areas of informal housing are likely to contribute to permanence by providing slum residents with security and confidence in the future (United Nations Commission on Sustainable Development, 2005: 3). Recent land tenure policies in Burkina Faso, Senegal, and Tanzania have been well received, with the number of evictions in these countries dropping (UNHSP, 2006: 219). Service provision may have a similar effect. If slum residents gain access to services such as potable water, solid waste removal, transportation, policing, and education, they will be better able to build safe and healthy communities. Democratic urban governance can help to counteract slum policies that generate negative social capital, such as clearances and demolitions. Such governments are more likely to protect vulnerable groups because they involve in decision making the very people they are trying to help. In Rio de Janeiro, the return to democratic governance in the 1980s corresponded with the development of projects such as Favela Bairo, a program that invested over US $600 million in the provision of public services including clean drinking water and roads to slums in the city (Foek, 2005: 32).

Involving the community in decision making and security provision is the operating principle behind community policing, a proactive approach to law enforcement that encourages closer bonds between security officers and the public.[19] Examples of community policing initiatives include staffing local policing units with community members as well as police officers; hiring more bilingual officers for neighborhoods with concentrated ethnic populations; encouraging police to patrol by foot rather than in vehicles; and establishing neighborhood advisory councils for community members to discuss specific concerns with police officers.

Community policing provides an encouraging example of how building positive social capital between community members and the state can help to peacefully resolve disputes, thereby potentially preventing violence from erupting in urban areas. In slum areas of Mumbai, community police stations, or *panchayats*, are staffed by local volunteers who work in

---

[19] In his writings on South Africa, Wilfried Schärf defines community policing as "any form of sustained partnership/consultation/liaison between local residents and the local state police," not to be confused with "community-generated policing" which "refers to civilian forms of policing (not for commercial gain) outside of a partnership with the state," including vigilantism, "and what became labelled in [South Africa] as 'urban terror.'" See Schärf, 2000: 5.

partnership with police officers to patrol the streets and resolve disputes (Roy et al., 2004: 135–138). This approach has helped community members resolve disputes in a fashion that is both cost-effective (community members volunteer their time) and empowering for women (who often make up a majority of *panchayat* members).

In Bogotá, community policing combined with educational programs, increased investment in policing, and innovations in street lighting have helped reduce the number of homicides per year by 75 percent since 1993 (Phillips, 2006). Community policing strategies in major Colombian cities are now being used by the United Nations Development Programme as models for other cities in Latin America such as in El Salvador and Ecuador (Phillips, 2006).

4.4. CAPACITY BUILDING FOR URBAN CONFLICT RESILIENCE

Capacity building for local governance and democratic development play important roles in peacebuilding in post-conflict environments. In the short term, a focus on the local level may be more practical and productive than a focus on state-level institutions. In the wake of violent conflict, citizens often no longer trust the state for protection and basic services, turning instead to community support networks. This is illustrated by the thousands of voluntary organizations that emerged in Belfast since the 1970s as a response to a breakdown in state legitimacy during periods of vicious sectarian fighting.[20] Municipal governments can cooperate with such local networks on urban reconstruction to reduce the likelihood of recurring violence and lubricate the rebuilding process (Bush, 2004: 14).

Urban centers have the potential to drive national post-conflict reconstruction efforts by projecting an image of recovery and peace that can foster confidence in larger peace processes. Sarajevo, for example, suffered a forty-three-month siege by Serbian forces in 1992–1995, resulting in an estimated 12,000 deaths (BBC News, 2002). However, the city's remarkable recovery —characterized by interethnic marriages and civic engagement among youth groups—served as a source of symbolic and practical peacebuilding for Bosnians (Arun, 2006). In the Middle East, the Municipal Alliance for Peace, working in partnership with city councilors from the UK, works with Israeli and Palestinian local governments to build capacity and foster peaceful dialogue through town-twinning arrangements (Brown, 2006).

---

[20] Ellis and McKay estimated that 5,500 such groups existed in 2000. See Ellis and McKay, 2000: 51.

4.5. ADVANCING HUMAN SECURITY IN THE BUILT ENVIRONMENT

The physical space where people live—their built environment—has an important, if often overlooked, impact on urban security. Some environments are more conducive to promoting social capital than others. Cities that are territorially segregated—those with hollowed-out cores and affluent suburbs, gated enclaves amidst sprawling slums, or physically divided or isolated ethnic or cultural groups—are less likely to enjoy the benefits of positive social capital.

On the other hand, urban policies that seek to promote interaction among groups, including the creation and maintenance of public spaces, can nurture diversity and integration, thereby building unity and supporting inter-group exchanges. Urban planning (by building diverse communities) and urban management (through effective representation and access to decision making) can be effective in reducing violence and insecurity that can stem from social stigma, discrimination, and isolation.

Effective local governance can promote urban conflict resilience by offering a path to physical improvements that encourage the growth of stable urban communities. Civil society also has a key role to play in enhancing safety and security in urban communities. Movements that empower marginalized groups to bring an end to forced evictions and ensure land security can help to prevent violent conflict that can flourish when people are excluded and discriminated against.

5. Safer Cities for an Urban Century

This paper has attempted to demonstrate the value of examining human security from an urban perspective. Rapid urbanization is setting in motion new dynamics in which organized gangs and transnational criminal networks are taking advantage of failed public security in sprawling urban spaces to generate new threats to people's safety and lives. From Cape Town to Cairo, Bangkok to Baghdad, Kingston to Kandahar, guns, gangs, and drugs are finding their way onto city streets with devastating consequences for civilian populations. In some cases, more deaths are being caused by armed violence in cities within countries formally at peace than within countries experiencing civil war.

Rapid urbanization is having a particularly profound impact in the developing world where many local governments lack the capacity to provide adequate public security for their ever-growing populations. In some cities, the inability or unwillingness by public security forces to provide public security is resulting in the progressive privatization of security. While elites are often able to hire private security forces, slum dwellers are increasingly being victimized by highly organized and heavily armed gangs who are filling the void left when public security fails.

Yet, out of these challenges emerge exciting opportunities for improving security, advancing human rights, and building dynamic, conflict-resilient communities. As the level of government closest to the people, inclusive, effective, and responsive local-level institutions can play a vital role in reducing tensions and resolving conflicts between groups, thereby preventing violence from erupting. Diverse, densely populated cities also have the potential to foster positive social capital, building a strong social fabric that can potentially help to buffer cities from conflict. Reconciling these positive and negative aspects of urbanization will be critical to building safer cities for the future.

5.1. EXISTING EFFORTS AND ACTORS IN URBAN SECURITY AND DEVELOPMENT

A variety of actors, ranging from UN bodies, to bilateral donors, to community-based organizations, have undertaken efforts to enhance the safety of people living in urban areas. These efforts signal a growing acknowledgement that cities are key entry points for programs that seek to enhance security and development.

In the UN system, there are a number of institutions that have been established to address the specific security and development needs of cities. UN-HABITAT's Safer Cities Programme has partnered with local governments in programs that seek to build good governance and prevent crime and violence in cities such as Dar es Salaam, Johannesburg, and São Paulo. Several other UN agencies, including the United Nations Development Programme and the United Nations Population Fund, have also undertaken city-based activities.

In recent years, city governments themselves have become actors in international diplomacy. Through what is known as "city diplomacy," cities are increasingly cooperating in international fora to play a role in strengthening each other's capacities to prevent and mitigate armed violence. United Cities and Local Governments, the so-called UN of cities, encourages this kind of diplomacy and conflict prevention at the urban level through its Committee on City Diplomacy.[21] Members of the Cities Alliance, a global coalition

---

[21] For more information, see http://www.cities-localgovernments.org/uclg/index.asp?L =,&ID=241&pag=newTemplate.asp.

of cities and development organizations that support slum-upgrading programs,[22] are also active in bridging development and security in cities.

At the local level, some programs, particularly those led by civil society, have been directed at improving human security in cities. For example, the Brazil-based NGO Viva Rio works in over 350 favelas, or shantytowns, on disarmament and gun awareness campaigns, with a focus on youth.[23] Examples of work being done to strengthen local-level democratic development include International IDEA's Local Democracy Assessment Guides and the National Democratic Institute and the World Bank's handbooks on best practices for local governance.[24] On the issue of urban forced evictions, the international NGO Human Rights Watch has spearheaded a campaign calling on the Cambodian government to strengthen its housing and resettlement policy to protect forced evictees in Phnom Penh (Human Rights Watch, 2006a).

While this evidence of increased international engagement on urban issues is welcomed, these efforts remain largely focused on developmental impacts rather than security objectives.[25] Nevertheless, there appears to be a clear and growing international recognition of the important synergies between security and development in the urban context.

5.2. TRANSNATIONAL DIMENSIONS OF URBAN ARMED VIOLENCE

The research presented in this paper suggests that much of the organized armed violence that takes place on city streets is perpetrated by groups that are linked to transnational criminal organizations or internationalized weapons flows and gang cultures. For example, international human trafficking is enabled by large criminal networks with operations in major cities throughout the world. These cities serve as major transit points or end destinations for the many thousands of women and children who are sold into the sex trade each year. It is also the case that American gang culture is being mimicked across the Americas, from music and fashion to a growing

---

[22] For more information, see www.citiesalliance.org.

[23] For more information, see www.vivario.org.br.

[24] For more information on International IDEA's work, see www.idea.int/news/local _ level_africa.cfm. For NDI's local governance programs, see www.ndi.org/globalp/localgov/ localgov.asp. For the World Bank, see info.worldbank.org/etools/mdfdb/Conf_Workshops _11.htm.

[25] Some recent examples include the Asian Development Bank's Country Assistance Plan for India, the Swedish International Development Agency's Urban Development Programs, and USAID's Local Governance Program in Iraq and its Haiti Transition Initiative.

willingness to engage in drive-by shootings and armed urban combat with rival gangs and state security forces (Strocka, 2006: 137). The widespread appeal of American gang culture, particularly its powerful bonding effect on marginalized urban youth, may help to account for the proliferation of organized gangs as well as why it is so difficult to reduce incidents of armed violence in many cities.

In this sense, urban gangs can be seen as a local manifestation of transnational crime. In Latin America, youth gangs are frequently involved in drug trafficking, one of the most challenging international illegal enterprises. Not only are there important links between urban gangs and transnational crime, but evidence also suggests that these links are becoming more prominent. As one recent study found, "youth gangs no longer operate only within the boundaries of a particular, relatively small neighborhood, but increasingly extend their sphere of influence across cities, regions and countries" (Strocka, 2006: 136).

Transnational criminal activities, such as trafficking in drugs and weapons, are often closely intertwined and mutually reinforcing. Cocaine, opium, guns, and even people are used as currency for illicit transactions between criminal groups.[26] Major criminal organizations, such as the Triads in Hong Kong and the Japanese Yakuza, profit from forging alliances with other criminal networks to expand their reach and profits (United Nations Department of Public Information, 2000). There are also demonstrated linkages between arms and drug trafficking and terrorism, with profits from illicit trade used to support terrorist networks (Monblatt, 2004).

The transnational dimensions of the gun trade are well documented, and the linkages to urban armed violence are frequently just as clear. The lethality of urban armed violence is often enabled by illegal arms trafficking. For example, many of Brazil's guns are smuggled illegally from Paraguay (Bacoccina, 2003), and many of West Africa's guns are imported through Warri, a port town in southern Nigeria (Ojudu, 2006). The high rates of firearm violence in South African cities today are assessed to be the result of the influx of guns from the civil war in Mozambique. The limited success

---

[26] Other examples include: (1) In 2001, Colombian officials arrested three members of the Irish Republican Army in Colombia, and later convicted them of teaching FARC militants bomb-making techniques. (2) Officials in several countries have documented complicated trade patterns involving illicit shipments of coca paste through the tri-border region of Argentina, Paraguay, and Brazil to the Bekaa Valley in Lebanon, the center of Hezbollah's influence. In May 2003, Paraguayan police arrested Hassan Dayoub while he was preparing to ship an electric piano containing more than five pounds of cocaine to Syria. See Monblatt, 2004.

of disarmament efforts following Angola's conflict also resulted in many guns being illegally smuggled across the border into South Africa.[27]

The links between actors that perpetrate violence—from members of international organized criminal networks, to armed insurgents waging combat against states, to paramilitary groups controlling regions within states, to violent gangs controlling city slums—support the argument that international actors have an important role to play in helping to combat urban violence. They suggest an explanation as to why efforts by local and national level governments to combat urban armed violence may have limited success: international issues are most effectively addressed by international actors and regimes.

Existing international norms and laws offer a platform upon which states can build to address the transnational dimensions of urban violence. The UN Convention against Transnational Organized Crime is one such example that recognizes the significant roles played by international drug cartels, human trafficking networks, and illicit arms traffickers in perpetuating armed violence in urban areas. The Convention has three protocols: one on human trafficking, one on the smuggling of migrants, and one on illicit small arms trafficking.

5.3. AN INTERNATIONAL AGENDA ON ORGANIZED ARMED VIOLENCE?

A central argument of this article is that the scale of organized armed violence in large urban areas frequently exceeds that of all but the most devastating of current wars. Research on contemporary armed conflicts frequently uses the threshold of 1,000, 100, or even 25 "battle-deaths" annually to define a civil war[28]—a comparatively low number compared with the scale of urban armed violence in numerous cities discussed in previous sections of this paper. Evidence from Colombia, a country experiencing an intense civil war, suggests that more people are dying from armed violence in urban areas than from the conflict between rebel groups and government forces.

Much has been done, particularly in the past fifteen years, to adapt the international laws and institutions originally designed to respond to

---

[27] For example, one report found that only about 10 percent of guns in circulation in Angola were collected by a government disarmament campaign. See "Angola: Widespread small arms could lead to increase in crime," IRIN, 2003.

[28] The Correlates of War Project, http://www.correlatesofwar.org, and the Political Instability Task Force's State Failure project, http://globalpolicy.gmu.edu/pitf, use the figure of 1,000 battle deaths per year.

the challenges of wars between states to address the challenges posed by a different kind of conflict—civil wars within states. The evidence is mounting, however, to suggest that the changing nature of organized armed violence may be more radical than many had imagined. Traditional definitions of war and armed conflict may be obscuring a crisis of armed violence within contemporary cities.

Take, for example, a central theme that has run through this article—the violent threats facing children in urban environments—and compare the international responses to this set of challenges with those devoted to the challenges facing child soldiers in armed conflicts.

In 1996, Graça Machel's ground-breaking report on the impact of armed conflict on children was released, noting that there were approximately 300,000 child soldiers worldwide. Since that time, the Optional Protocol to the Convention on the Rights of the Child on the involvement of children in armed conflict has been negotiated, raising the minimum age of soldiers; the Rome Statute establishing the International Criminal Court has defined the conscription, enlistment, or use in hostilities of children under the age of fifteen as a war crime; the UN Secretary-General has created a Special Representative for Children and Armed Conflict; 180 countries endorsed a global action plan entitled A World Fit for Children at the UN Special Session on Children; and the UN Security Council has adopted seven thematic resolutions devoted to children and armed conflict. Agencies and NGOs in the field have responded by ensuring that child protection is included in peacekeeping mandates, targeting disarmament, demobilization, and reintegration (DDR) programs to children (including specific emphasis on girls), and monitoring and reporting on persistent violators as listed by the Secretary-General in his annual list of armed groups who recruit and/or use children.

Children fighting in urban gangs experience violence comparable to that faced by child soldiers. This is particularly true when children in urban gangs are given military-grade weapons and put on the front lines of armed combat against enemy gangs or state security forces. Beyond the direct violence they face, there are other important similarities: aggressive recruitment strategies, the widespread use of drugs, the prevalence of sexual violence, social stigmatization, and the likelihood of long-term psychological effects caused by exposure to violence. And the millions of children living in the streets and fighting in urban armed gangs far outnumber the estimated 300,000 child soldiers fighting in war zones throughout the world.

The evidence of alarming levels of violence faced by children living in some urban areas makes a compelling case for enhancing protection of these young people. But where is the international response on behalf of children facing the brunt of organized armed violence in cities supposedly at peace?

Or consider another key theme that reoccurs throughout this paper: that the aggressive, even repressive tactics employed by law enforcement officials in situations where urban security has broken down are frequently counterproductive. These harsh tactics frequently exacerbate levels of violence, harden the attitudes of communities against law enforcement, and do nothing to address the underlying causes of insecurity. Here, an internationally agreed-upon set of standards that are directly applicable already exists.

The Basic Principles on the Use of Force and Firearms by Law Enforcement Officials, adopted by consensus at the UN in 1990, establish clear and detailed guidelines for law enforcement to ensure that the use of force is necessary, proportionate, and accountable within a legal system (United Nations High Commissioner for Human Rights, 2006a). Although the principles are nonbinding and not widely known, their application could assist in the rebuilding of failed public security systems. The Basic Principles also demonstrate the potential role of international standards of conduct in promoting the rule of law and preventing human rights violations. To help address the crisis of insecurity in large urban areas around the world, the establishment of similar standards might be warranted with respect to the promotion of community-based policing and the regulation of private security companies.

Clearly these are only tentative suggestions. Much more research and analysis is needed in order to know how and where international efforts could contribute to addressing urban insecurity. One thing, however, is clear. International efforts to promote human security—to enhance the safety of people and their communities from the threat of physical violence—must respond to the real insecurities people face in their daily lives. The fact that much organized armed violence takes place outside situations defined as armed conflicts should lead to a systematic examination of whether the international normative, institutional, and legal framework constructed in the twentieth century to respond to the predominant form of organized armed violence of that era—international and intra-state armed conflict—can be adapted to the urban insecurity realities of the twenty-first century.

References

Ali, I., 2006, *Human (In)Security and Cities: Case Study—Mogadishu, Somalia*, Centre for Research and Dialogue, Mogadishu, p. 16.

Arun, N., 2006, Sarajevo finds love after the war, BBC News, 28 February; http://news.bbc.co.uk/2/hi/europe/4746082.stm, accessed 1 November 2006.

Asian Coalition for Housing Rights, 2003, Thailand to build 1 million low-income housing units, 13 January; http://www.achr.net/country_news.htm, accessed 1 November 2006.

Bacoccina, D., 2003, Brazil seeks to curb gun crime, BBC News, 24 July; http://news.bbc. co.uk/2/hi/ americas/3089417.stm, accessed 1 December 2006.

BBC News, 2002, Bosnia marks war anniversary, 6 April; http://news.bbc.co.uk/2/hi/europe/ 1914133.stm, accessed 1 November 2006.

Beall, J., 2005, Exit, voice, and tradition: Loyalty to chieftainship and democracy in metro-politan Durban, South Africa, London School of Economics and Political Science, Crisis States Programme, Working Paper 59, p. 20.

Blair, D., 2006, Children of the streets feel wrath of Mugabe, *The Telegraph,* 16 May; http:// www.telegraph.co.uk/news/main.jhtml?xml=/news/2006/05/16/wzim16.xml, accessed 1 November 2006.

Botha, A., 2001, Fear in the city: Urban terrorism in South Africa, in: *The Multi-Headed Monster: Different Forms of Terrorism,* Institute for Security Studies, Monograph no. 63.

Brennan-Galvin, E., 2002, Crime and violence in an urbanizing world, *Journal of International Affairs* **56**(1):123–145.

Brown, A., 2006, Peace in our time?: Local government offers hope for the future in the Arab-Israeli conflict, UK Local Government Alliance for International Development, 4 April; http://www.lgib.gov.uk/lg-alliance/features/features/2006/Peace_in_our_time.html, accessed 1 November 2006.

Bush, K., 2004, *Building Capacity for Peace and Unity: The Role of Local Government in Peacebuilding,* Federation of Canadian Municipalities, Ottawa, p. 5.

Canadian International Development Agency (CIDA), 2006, Street children; http://www.acdi-cida.gc.ca/CIDAWEB/acdicida.nsf/En/REN-218125542-Q3B?OpenDocument, accessed 22 September 2006.

Casa Alianza UK, 2005, Casa Alianza UK Newsletter, May 2005; http://www.casaalianza. org.uk/northsouth/CasaWeb.nsf/Resources/CB061899BBD685738025715A003D00C5/ $FILE/ may_2005.pdf?openElement, accessed 1 November 2006.

Casa Alianza UK, 2006, Violent deaths of children and youth in Honduras continue to increase; http://www.casa-alianza.org.uk/northsouth/CasaWeb.nsf/CasaNews/Children_ Honduras_Violence?OpenDocument, accessed 1 August 2006.

Ceballos Melguizo, R., 2001, The evolution of armed conflict in Medellín: An analysis of the major actors, *Latin American Perspectives* **28**(1):110–131.

Child Rights Information Network, 2005, Nigeria: 15 Million children toil in slavery, 18 November; http://www.crin.org/violence/search/closeup.asp?infoID=6608, accessed 1 November 2006.

Dowdney, L., 2003, *Children of the Drug Trade: A Case Study of Children in Organized Armed Violence in Rio de Janeiro,* Viveiros de Castro Editora Ltda, Rio de Janeiro, p. 130.

Dowdney, L., 2004, *Neither War Nor Peace: International Comparisons of Children and Youth in Organised Armed Violence,* Viva Rio/Instituto de Estudos da Religião, Rio de Janeiro, p. 12.

Eberwine, Donna, 2003, The violence pandemic: How public health can help bring it under control, *Perspectives in Health* **8**(2).

Ellis, G., and McKay, S., 2000, Belfast: City management profile, *Cities* **17**(1):51.

Faiola, A., 2002, Brazil's elites fly above their fears; Rich try to wall off urban violence, *Washington Post,* 1 June, p. A01.

Foek, A., 2005, Rio de Janeiro: Microcosm of the future, *Humanist* **65**(4):32.

Goldfrank, Benjamin, 2002, The fragile flower of local democracy: A case study of decentrali-zation/participation in Montevideo, *Politics and Society* **30**(1):51–83.

Graham, S., 2002, Special collection: Reflections on cities, September 11th and the "War on Terrorism"—One year on, *International Journal of Urban and Regional Research* **26**(3):589–590.

Gutiérrez Sanín, F., and Jaramillo, A. M., 2004, Crime (counter-)insurgency, and the priva-tization of security—The case of Medellín in Colombia, *Environment and Urbanization* **16**(2):17–30.

Hodgson, Martin, 2001, Reportage—Guns for hire, *The Independent,* 17 March.

Human Rights Watch and the Centre for Law Enforcement Education, 2002, The Bakassi boys: The legitimization of torture and murder; http://www.hrw.org/reports/2002/ nigeria2, accessed 1 November 2006.

Human Rights Watch, 2002, Kenya: Crackdown on Nairobi's refugees after Mombasa attacks, 6 December; http://www.hrw.org/press/2002/12/kenya1205.htm, accessed 1 November 2006.

Human Rights Watch, 2006a, Cambodia: Phnom Penh's poor face forced evictions; http:// hrw.org/english/ docs/2006/08/02/cambod13889.htm, accessed 1 December 2006.

Human Rights Watch, 2006b, Vietnam: Street children at risk before APEC summit, 13 November; http://hrw.org/english/docs/2006/11/13/vietna14543.htm, accessed 20 November 2006.

Human Security Centre, 2005, *Human Security Report 2005: War and Peace in the 21st Century,* Oxford University Press, New York, p. 125.

Integrated Regional Information Networks (IRIN), 2003, Angola: Widespread small arms could lead to increase in crime, 7 February; http://www.irinnews.org/report.asp?ReportID =32179&SelectRegion=Southern_Africa&SelectCountry =ANGOLA, accessed 1 December 2006.

Internal Displacement Monitoring Centre, 2006, *Internal Displacement: A Global Overview of Trends and Developments in 2005,* Internal Displacement Monitoring Centre, Geneva.

International Action Network on Small Arms, 2006, 2006: Bringing the global gun crisis under control; http://www.iansa.org/campaigns_events/gun-control-2006.htm, accessed 1 November 2006.

International Committee of the Red Cross (ICRC), 2005, Dossier de presse: Le travail du CICR et de la Croix-Rouge Haïtienne à Cité Soleil, 11 August; http://www.icrc.org/ web/eng/ siteeng0.nsf/iwpList86/6D27B1CC81B58B22C12570EB00456534, accessed 27 June 2006.

International Crisis Group, 2005, Spoiling security in Haiti, 31 May; http://www.crisisgroup. org/home/index.cfm?id=3485&l=1, accessed 1 November 2006.

Interpeace, 2006, *Human (In)Security and Cities: Summary of a Rapid Research Project,* Interpeace, Geneva, p. 9.

Jacobsen, K. et al., 2003, The Sudan: The unique challenges of displacement in Khartoum, in: *Caught between Borders: Response Strategies of the Internally Displaced,* M. Vincent and B. Refslund Sorensen, eds., Pluto, London.

Judge, R., 2003, Social capital: Building a foundation for research and policy development, *Horizons* 6(3); http://www.policyresearch.gc.ca/page.asp?pagenm=v6n3_art_03, accessed 1 November 2006.

Juma, M. K., and Kagwanja, P. M., 2003, Securing refuge from terror: Refugee protection in East Africa after September 11, in: *Problems of Protection: The UNHCR, Refugees, and Human Rights,* N. Steiner et al., eds., Routledge, New York.

Karatnycky, A., 2005, Ukraine's Orange Revolution, *Foreign Affairs* 84(2):43.

Kuzio, T., 2005, Ukraine's Orange Revolution: The opposition's road to success, *Journal of Democracy* 16(2):128.

LeGates, R. T., and Stout, F., eds., 2003, *The City Reader,* Routledge, New York, p. 31.

McCarney, P., 2006, Our future: Sustainable cities—Turning ideas into action, World Urban Forum III, Background Paper, p. 4.

McKinley, T., 2001, Kenya's slum war, BBC News, 7 December; http://news.bbc.co.uk/ 2/hi/ africa/1697809.stm, accessed 1 November 2006.

McRae, R., 2001, Human security in a globalized world, in: *Human Security and the New Diplomacy,* R. McRae and D. Hubert, eds., McGill-Queen's University Press, Montreal, pp. 14–27.

Miller, J., 1998, Gender and victimization risk among young women in gangs, *Journal of Research in Crime and Delinquency* 35(4):445.

Mitlin, D., 2001, Civil society and urban poverty—Examining complexity, *Environment & Urbanization* 13(2):162.

Monblatt, S., 2004, Terrorism and drugs in the Americas: The OAS response, Organization of American States, February; http://www.oas.org/ezine/ezine24/Monblatt.htm, accessed 1 December 2006.

Morrison Taw, J., and Hoffmann, B., 1994, *The Urbanization of Insurgency*, RAND Corporation, Santa Monica, CA.

Moser, C., 2004, Urban violence and insecurity: An introductory roadmap, *Environment & Urbanization* 16(2):6.

Moser, C. O. N., and Rodgers, D., 2005, *Change, Violence and Insecurity in Non-Conflict Situations*, Overseas Development Institute, London, p. 23.

O'Meara Sheehan, M., 2002, Where the sidewalks end, *World Watch* 15(6):32.

Ojudu, B., 2006, Gun smuggling in the Niger Delta, World Press Review Online, 1 December; http://www.worldpress.org/cover7.htm.

Orhant, M., 2001, Trafficking in persons: Myths, methods, and human rights, Population Reference Bureau; http://www.prb.org/Template.cfm?Section=PRB&template=/ContentManagement/ContentDisplay.cfm&ContentID=5261, accessed 1 November 2006.

Pan American Health Organization, 2000, *A Portrait of Adolescent Health in the Caribbean: 2000*, Pan American Health Organization, Washington, p. 17.

Phillips, T., 2006, Rio turns to Colombia for help in curbing violence, *Guardian Unlimited*, 23 November; http://www.guardian.co.uk/brazil/story/0,,1955486,00.html, accessed 1 December 2006.

Rapley, J., 2006, The new middle ages, *Foreign Affairs* 85(3):95–103.

Rodgers, D., 2003, Dying for it: Gangs, violence and social change in urban Nicaragua, London School of Economics and Political Science, Crisis States Programme, Working Paper 1, p. 2.

Roy, A. N. et al., 2004, Community police stations in Mumbai's slums, *Environment & Urbanization* 16(2):135–138.

Save the Children Sweden, 2005, *Urban Flight and Plight of War Affected Children in Africa*, Save the Children Sweden, Copenhagen, p. 16.

Schärf, W., 2000, Community justice and community policing in post-apartheid South Africa: How appropriate are the justice systems of Africa? Paper delivered at the International Workshop on the Rule of Law and Development: Citizen Security, Rights and Life Choices in Law and Middle Income Countries, Institute for Development Studies, University of Sussex, p. 5.

Small Arms Survey, 2005, *Small Arms Survey 2005: Weapons at War*, Oxford University Press, Oxford, p. 232.

Small Arms Survey, 2006, *Small Arms Survey 2006: Unfinished Business*, Oxford University Press, Oxford, p. 295.

Snoxell, S. et al., 2006, Social capital interventions: A case study from Cali, Colombia, *Canadian Journal of Development Studies* 27(1):68.

Sommers, M., 1999, Urbanisation and its discontents: Urban refugees in Tanzania, *Forced Migration Review* 4:22–24.

South African Press Association (SAPA), 2006, Security guards outgun cops 3 to 1, 4 April; http://iafrica.com/news/sa/170380.htm, accessed 1 November 2006.

Strocka, C., 2006, Youth gangs in Latin America, *SAIS Review* 26(2):137.

Transitional Islamic Government of Afghanistan, 2004, Securing Afghanistan's future: Accomplishments and the strategic path forward—National police and law enforcement technical annex; http://www.af/resources/mof/recosting/draft%20papers/Pillar%203/ Nati onal%20Police%20&%20Law%20Enforcement%20-%20Annex.pdf#search=%22 kabul% 20%2B%20police%20%2B%20wage%22, accessed 1 November 2006.

U.S. Agency for International Development (USAID), 2006, USAID/OTI field report: January–March 2006; http://www.usaid.gov/our_work/cross-cutting_programs/transi-tion_initiatives/ country/haiti/rpt0306.html, accessed 1 November 2006.

U.S. Citizenship and Immigration Services, 2001, Colombia—RIC query, 11 July; http://www.uscis.gov/graphics/services/asylum/ric/documentation/COL01001.htm, accessed 1 November 2006.

U.S. Department of State, 2004, Trafficking in persons report; http://www.state.gov/ docu-ments/organization/34158.pdf, accessed 1 November 2006.

U.S. Department of State, 2005, Fact sheet: Sudan: Death toll in Darfur, 25 March; http://
www.state.gov/s /inr/rls/fs/2005/45105.htm, accessed 1 November 2006.

UN-HABITAT, 2001, Urbanization: Facts and figures, in: *Urban Millennium,* UN-HABITAT,
New York, p. 9.

United Nations Children's Fund (UNICEF), n.d., Egypt—Child protection—Street children: Issues and impact; http://www.unicef.org/egypt/ protection_144.html, accessed
22 Sept. 2006.

United Nations Commission on Sustainable Development, 2005, *Human Settlements: The
Growing Problem of Urban Slums,* Backgrounder, April 2005, United Nations Department
of Public Information, New York, p. 3.

United Nations Department of Public Information (UNDPI), 1995, Urban crime: Policies
for prevention; http://www.un.org/ecosocdev/geninfo/crime/dpi1646e.htm, accessed 29
June 2006.

United Nations Department of Public Information (UNDPI), 2000, Fighting transnational
organized crime—Press Kit Backgrounder No.1, Tenth United Nations Congress on
the Prevention of Crime and the Treatment of Offenders; http://www.un.org/events/
10thcongress/ 2088f.htmaccessed 1 December 2006.

United Nations Development Programme (UNDP), 1994, *Human Development Report 1994,*
Oxford University Press, New York.

United Nations High Commissioner for Refugees, 2005, Radio campaign informs displaced
Colombians of their rights, 28 February; http://www.unhcr.org/cgi-bin/texis/vtx/ news/
opendoc.htm?tbl=NEWS&page=home&id=42234ad94, accessed 1 November 2006.

United Nations High Commissioner for Human Rights, 2006a, Basic principles on the use
of force and firearms by law enforcement officials; http://www.unhchr.ch/html/menu3/b/h
_comp43.htmaccessed 1 December 2006.

United Nations High Commissioner for Refugees, 2006b, *2005 Global Refugee Trends,* United
Nations High Commissioner for Refugees, Geneva.

United Nations High Commissioner for Refugees, 2006c, Refugee livelihoods network—
September 2005; http://www.unhcr.org/cgi-bin/texis/vtx/research/opendoc.pdf?tbl=
RESEARCH&id=434d2ce52, accessed 1 November 2006.

United Nations Human Settlements Programme (UNHSP), 2004, Youth employment and
urban renewal—Discussion paper for expert group meeting, 22–24 June, 2004; http://
www.unhabitat.org/downloads/docs/274-Youth_employment_and_urban_renewal-Youth
employmentHabitat%20v1.pdf, accessed 11 September 2006.

United Nations Human Settlements Programme (UNHSP), 2006, *State of the World's Cities
2006/7,* UN-HABITAT, Nairobi, p. 4.

United Nations Office for the Coordination of Humanitarian Affairs (UNOCHA), 2005,
Humanitarian Situation Report: November 2005, http://www.hahin.org/Document
%20Library/10%20%20HumSitrep%20November.pdf, accessed 1 November 2006.

United Nations Sub-Committee on Nutrition, 2001, *Report on the Nutrition Situation of
Refugees and Displaced Populations* 34:36.

Varshney, A., 2002, *Ethnic Conflict and Civic Life,* Yale University, New Haven, CT, p. 228.

Viva Rio, 2005, *Women and Girls in Contexts of Armed Violence: A Case Study on Rio de
Janeiro,* Viva Rio, Rio de Janeiro.

vom Hove, T., 2006, More than one billion people call urban slums their home, City Mayors,
25 August; http://www.citymayors.com/report/slums.html.

Weeks, C., 2006, Canada to equip Afghan police, *Ottawa Citizen,* 30 October.

URBANIZATION AND ENVIRONMENTAL SECURITY

Infrastructure Development, Environmental Indicators, and Sustainability

NIKOLAI BOBYLEV*

Research Center for Interdisciplinary Environmental Cooperation of the Russian Academy of Sciences St. Petersburg, Russia

Abstract: Urban infrastructure is one of the key elements in the struggle for making cities more sustainable and ensuring the environmental security of the population. This paper will analyze some patterns in urban infrastructure development, focusing on environmental impacts of different types of infrastructure. Urban infrastructure and environmental indicators are discussed, as well as application of multiple criteria decision analysis to urban infrastructure environmental assessment on the basis of the above-mentioned indicators. Environmental performance of underground infrastructure is compared in three cities: Paris, Stockholm, and Tokyo. Some important issues for the development of sustainable urban infrastructure have been identified: functionality, land use, integrality, flexibility, rationality, vulnerability.

Keywords: Urban environment; urban infrastructure; environmental security; environmental assessment

1. Introduction

The demographic conentration of people in urban areas and their consumption patterns and economic activities have a large impact on the cities' internal and external environment. Big urban agglomerations, seen as an absolute disaster for the environment in the past, now offer potential opportunities for managing world population growth in a sustainable way. Innovative compact urban communities can combine friendly living standards for their citizens and maintain a high efficiency in the use of land, water, and energy resources.

* Address correspondence to Nikolai Bobylev, P.O. Box 45, 195267, St. Petersburg, Russia; e-mail: nikolaibobylev@yahoo.co.uk; nikolaibobylev@gmail.com

P.H. Liotta et al. (eds.), Environmental Change and Human Security, 203–216.

Urban infrastructure is one of the key elements in a struggle for making cities more sustainable and ensuring environmental security of the population. Different types of urban infrastructure include transport facilities, resource supply networks, and waste management systems.

Experiences in the development of urban infrastructure have not always been positive. Some examples include hindrances for further development due to inappropriate location of facilities and the destructive impact of motorization on cities' environment. New megacities in developing countries are growing, and there is a unique chance to use the experiences of the developed world and learn from the successes and mistakes of urban infrastructure development.

2. Human Security

Several concepts for the future well-being of humanity have been developed; the most famous among them is the sustainable development concept proposed by the Bruntland Commission in 1987 (Bruntland, 1987). During the last decades a concept of human security emerged in opposition to traditional security concepts. Human security is viewed as a paradigm shifting from military, political, and economy-related issues to environmental, public health, and quality of life concerns.

2.1. ENVIRONMENTAL SECURITY

The environmental component of human security can be described as the state of protection of vital interests of the individual and society from anthropogenic and natural threats. Belluck et al. (2005) defined environmental security as actions that guard against environmental degradation in order to preserve or protect human, material, and natural resources at scales ranging from global to local. The concept of environmental security addresses key challenges to humanity posed by depletion of critical resources and environmental change and degradation. Glenn et al. (1998) define five key elements of environmental security: public safety from environmental dangers caused by natural or man-made processes due to ignorance, accident, mismanagement, or unsound design; amelioration of natural resource scarcity; maintenance of a healthy environment; amelioration of environmental degradation; and prevention of social disorder and conflict (i.e., promotion of social stability). Arguably, social conflicts can be a cause or consequence of environmental insecurity, and should be considered within a broader concept of human security, rather than diverting focus from the environmental issues in the environmental security concept. The most important components of an environmental security concept are equitable distribution of environmental resources and uninterrupted provision of environmental services.

3. Urbanization Trends

According to the United Nations Population Division, the world continues to urbanize quickly. Urban dwellers represented 49 percent of the global population in 2005, and half of the world's population is projected to be urban by 2008. Virtually all the population growth expected during 2005–2030 will be concentrated in urban areas (Population Newsletter, 2006). In 1950 there were just two megacities, New York and Tokyo; their number had reached twenty by 2005 and is expected to rise to twenty-two by 2015, of which seventeen will be located in developing countries.

Urbanization is inherent for both developed and developing states. Yet it is surprising that states with declining populations continue to urbanize, and urban areas are growing faster than their population. There has been a 20 percent expansion in Europe in the last twenty years with only a 6 percent increase in population over the same period (Eurostat, 2007).

4. Challenges to the Urban Environment

Cities, being large resource consumers and pollutants, have a significant impact on the surrounding areas and beyond. Apart from this external to urban area impacts, there are serious internal impacts on the urban environment. There are major adverse environmental impacts in urban areas on human health and artificial environment. Natural environment, except for soils, is very limited in urban areas. Artificial environment, presented by man-made structures such as historic buildings, often suffers from air and groundwater pollution.

In general, environmental threats to human security in urban settings are posed by natural hazards (earthquakes, floods, storms) and air, water, and soil pollution. It is difficult to classify adverse environmental impacts by any criteria, such as origin (natural, artificial), impact on a particular biosphere component, or consequences. It should be recognized that in urban areas, given the concentration of human and industrial resources, most of those impacts have complex, multi-origin causes and their consequences touch many components of the environment. It would be more precise to talk about adverse environmental processes or events.

Floods are a good illustration of the above described complexity of adverse urban environmental events. Floods, originally a natural phenomenon that brings no harm but benefits the ecosystems in many areas, now have a severe adverse impact on the urban environment. Flood waters are also rising higher due to low soil permeability and high surface runoff in urban areas. Toxic substances in water and soil make floodwater dangerous to flora, fauna, and humans. Many floods originate purely from the consequences of urban development and have a grave negative impact on urban areas.

Some other, often overlooked, examples of complex challenges to the urban environment are the construction industry, the solid waste management process, and intensive tourism.

It is widely recognized that urban traffic and the use of private cars is the major threat to human security in big cities and especially in megacities. The most remarkable impacts are routine air pollution, traffic accidents, and noise.

Pollution generated by urban transport has a proven impact on public health. Nearly all (97 percent) of Europe's urban citizens are exposed to air pollution levels that exceed EU standards of quality for particulates, 44 percent for ground-level ozone and 14 percent for nitrogen dioxide (SUD-LAB, 2007). It is recognized worldwide (SUD-LAB, 2007; Ingram and Liu, 1980; World Development Indicators, 2007; WHO, 1999) that particulate matter poses the key challenge. Long-term exposure to high doses of small particles in the air contributes to a wide range of health effects, including respiratory diseases, lung cancer, and heart disease (World Development Indicators, 2007). The construction and waste management industries are also significant sources of particulate matter emissions.

Other common urban air pollutants threatening human health are volatile organic compounds, sulfur oxides (which also contribute to the acid contamination of water reservoirs, buildings, and flora), nitrogen oxides (which also create smog and acid precipitation), carbon oxides, and lead.

According to the World Bank Thirty-seven-city Study (Newman and Kenworthy, 1999), death resulting from transport accidents accounts for 2 percent of total deaths in the United States, 1 percent in Europe, and 4 percent in Asia. Infrastructure functional composition and design are the key elements in managing transport safety.

In Europe at least 100 million people in urban agglomerations are exposed to road traffic noise levels above the WHO recommended level of 55 dBA, and about 40 million are exposed to levels above 65 dBA, a level at which noise is seriously detrimental to health (WHO, 2000).

The noise problem should be addressed at an early stage of decision making about types of infrastructure to be installed. Low-noise infrastructure options such as trains on tiers have been gaining popularity among city planners. Adjusting existing infrastructure to low-noise requirements is a rather difficult task. Noise mitigation measures, such as noise breaking screens, have negative impact on landscape and on-road air quality.

5. Urban Infrastructure and the Environment

5.1. URBAN INFRASTRUCTURE

Urban infrastructure plays one of the key roles in the sustainability of cities. The impact of urban infrastructure composition and arrangement can be compared with that of a city master plan. Different types of urban infrastructure include transport facilities, resource supply networks, and waste management systems.

Infrastructure provides certain services at a certain environmental cost. Infrastructure can be an instrument for policy implementation. It can encourage or discourage citizens to behave in a certain way and affect their lifestyle and consumption patterns.

For instance, poorly maintained water supply systems that have a high percentage of loss due to leakage discourage consumers from saving water. Authorities in many Russian cities have for many years had problems in persuading people not to waste water when about 10 percent of the total supply was lost due to technical mismanagement.

Another example of how urban infrastructure affects citizens' lifestyle is road networks. In an attempt to confront traffic jams in city centers, planners increased road networks and built new parking facilities. Apparently this policy had an opposite effect, encouraging people to use private cars, thereby aggravating traffic and air quality problems.

The above examples illustrate a strong link between urban environmental security and the shape of urban infrastructure. Indicators of infrastructure development and the environmental gauges can help to study the environmental impacts of urban infrastructure in depth.

5.2. URBAN INDICATORS

Indicator can be defined as a piece of information that, derived from a series of observed facts or expert judgments can reveal relative changes in a complex system. Urban data, presented also in a form of indicators, is monitored by local authorities and city administrations. Cities provide urban data to national governments and international organizations concerned such as Eurostat, Metropolis, or UN HABITAT. However, comparative analysis of urban data is not an easy task for several reasons, which are summarized below.

- Many indicators addressing the same issue are formulated in different ways.

- Different methodologies are used for collecting data on the same indicator.

- To analyze comparative performance of a certain indicator, the indicator should be viewed in a certain setting, represented by relevant city data like climate, landscape, and GDP.
- Data refers to different time periods.

The above observations suggest that there is a lack of common frameworks for urban indicators. This problem has been recognized and addressed at the international level by such organizations as World Bank and Eurostat. Obviously some progress has been made, especially in establishing methodologies for indicator data collection (*Indicators for Sustainable Development,* 2007).

There are many descriptions of urban problematics on the indicators basis, and generic indicator sets.

Many urban indicator sets are very well structured and are accompanied by clear methodologies for data collection. Often the major problem arises with data compatibility, especially if the data is provided by many international cities. Local city and town authorities often have limited resources to follow the required guidelines and present data in a required format. The World Bank acknowledges that indicator data on urban environment "should be used with care; and because different data collection methods and definitions may have been used, comparisons can be misleading" (World Development Indicators, 2007). However, comparative data analysis represents the major interest for judging the success of urban development.

5.2.1. Infrastructure Indicators

Traditionally, data on a particular infrastructure type has been collected by respective professional organizations, e.g., data on a road network would be managed by construction companies, maintenance organizations, and automobile associations. Usually these indicators of infrastructure development are very much focused on the needs of respective industries, and for that reason are difficult to use in comparative analysis of the urban environment. Infrastructure indicators provided by industries usually reflect those industries' development achievements. Examples of these indicators are annual industry investments, kilometers of roads built per year, tons of raw materials consumed. However, for the purpose of analyzing infrastructure impact on urban environment, interdisciplinary indicators would be the most useful. An example of an environmentally reflective infrastructure indicator is the ratio between total road cover and road cover actually used by traffic. This indicator would be important to analyze the possibility of increasing rainwater infiltration and reducing surface water runoff.

Some other examples of infrastructure indicators that can be used in environmental assessment are the share of transport means in a city, energy consumption, and stock density.

5.2.2. *Environmental Indicators*

Environmental indicators can be divided into two categories: state and pressure indicators. Table 1 is composed of commonly used urban indicators (World Development Indicators, 2007; *Indicators for Sustainable Development*, 2007; CEROI, EUROSTAT; SUD-LAB) and gives some examples of indicators that fall into either of the above categories.

A separate group of indicators that fall into neither of the categories in Table 1 can be identified as issues relevant to urban environmental quality. Examples of indicators that fall into this category are

- Existence of local environmental plans
- Citizens who have less than 15-min access to green areas
- Citizens' awareness of specific environmental challenges their city faces

The incorporation of all three groups of environmental indicators (state, pressure, and those relevant to urban environmental quality issues) is important for a comprehensive understanding of the status of urban environment.

TABLE 1. Urban Environmental State and Pressure Indicators.

State environmental indicators	Environmental pressure indicators
Air pollutants: • Benzene • Carbon monoxide • Nitrogen dioxide • Ozone • Sulfur dioxide • Suspended particulate matter	• Population of urban agglomeration • Population density • GDP per capita • Transport modes used (private car, bus, electric transport, underground, taxi, bicycle) • Passenger cars per 100 inhabitants
Surface and ground water pollutants: • Heavy metals • Nutrients • Oils • Organic toxins • Suspended matter	• Percentage of journeys to work by car • Road network density • Water consumption • Energy consumption • Energy supply type (nuclear, coal, hydropower, solar, waste incineration, biogas, wind, thermal)
Soil pollutants: • Heavy metals • Oils	• Solid waste treatment (open damp, landfill, incineration, composition, recycling)
Municipal waste: • Percentage of waste collected • Percentage of wastewater treated	

5.3. URBAN UNDERGROUND INFRASTRUCTURE

Urban underground infrastructure (UUI) can serve as an interesting illustration of how infrastructure can affect urban environment, citizens' security, and cities' sustainability. Naturally, infrastructure can have both positive and negative impacts on different components of the environment and aspects of urban life. Overall judgment on benefits and drawbacks of infrastructure use is made on the basis of integrated interdisciplinary assessment, using, for instance, methods of multiple criteria decision making.

Examining the relation between infrastructure indicators, city characteristics, and environmental indicators represents the emphasis context of this paper. Another pressing research question is to ask what are the benefits of a certain type of infrastructure (e.g., UUI) to the urban environment, i.e., making a judgment about the environmental impact of infrastructure on the basis of its performance in different cities.

Underground infrastructure can be defined as a set of subterranean structures, interconnected physically or functionally. Urban underground space encompasses structures of various functional purposes: storage (e.g., food, water, oil, industrial goods, and waste, including hazardous waste); industry (e.g., power plants); transport (e.g., railways, motor roads, and pedestrian tunnels); utilities and communications (e.g., water, sewerage, gas, electric cables); public (e.g., shopping centers, hospitals, civil defense structures); and private and personal (e.g., car garage).

Underground space in urban areas has been extensively utilized. In many cities underground space at the depth of 5–10 m is very congested: it is a mixture of old—in many cases unused—service tunnels, forgotten cables, and recently installed utilities. Among underground structures that have significant volume are multipurpose urban public structures and underground public transportation systems. Multipurpose structures most often encompass motorcar parking and shopping areas and have depths up to 20 m. Metro tunnels are found at a depth of up to 60 m.

A pilot study on UUI and the environment analyzes the relation between UUI and cities' indicators in Paris, Stockholm, and Tokyo. These three cities were selected on the basis of availability of data on UUI. Collecting and aggregating data on UUI represented the major challenge for research.

5.3.1. Considered Indicators

UUI is presented by its main indicator—total volume of underground space used in cubic meters. The indicator, which derives from the UUI volume, is UUI thickness—UUI volume divided by city area (cubic meters of underground space per square kilometer area). The third UUI indicator considered is urban underground space use by function (utilities, industry, transport, public use), in percents.

City indicators include population, land area, and population density (inhabitants per square kilometer). City environmental indicators are presented by pollutant concentrations (SOx, NOx, SPM). Relevant to the environment infrastructure indicators are the average speed of the journey to work (km/h), and road network density (km/km²).

Figure 1 indicates that use of urban underground space becomes increasingly important in densely populated urban areas.

Naturally, UUI was developed to tackle lack of free space, especially in city centers. There are also strong environmental reasons for developing UUI—motor traffic streams optimization, travel time and emission reduction, combating noise, and promoting green areas (Bobylev, 2006).

Figure 2 shows the relation between volume of UUI and density of road network. A significant percent of UUI and a road network has the same transport functions. These are traffic and parking of motorcars. The underground location of transport functions allows the gain of environmental benefits: emissions and noise reduction, landscape preservation, and less surface runoff disturbance. A UUI substitute road network provides city transportation functions in a more environmentally friendly way (Figure 2). It is possible to reduce road density by using underground space.

Major use of urban underground space is related to transportation (Figure 3). Variations in other UUI functions suggest that there is a high potential in multipurpose utilization of underground space. Cities use over

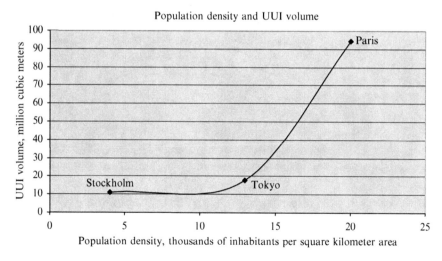

Figure 1. Relation Between Population Densities in Urban Areas and Volume of UUI.

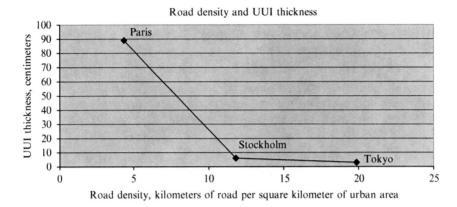

Figure 2. UUI Substitutes for a Road Network.

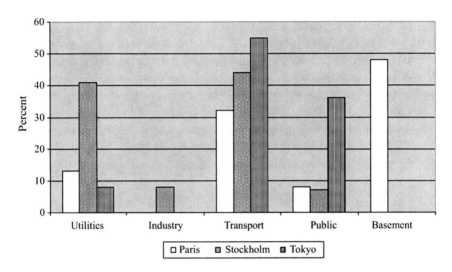

Figure 3. UUI Usage by Function.

30 percent of UUIs for transportation and over 5 percent for utilities. The significant volume of utilities infrastructure in Stockholm may reflect severe climate.

The indicator "speed of journey to work" is a good example of how cities can benefit from UUIs (Figure 4).

In spite of expectations, the comparison of UUI indicators and urban air quality indicators has not revealed any clear correlation. However, analytic research shows a positive impact of UUI on urban air quality. UUIs combat air pollution by reducing road traffic congestion (Bobylev, 2005).

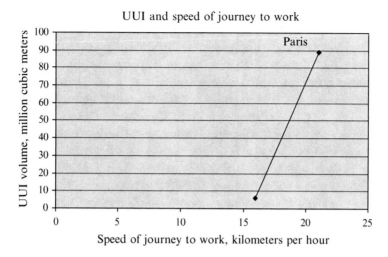

Figure 4. UUI Utilization Clearly Contributes Toward More Effective Transport System.

5.3.2. *Future Research*

Data collection represents a major challenge for future research on the urban environment and UUI. It is desirable to expand the number of cities involved in the analysis, selecting some representative cities and taking into account climate, ground conditions, GDP, and build stock density. A greater number of "direct environmental indicators" and "indicators relevant to the environment" should be considered. UUI use by function represents the most interesting indicator, especially in terms of studying functions' alteration in time.

6. Environmental Assessment

Evaluation of environmental impact of urban infrastructure requires analysis of many different multidisciplinary and often conflicting criteria. A number of methods or processes have been developed to address environmental concerns in development projects. These processes are listed as they evolved in historic order: environmental impact assessment (EIA), strategic environmental assessment (SEA), and sustainability appraisal (SA). The key difference between these processes is a scale of consideration of the impact. It has been increasingly recognized that not only on-site works, but perspective plans and policies need to be assessed from the environmental standpoint, preferably at a stage of their elaboration. Usually EIA addresses impacts at a project and program level, SEA at a plan and policy level. SA includes social and economic considerations on a par with environmental ones. All of these processes have been utilized for urban infrastructure planning and development. A particular interest represents environmental assessment of existing infrastructure and review of policies under implementation. SEA should be a process embedded into the development of a city master plan.

Methods, tools, and techniques for environmental assessment constitute a range from experts' workshops and public involvement to the use of computer models and scenario building. Multiple criteria analysis (MCA), or multiple criteria decision analysis (MCDA) is one the methods that is used in environmental assessment and can be particularly useful for urban infrastructure development assessments. MCDA is a method of decision analysis that involves the use of scoring and weighting systems based on criteria, in order to test and compare impacts of alternatives. MCDA addresses situations where decisions are needed to handle complex problems involving alternative options. A city policy to promote bus or tram public transportation systems is an example of the alternatives in MCDA. Alternative policies, plans, programs, or projects are evaluated along many conflicting criteria. These criteria can be formed on a basis of infrastructure and environmental indicators (see section 5). MCDA can be conducted using different methods and classifications, descriptions of which can be found in Getzner et al. (2005). MCDA results provide a clear understanding on the contribution of each criterion into the overall assessment goal, which can be urban environmental quality or environmental impact. Alternative urban infrastructure options will be rated according to their relation to the assessment goal. One of the most important reason for using MCDA is arriving at a clear understanding of how criteria impact the assessment goal under different policy options and therefore which trade-offs should be made.

Comparative assessment of different urban indicators, including retrospective assessment, is an important component of environmental assessment process. This can be especially beneficial at a policy level, where it is difficult to identify sustainable infrastructure solutions on a basis of design parameters or environmental characteristics forecast.

7. Sustainable Development of Urban Infrastructure

Urban infrastructure is to be utilized for many years; therefore the use of sound infrastructure policy and design solutions makes it possible to provide cities with long-term sustainability impetus. In this view, careful planning and environmental assessment of infrastructure is an important issue for urban environmental security. Some criteria that resulted from observations and analysis of urban infrastructure performance are presented below.

The functionality criterion is important to understanding whether an infrastructure can be multifunctional, and adjusting to new functions that could possibly be required in the future. For example, service tunnels should accommodate new and increased numbers of cables, and underground parking should be adjustable to accommodate different types of vehicles.

Land use criteria should be considered to assess whether an infrastructure is likely to require more space or less in the future, and how the scarcity of land can be tackled and excess utilized.

The integrality criterion judges how cohesive the infrastructure is internally and externally and how well it is integrated with other city structures. Singular underground and aboveground pedestrian crossings can be an example of disintegrated structures. They pose maintenance problems, including public security issues, and are inconvenient for pedestrians. On the other hand, above or under road passages connected to adjacent buildings take less time to pass and create multifunctional environments that are enjoyable for pedestrians.

The flexibility criterion reflects the possibility to upgrade and renew infrastructures. Installation of new equipment into structures and changing their inner spatial design can be examples of flexibility requirement.

The rationality criterion is important to use of natural resources, including urban space. The rationality criterion also reflects pressures imposed by an infrastructure on other natural and artificial components of urban environment. For example, the expansion of transport infrastructure may dramatically increase tourism and put unwanted pressure on urban historic and cultural environment, endangering intangible city assets.

The vulnerability criterion is directly related to provision of environmental security: vital vulnerable infrastructure significantly increase overall city vulnerability. For example, UUI is vulnerable to floods, so the underground location of vital urban services like transport and emergency response can significantly increase urban area vulnerability.

These criteria can be used to aid infrastructure and environmental indicators (see section 5) for environmental and environmental security assessments.

8. Conclusions

Environmental security, as an essential component of human security, has a particular importance in densely populated urban areas, where environmental problems are aggravated. Different types of urban infrastructures can have positive as well as negative impact on the urban environment. Infrastructure performance should be assessed using many criteria, including infrastructure and environmental indicators; MCDA methods are helpful in conducting the assessment process.

Given the formative role of urban infrastructure, sustainable infrastructure decisions serve as a basis to ensure sustainable development of cities.

Acknowledgements

The author would like to acknowledge the contribution of

1. The following persons, who helped immensely in collecting the data on the use of urban underground space: Daniel Morfeldt (Sweden), and Pierre Duffaut (France).

2. The following organizations, which provided the opportunities for conducting this research: Japan Society for Promotion of Science; United Nations University.

References

Belluck, D. A., Hull, R. N., Benjamin, S. L., and Linkov, I., 2005, Environmental security, critical infrastructure and risk assessment: Definitions and current trends, in: *Environmental Security and Risk Assessment*, I. Linkov and G. Kiker, eds., Kluwer, Dordrecht.

Bobylev, N., 2005, Environmental assessment of urban underground infrastructure, in: *New Technologies for Urban Safety of Mega Cities in Asia*, T. Pan and K. Meguro, eds., proceedings of the fourth international symposium, 18–19 October, Nanyang Technological University, Singapore, pp. 499–508.

Bobylev, N., 2006, Strategic environmental assessment of urban underground infrastructure development policies, *Tunn Undergr Sp Tech*. **21**(3–4), Safety in the underground space—Proceedings of the ITA-AITES 2006 World Tunnel Congress and 32nd ITA General Assembly, Elsevier, pp. 469–479.

Bruntland, G., ed., 1987, Our *Common Future: The World Commission on Environment and Development*, Oxford University Press, Oxford.

Eurostat, 2007; http://epp.eurostat.ec.europa.eu.

Cities Environment Reports on the Internet (CEROI), 2007; http://ceroi.net.

Getzner, M., Spash, C. L., Stagl, S., 2005, *Alternatives for Environmental Valuation*, Routledge, Abingdon.

Glenn, J. C., Gordon, T. J., and Perelet, R., 1998, *Defining Environmental Security: Implications for the U.S. Army*, ed. Molly Landholm, AEPI-IFP-1298.

Indicators for Sustainable Development: Guidelines and Methodologies, 2007, United Nations.

Ingram, G., and Liu, Z., 1980, *Motorization and Road Provision in Countries and Cities*, World Bank.

Metropolis, 2007; http://www.metropolis.org/index.htm.

Newman, P., and Kenworthy, J., 1999, *Sustainability and Cities: Overcoming Automobile Dependence*, Island, Washington, DC.

Population Newsletter, no. 81, June 2006, United Nations Population Division Secretariat Department of Economic and Social Affairs; www.unpopulation.org.

Sustainable Urban Development Laboratory (SUD-LAB), 2007; http://www.sud-lab. com/ src/abt01.asp.

World Development Indicators, 2007, The World Bank; http://go.worldbank.org/ 3JU2HA60D0.

World Health Organization (WHO), 1999, *Monitoring Ambient Air Quality for Health Impact Assessment*, WHO Regional Publications, European Series, no. 85.

World Health Organization (WHO), 2000, *Human Exposure Assessment*, Environmental Health Criteria Document 214, Programme of Chemical Safety.

APPROACHING ENVIRONMENTAL SECURITY

From Stability to Sustainability

STEVEN R. HEARNE*

Army Environmental Policy Institute

Abstract: Environmental security implies a freedom from environmental threats that can contribute to instability and the outbreak of conflict. The complex and long-term nature of such threats often makes them difficult to identify and act upon; however, it is widely recognized that such threats need to be addressed in the next U.S. National Security Strategy. Countries that are less environmentally sustainable are more prone to instability. Early identification of instability is necessary in structuring successful interventions, and a simple framework and set of indicators are presented as a means to better identify and communicate this instability. Defense environmental cooperation can be an effective tool to promote sustainability overseas. The U.S. Army has demonstrated its leadership in sustainability, and can leverage its experience to help implement related defense cooperation activities in support of the new Africa Command.

Keywords: Cooperation; conflict; engagement; environment; resource scarcity; security; stability; stress; sustainability; threats; triple bottom line; vulnerabilities

1. Introduction

Environmental stress is a contributor to instability and potential conflict and, therefore, is important to national security. Early identification of instability is necessary in structuring appropriate and successful interventions. Environmental cooperation on issues of common interest has been used effectively by the U.S. Department of Defense (DoD) and its geographic combatant commands as a nonthreatening means of engagement to encourage sustainable practices that help promote stability.

---

* Address correspondence to Steven R. Hearne, P.E., Senior Fellow, Army Environmental Policy Institute, Suite 1301, 1550 Crystal Drive, Arlington, Virginia, 22202-4144, U.S.A.; tel.: 703-602-0191; e-mail: steven.hearne@hqda.army.mil. The views expressed in this paper are those of the author and do not necessarily reflect the official position of the Department of the Army, Department of Defense, or the U.S. government.

P.H. Liotta et al. (eds.), Environmental Change and Human Security, 217–251.
© 2008 *Springer Science + Business Media B.V.*

These efforts, however, have been limited because of a poor understanding of the nature of environmental threats and vulnerabilities, inadequate resources, and a failure to properly address the important relationship of environment to regional stability in more recent national security strategies. For comparison, excerpts from earlier national security strategies are offered throughout this paper to highlight the importance given to environmental threats and vulnerabilities by prior administrations.

> *Environmental threats do not heed national borders and can pose long-term dangers to our*
> *security and well-being.*
> (U.S. National Security Strategy, 1997)

It is the objective of this paper to provide a synthesis of a large body of relevant research on the important relationship of the environment to security, stability, and sustainability. These relationships also are valuable in developing a succinct definition for environmental security that will help garner support for future defense environmental cooperation activities promoting regional stability.

A simple framework of stability, with accompanying indicators, also is presented as a means to help identify and communicate national and regional instability. A basic correlation analysis is used to determine the relationship between a key index of environmental sustainability and an index representative of nation-state failure. The concept of sustainability, with its focus on shared values and equity, is introduced, as is its implementation within the U.S. Army, and its viability as a tool for early intervention to address underlying causes of regional instability.

2. Environment and Security

The relationship of environment to security has been intensively studied and hotly debated for nearly two decades, yet the term "environmental security" has remained ill defined. A broad interpretation of the term views severe environmental degradation and stressors as presenting a security threat as serious as war. Those supporting a narrower definition focus on the protection of natural (domestic) resources from both external and internal environmental threats and disasters. The term has also been viewed from the reverse perspective of how violent conflict might have an adverse impact on the natural environment. Yet another variant views environmental security as protection from eco-terrorism, a term that is also not well defined, and therefore subject to some misunderstanding and debate. Past attempts to provide a consistent definition of environmental security generally have found little consensus, although the term appears in common use (Glenn et al., 1998). Not surprisingly, traditional policy and security analysts continue to view this concept with some skepticism.

The term *security* itself, from the Latin securitas, also is viewed as a rather vague concept, without a commonly agreed-on definition (Renner, 2006). It commonly is defined as "freedom from danger and freedom from fear or anxiety" (Merriam-Webster, 2003). The term *environment* is not without its differing interpretations, but generally is defined as the complex of physical, chemical, and biotic factors that act upon an organism or an ecological community at any point in the life cycle and ultimately determine its form and survival (Botkin and Keller, 1995). It is the very nature of the complexity of these living and nonliving factors, in combination with socioeconomic and political factors, that determine the impact the environment has on national or regional security. Two conceptual models will be introduced in the next section to address and simplify this complex interdependence of factors.

In an early interview, General Anthony Zinni, the former combatant commander of the U.S. Central Command (CENTCOM), was asked to provide a definition that reflected the importance he placed on the environment as a security issue. He responded, "when environmental conditions, either natural or man-made, are destabilizing a region, a country, or have global implications, then there are major security implications. Those problems could affect stability, could result in conflict, could threaten our U.S. interests or other vital interests, and could impact our welfare" (Manwaring, 2002). It was suggested that their resulting impact can be felt well beyond where they originate, as with fisheries resource exploitation.

2.1. ENVIRONMENTAL STRESS AND CONFLICT

Research has provided unique insights into how to better address the environment in national security policy, focusing on the linkages between social, economic, political, and environmental effects, and how their interaction might lead to increased instability and conflict. Among the more quoted—and debated—research in this area is that from the Peace and Conflict Studies Program at the University of Toronto. In the seminal book *Environment, Scarcity, and Violence,* the author examines these complex linkages and proposes a Core Model of causality based on national and regional case studies (Homer-Dixon, 1999). Mainly, environmental scarcity was found to be an indirect cause of violent conflict contained within national borders. Furthermore, it was suggested that when environmental scarcity plays a role, it always joins with other economic, political, and social (contextual) factors to produce its effects. These causal relationships have important consequences to policy makers considering how and when to intervene to resolve a conflict, ideally in its early stages before it escalates to a point where peacekeeping forces may be needed to provide stability and security.

Natural resource scarcities can trigger and exacerbate conflicts.
(U.S. National Security Strategy, 2000)

The North Atlantic Treaty Organization (NATO) has been increasingly concerned with nontraditional threats to security, including the consequences of environmental change. In 1999, NATO directed a pilot study with the objective of better integrating environmental considerations in future security deliberations (NATO, 1999). Environmental change was conceived in terms of the nature and extent of environmental stress that results from the scarcity of renewable natural resources and environmental degradation. A conceptual model was developed by the NATO pilot study that is a simpler representation of the Toronto Core Model. It highlights the importance of contextual factors (e.g., perception of key actors, political stability, economic vulnerability, and resource dependency) in understanding why one nation may be more effective than another in dealing with environmental stress, limiting the vulnerability of its resources to such stressors, and resolving conflict.

2.2. THE NATURE OF ENVIRONMENTAL STRESS

The nature of environmental stressors generally is subtle and long-term, and often takes decades to emerge as a recognized threat. These stressors have been characterized as "creeping vulnerabilities," given that they may not be easily identifiable, are often linked to the complex interdependencies discussed previously, and do not always suggest an appropriate response (Liotta, 2005). Table 1 provides a summary of major global environmental threats and vulnerabilities, reflecting findings from several assessments, all of which reach similar conclusions as to key environmental stressors (Speth and Haas, 2006; Abbot et al., 2006; OECD, 2001; King, 2000, 2006). Importantly, environmental resource degradation is closely connected with scarcity because it can have an impact on the availability of renewable natural resources.

Not all security risks are immediate or military in nature. Transnational phenomena such as ... environmental degradation, natural resource depletion, rapid population growth, and refugee flows also have security implications for both present and long term American policy.
(U.S. National Security Strategy, 1995)

Interestingly, in what was admittedly a nontraditional approach, the National Intelligence Council (NIC), in collaboration with outside experts from academia, think tanks, and business, issued an assessment in 2000 on global trends through 2015 that identified natural resources and environment as among the seven "global drivers" shaping the next fifteen years (NIC, 2000). The most recent NIC assessment (2004) lists environmental issues,

TABLE 1. Global Environmental Threats and Vulnerabilities.

Resource scarcity	Resource degradation	Demographic change	Shocks
Deforestation	Climate change	Population growth	Natural disasters
Desertification	Ozone depletion	Rapid urbanization	Industrial disasters
Soil erosion	Air pollution	Growth—youth bulges	Infrastructure failures
Loss of arable land	Acidification	Globalization/equities	Pandemics
Fisheries decline	Water pollution	Population migration	Conflicts/wars
Water availability	Eutrophication	Population flight	
Loss of wetlands	Pesticides		
Biodiversity loss	Solid waste		
Invasive species	Toxic waste		

such as climate change, as continuing to come to the forefront of the 2020 global landscape, exacerbated by globalization in developing economies and the integration of migrant populations to compensate for changing demographics in established, but aging, powers.

Among the listed threats and vulnerabilities in Table 1, an earlier study suggested that land degradation, deforestation, and freshwater scarcity, alone and in combination with high population density, have the potential to increase the risk of internal low-level conflict (Haugue and Ellingsen, 1998). The researchers, however, found economic development and type of political regime as more decisive predictors of conflict. A critique of this study suggested that linkage might be better viewed as a development issue (Gleditsch, 1998). This critique involved a review of other empirical studies and found only weak evidence directly linking environment and conflict. This same critique found many such empirical studies to be suspect because of the problem of obtaining valid and reliable environmental data at both the national and subnational level, and recommended major improvements in systematic data collection.

Freshwater availability remains an important global environmental issue (Gleick, 2006). Water scarcity has been suggested as an emerging threat that could lead to increasing conflict in this century. However, based on a review of approximately 400 cases of interstate conflicts between 1918 and 1994, only a few were found to involve conflict over water (Khagram et al., 2003). While water wars should not be ruled out, it is suggested that future conflicts over water likely will be limited to intrastate versus interstate, and not necessarily involve violent intervention. Water also is viewed as a "pathway to peace, not war" because it is so important a resource that nations simply "cannot afford to fight over it," as evinced by the number of water treaties among countries that share a common river basin (Wolf et al., 2006: 1). Others argue that water scarcity is the emerging threat of the twenty-first century (Pearce, 2006), especially among those water-limited nations that extend across an arc of instability from North Africa to China (Reed, 2004).

Environmental threats such as climate change, ozone depletion, and the transnational movement of hazardous chemicals and wastes directly threaten the health of U.S. citizens.
(U.S. National Security Strategy, 1998)

Climate change is singled out for discussion because of the recent international consensus in the scientific community that "the greenhouse effect is real" (NIC, 2004), irrespective of its causation and possible remedies, and given its potential impact on global economic stability (Stern, 2007). The U.S. National Security Council recently has put climate change on its agenda for the first time, given its potential to serve as a catalyst for new conflicts. Proposed legislation also has been introduced by the U.S. Congress to elevate climate change to a national defense issue because of concerns that it could be an additional stressor that could further destabilize already vulnerable and fragile states near a tipping point. This would require the first-ever national intelligence estimate on this issue.

A blue-ribbon panel of retired U.S. generals and admirals recently released a report on climate change, demanding an urgent global response. They warn that "climate change can act as a threat multiplier for instability in some of the most volatile [and vulnerable] regions of the world" (CNA, 2007: 1). They further suggest that climate change may even threaten the most stable of the African governments, drawing the United States into regional politics to a much greater extent than in the past to thwart attempts by extremists to exploit the resulting instability.

Of particular concern is that climate change, in combination with human-induced environmental degradation, has been suggested as being at the real roots of recent violence in Darfur and other parts of the Sahel region in Africa (Faris, 2007). Notably, the European Union (EU) High Representative for Common Foreign and Security Policy, Javier Solana, challenged traditional ways of dealing with complex security challenges, recognizing that, "in many ways, Darfur is the first time we are aware that a war is caused by climate change," further cautioning, "it will not be the last" (Solana, 2007). Weather shock is closely correlated with civil conflict in sub-Saharan Africa, with conflict expected to worsen even under the most positive climate projections (Miguel et al., 2004, 2007).

There are also conflicting visions of how the global environment will be affected by these environmental threats and vulnerabilities. Optimists who suggest technological advances, coupled with the resilience and adaptability of society as a panacea, may be placing future generations at risk. Many contest that Earth's capacity to meet humanity's needs has been exceeded, and recommend a rapid transition to sustainability to offset the current ecological deficit, or debt (WWF, 2006). They are especially alarmed by increasing Western-like consumption patterns and unbridled economic growth among less-developed and transitional economies, many with burgeoning populations vying to improve their standards of living and to share in any future distribution of wealth. In response to the complexity and uncertainty of such stressors, the *precautionary principle* has emerged and increasingly been adopted in national and international law. This principle requires measured action be taken to address serious environmental threats and preclude irreversible damage before full scientific proof is available (Speth and Haas, 2006).

2.3. ENVIRONMENTAL SECURITY DEFINED

The term *environmental security,* while in common use today, remains poorly defined. The previous section discusses the potentially destabilizing effects that environmental stressors can have on vulnerable nations and regions, which, if not addressed, could lead to conflict. A succinct definition is offered that accounts for the importance of such stressors on security interests and stability. This definition builds upon earlier efforts by respected academics and practitioners to define the term (Butts, 2007; Liotta, 2003; Manwaring, 2002; King, 2000; Glenn et al., 1998).

Environmental security is defined and applied in this paper as: *the freedom from natural and anthropogenic environmental threats and vulnerabilities that have the potential to adversely impact on national security interests and, if left unchecked, could contribute to increasing intrastate or broader regional instability and to the outbreak of conflict.*

2.4. TRADITIONAL AND NONTRADITIONAL SECURITY

The traditionalist security mindset typically has focused attention on the more immediate and identifiable "threats" that are believed to require a more state-centered, and often hard-power, response. Unfortunately, given that environmental stressors typically are subtle and long-term in nature, they generally are given little attention by comparison, making them difficult to address later. It is the nature of these and other threats and vulnerabilities that characterize the major type of security challenges and responses summarized in Table 2.

In addition to environmental security, an important, nontraditional, security concept that has been heavily researched and debated since the mid-1990s is that of *human security.* This concept suggests the focus of

TABLE 2. Comparison of Traditional and Nontraditional Security Types.

Type	Focus	Concerns	Threats/vulnerabilities	Responses
Traditional security	The state	Sovereignty and territorial integrity	Challenges from other states and stateless actors	Diplomatic intervention Economic crisis response Military intervention Humanitarian support
Environmental security	The ecosystem	Protection of natural infrastructure	Resource scarcity/depletion Resource degradation—pollution/waste Demographic changes Shocks—natural, manmade	Multinational governance Conflict prevention Conflict resolution
Human security	The individual	Integrity of the individual (Freedom from fear) (Freedom from want)	Personal security—violence, hazards Political security—repressive state - - - - - - - Economic security—poverty Food security—famine, contamination Health security—injury, disease Environmental security—scarcity, waste Community security—cultural integrity	Preventive diplomacy Disaster planning Humanitarian support Aid investment
Sustainable security	The generational quality of life	Global sustainability	Non-sustainable environmental, social, economic, and institutional systems Inequitable resource distributions Marginalization of the majority world Uncontrolled development/ urbanization Non-adaptive and non-resilient systems	Institution building Security sector reform Military professional-ism Energy sector efficiencies Economic investment Adaptive management of natural resources and land uses

Sources: Adapted from Liotta (2005), Liotta and Owen (2006a), UNDP (1994), Khagram et al. (2003), Abbott et al. (2006).

traditional state-centric security be redefined or redirected to the individual, with an emphasis placed on underlying social, political, and economic reforms (Richmond and Franks, 2005). As with environmental security, human security is interpreted both narrowly and broadly. While not a new concept, human security is believed to have been first used in an early 1994 United Nations (UN) report that focused on seven main categories of threats or deprivation (UNDP, 1994). These seven categories of threats are listed in Table 2.

The narrower definition of human security, often termed the Canadian Approach, is seen as providing more of a "freedom from fear," i.e., violent threats from state and non-state actors (Renner, 2006; Liotta and Owen, 2006a). The first two categories of threats listed in Table 2 for human security are representative of this "narrow" interpretation. The European Union recently developed a Human Security Doctrine for Europe primarily focusing on response to violent threats, thus also addressing the narrower definition of human security (EU, 2004; Liotta and Owen, 2006b). The last five categories of listed threats are more representative of the broader interpretation, which is seen as providing more of a "freedom from want."

The human security construct generally treats environmental security as one of its seven main categories of threats. More often, environmental security is treated as a separate security type in the literature and in common security discourse and practice, as seen in Table 2. Sustainable security is a relatively new concept and will be addressed later.

3. Environment and Stability

Several frameworks for instability have been suggested in the literature. This paper discusses two simplified approaches that address or consider the environmental threats and vulnerabilities introduced in Table 1. The first such framework is described in the United Kingdom Prime Minister's Strategy Unit Report to the government in early 2005 (UK, 2005). In this systems-oriented framework, climate change and natural resource dependency are viewed—among other risk factors—as key external and internal risk factors for instability, respectively. Natural disasters are also seen as having the potential to further shock (or exacerbate) already unstable national regimes. This framework presents a number of external stabilizing factors, such as security guarantees, political relationships, and economic associations— all of which are important to a country's capacity and resilience to address the risks posed by environmental threats and vulnerabilities. The framework suggests that where few of these external stabilizing factors are present, a country can easily tip into violent conflict, political instability, and economic crisis.

The range of environmental risks serious enough to jeopardize international stability extends to massive population flight from man-made or natural catastrophes ... large-scale ecosystem damage caused by industrial pollution, deforestation, loss of biodiversity, ozone depletion, desertification, ocean pollution, and ultimately climate change.

(U.S. National Security Strategy, 1995)

3.1. THE STABILITY PYRAMID—A SIMPLIFIED FRAMEWORK

The second framework was proposed by this author as part of earlier research conducted at the U.S. Naval War College (Hearne, 2001). This research examined an extensive body of then recently completed work on state stability and failure, and efforts to develop improved indicators for use in related models. This synthesis was necessary in addressing two questions of concern to U.S. policy makers: (1) can a set of indicators and a simple framework be constructed to help identify instability; and (2) how might a U.S. geographic combatant command, government agencies, and others employ such a framework to better prioritize engagement activities and leverage resources to address instability?

Previous conflict and state-failure research that investigated the complex and multidimensional relationships, or linkages, among the three major environmental, socioeconomic, and institutional-governance dimensions concluded that environmental stress is an important, albeit indirect, contributor to state stability. Environmental stress alone was seen as neither a necessary nor sufficient cause leading to state weakening or failure. Thus, environmental stress is best viewed as joining with other socioeconomic and institutional factors to produce an impact on state stability. It is in this context that a simpler framework of stability—the Stability Pyramid—was proposed, and a Core Set of indicators identified.

The Stability Pyramid and Core Set of indicators are shown in Figure 1. The linkage between each pairing of dimensions is depicted by the two-way arrows at each corner of the pyramid. The relative strength of relationships between each pairing of dimensions generally was found to be positive based upon the aforementioned research comprising a detailed review of the literature and correlation analysis of key indicators (Hearne, 2001).

3.2. INDICATORS OF INSTABILITY AND INSECURITY

This Core Set of indicators was developed based on an exhaustive review of ongoing efforts to develop indicators of environmental performance, state failure, and sustainable development, as well as initiatives related to aggregating these indicators further into indices, or a single index. A concerted

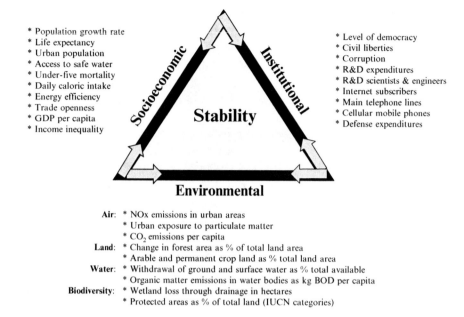

* Population growth rate
* Life expectancy
* Urban population
* Access to safe water
* Under-five mortality
* Daily caloric intake
* Energy efficiency
* Trade openness
* GDP per capita
* Income inequality

Socioeconomic

Institutional

Stability

* Level of democracy
* Civil liberties
* Corruption
* R&D expenditures
* R&D scientists & engineers
* Internet subscribers
* Main telephone lines
* Cellular mobile phones
* Defense expenditures

Environmental

Air: * NOx emissions in urban areas
             * Urban exposure to particulate matter
             * CO_2 emissions per capita
Land: * Change in forest area as % of total land area
             * Arable and permanent crop land as % total land area
Water: * Withdrawal of ground and surface water as % total available
             * Organic matter emissions in water bodies as kg BOD per capita
Biodiversity: * Wetland loss through drainage in hectares
             * Protected areas as % of total land (IUCN categories)

Figure 1. A Simplified Framework—The Stability Pyramid and Core Set of Indicators.

effort was made to provide implicit weighting by ensuring a balance in the number of key indicators selected for each of the three dimensions. It was important to select indicators reflective of the major components of the environmental dimension (air, land, water, and biodiversity) to capture the most salient aspects of environmental quality (Yu et al., 1998). Every effort was made to select indicators that were properly balanced across and within the environmental, socioeconomic, and institutional dimensions; widely accepted and in use by recognized international agencies and academia; and readily available.

A comparison of commonly used indicators is provided at Appendix Not surprisingly, the concentration of indicators across each of the three dimensions (environmental, socioeconomic, and institutional) was reflective of an agency's focus and mission. For example, the European Environmental Agency (EEA) was heavily involved in monitoring primarily environmental performance, whereas the United Nations Commission for Sustainable Development (CSD) pursued harmonized indicators of sustainability across all three dimensions. The utility of the Stability Pyramid framework and Core Set of stability indicators was demonstrated for several reference countries. The results suggested that this framework could provide an important and relatively simple tool to quickly identify and respond to situations of instability. Given the long-term nature of many environmental stressors, it was not surprising that the Core Set of socioeconomic and institutional indicators may provide a better early warning of instability in those situations where environmental stress is not (yet) highly pronounced.

The need for a simplified framework, such as the Stability Pyramid, has support from several independent research efforts in other disciplines. One of the more relevant, and parallel, efforts was conducted by the Global Environmental Change and Human Security (GECHS) Project. Its purpose is to better recognize the "inter-linkages of environment and society" and the way humans' interactions with the environment are "historically, socially, and politically constructed" (GECHS, 2000: 1). Employing a categorization similar to that of the Stability Pyramid, the GECHS project developed sixteen standard indicators of human security, or, more appropriately, human insecurity. Interestingly, the majority of these indicators of human insecurity are very similar to those found in Appendix.

A single Index of Human Insecurity (IHI) was further constructed by the GECHS project using a complete time series for all indicators and countries over a twenty-five-year period. Data were standardized and index classified, and an IHI average value assigned between one and ten for each country. The IHI was proposed as a simple and inexpensive means—a "first cut"—to identify where human security is most threatened, and to provide some underlying rationale for this assessment.

Recently, it has been questioned whether it is possible, or even desirable, to attempt to produce a reliable human security index. The Human Security Centre (2005), in their *Human Security Report 2005*, suggests that prior attempts, such as the aforementioned IHI, have focused too much on development, have ignored any measure of violence, and have not been updated regularly. Single indices also raise questions as to how supporting indicators were weighted at arriving at the composite value.

Such composite indices could mask more regional and sub-regional insecurities. The Centre's report proposed three simpler core indicators from well-recognized sources to develop a list of what they believe are the world's least secure countries: (1) fatalities from political violence; (2) core human-rights abuses; and (3) political instability and violence. The World Bank's *Political Instability and Absence of Violence* indicator is used to provide the measure for the last indicator to gauge the likelihood of government destabilization or overthrow (World Bank, 2004).

A common challenge to the many existing and proposed frameworks, indicator sets, and indices is the very nature of their underlying datasets. Data analyzed by researchers today primarily remains nationally based. Unfortunately, the quality and availability of these data can be held hostage by politics. Time-series data can be problematic and often incomplete, with researchers filling missing data by content analysis and coding. The seasonality of environmental data also may not be reflected in data collected by many governmental agencies and nongovernmental organizations. Trans-boundary environmental threats also have an impact on ecosystems that are not contiguous with political borders. Notwithstanding the above, "complex large-N empirical studies still have a future in the discipline, however, they must be based on high quality data" (Matthew and Fraser, n.d.: 29).

To address the problems of data availability, integrity, and aggregation, an alternative methodology has been suggested that would maximize the acquisition of local and subnational data and expertise in order to develop a manageable list of relevant threats, and to ensure that all data, whether quantitative or qualitative, are spatially referenced for improved visualization and analysis (Owen, 2003, 2004). This approach provides for the identification of "hotspots" of high-threat human insecurities within a specific country or region that is not possible when using only national aggregated indicators or indices. The geospatial approach also provides for better visualization and analysis of overlapping (i.e., multiple) threats in one specific area, and reportedly has the added potential to facilitate interdisciplinary analysis.

Geographic information systems (GIS) provide powerful tools for identifying trends and cumulative effects by allowing for the overlay of relevant spatial data for a particular country or region, and for examination of correlations between relevant datasets (King, 2000). Strong correlations between environmental and related human-security risks and regional instability have been reported using geospatial approaches; however, such correlations remain very dependent upon national-level data (King, 2006). Continuing progress in GIS software and related analytical tools is having the obvious effect of increasing demand for quality geo-referenced data at the local, subnational, ecosystem, and other regional levels.

3.3. INSTABILITY AND CONFLICT

Instability and conflict likely will be an enduring part of an increasingly complex and unpredictable twenty-first-century security environment. Many assumptions and institutions that governed international relations before the end of the Cold War are a poor fit in today's realities. Understandably, the terrorist attacks of September 11, 2001 refocused U.S. security strategy on the asymmetric threats posed by state and non-state actors. Increasingly, it is urged that a more balanced approach be placed on other destabilizing forces and nontraditional threats in an effort to improve the human condition that, if unabated, will continue to reinforce the root causes that support and favor terrorism as a tactic of choice (Richmond and Franks, 2005; Mueller, 2004).

Interstate war has remained at a relatively low level since the end of World War II, due in large part to successful United Nations intervention in defusing potential conflicts. However, the end of the Cold War and dissolution of the Warsaw Pact sparked a number of state failures when many of the former Soviet Union's unwilling satellites were freed. The end of Soviet hegemony in Europe created a security vacuum and also provided the United States with much more latitude to address regional instability as it felt its interests were threatened. The reunification of Germany also changed the security dynamics in Europe, which led to the Maastricht Treaty and a number of other treaties that further defined the European Union and ultimately shaped their Common Foreign and Security Policy.

Since 1989, the United Nations has conducted over forty peace operations worldwide, as compared to only fourteen conducted from 1945 to 1989 (Dobbins et al., 2003). Lessons learned are that the costs and risks associated with stability operations remain high. It is often said that it is easier to win the war than to secure the peace, as reflected by those killed in high-casualty terrorist bombings, which totaled 15,614 for the last eleven years, with over one-third attributable to terrorist incidents in Iraq since 2003 (Marshall and Goldstone, 2007). Casualties are continuing to climb.

3.4. STABILITY OPERATIONS IN PEACE AND CONFLICT

Stability operations are defined as those military and civilian missions and tasks that provide security, restore essential services, and meet humanitarian needs (JP, 2006). On a more long-term basis, stability operations are seen as important to helping develop indigenous capacity for securing essential services, rule of law, democratic institutions, and a robust civil society. The current expectation is that the military must assume a primary execution role when civilian institutions cannot (DoDD, 2005).

Historically, stability operations have stressed the U.S. military at a rate that far exceeds combat operations, accounting for six times the cost in lives and five times the cost in dollars (DSB, 2004). The United States will continue to become increasingly embroiled in stability operations overseas. Many will be of longer duration than planned. This is important, given that "the record suggests that while staying long does not guarantee success, leaving early ensures failure" (Dobbins et al., 2003: xxiv). There is a growing recognition of the importance of stability operations, across what is termed *full-spectrum operations* in military vernacular.

Full-spectrum operations combine three primary components: offensive, defensive, and stability operations, which are seen as being conducted simultaneously to differing degrees (AUSA, 2006). Figure 2 illustrates the relative

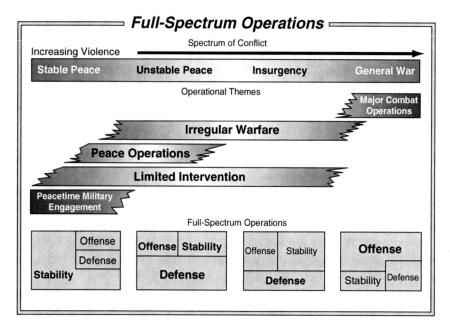

Figure 2. Importance of Stability to Peacetime Military Engagement and in Conflict
Source: Association of the U.S. Army.

importance that stability operations are afforded, from stable peace, through varying forms of intervention and conflict, to war.

While stability operations are not new, recognizing and proactively addressing the underlying causes of instability is a relatively new concept within DoD. The military has found itself increasingly involved in having to address complex socioeconomic, institutional, and other problems of affected states in longer deployments. Responding to such problems and related taskings have not traditionally been viewed as inherently military.

Regional instability typically is a "pay now or pay later" proposition. Refocusing national military and civilian engagement activities on root causes of instability may preclude costly state failure and conflict and help to legitimize unstable regimes that otherwise might be more susceptible to influence from extremist groups. Stability and reconstruction needs must be planned for in the earliest phase of an operation to optimize military and other stakeholders', e.g., nongovernmental organization, assets (JP, 2006).

DoD leadership recognized the need to improve its capability to conduct stability operations, and signed DoD Directive 3000.05, which assigns specific responsibilities within the department for planning, training, and preparing to conduct and support stability operations (DoDD, 2005).

This directive also places increased emphasis on building capabilities and sensitivity to ethnic, cultural, and religious issues. Stability operations were designated as a core U.S. military mission and given a priority comparable to combat operations. The U.S. Army views the tasks and responsibilities associated with stability operations as including civil security, control, and support; restoration of essential services; support to governance; and support of economic and infrastructure development.

This broad listing of responsibilities expands the military's role into areas more typically reserved for other U.S. government agencies and non-governmental and international governmental organizations. The directive also requires the development of pre-conflict indicators of instability and a list of countries with the potential to embroil the U.S. military in stability operations. Efforts currently are under way to develop an interagency metrics-analysis capability that would be applicable to any stability operation of interest. A Defense Science Board study suggested that open sources can provide much of the required intelligence (DSB, 2004).

Recognizing past problems with interagency coordination and cooperation, President Bush signed National Security Presidential Directive Number 44 in an effort to improve "coordination, planning, and implementation for reconstruction and stabilization assistance for foreign states and regions at risk of, in, or in transition from conflict or civil strife" (NSPD, 2005). NSPD-44 appoints the Secretary of State as the lead integrator for U.S. government efforts, and requires the department to work with the Secretary of Defense to "ensure harmonization with any planned or ongoing U.S. military operations across the spectrum of conflict."

Both the DoD directive and NSPD-44 are noticeably silent on environmental threats and responsibilities. Neither addresses basic human needs or sustainability considerations as part of stability or reconstruction operations and assistance. More recently, a draft Army plan to implement DoDD 3000.05 reportedly will require the development of standards that will address human health, environmental considerations, and sustainability, while providing flexibility commensurate with the operational situation.

4. Environment and Sustainability

Sustainability is not a new concept. Its basic tenet is relatively simple. It strives to satisfy the legitimate interests of different stakeholders, not only by providing an equitable distribution of resources today, but by ensuring that these resources will be available to support future generations.

Early Native American cultures recognized the importance of the environment and its natural resources to future generations, as did the nation's founding fathers. However, as the United States made the transition from a primarily agrarian to an increasingly urban and industrial society, the importance of, and human connection with, the environment waned. A major impetus beginning in the latter half to the end of the nineteenth century was the popular, but unofficial, policy of Manifest Destiny, which was used as justification for an acceleration in westward expansion and territorial acquisition, often with little regard for the environment, natural resource equities, or impact on future generations (Hallmark, 2006).

4.1. SUSTAINABILITY—A SHARED VALUE SYSTEM

The modern rebirth of sustainability has it roots in contemporary environmentalism. Environmental activism in the United States gained prominence in the 1960s, with a toxins alarm signaled by Rachel Carson in *Silent Spring*, and later fueled by outcries over environmental disasters that focused national attention on the lack of industry oversight and disregard for community. The response was an exponential growth in environmental legislation. Internationally, interest in simple regulatory compliance has shifted to embrace the enduring principles of sustainability.

These principles of sustainability were first articulated on the global stage during the United Nations Conference on the Human Environment in 1972. The United States officially embraced the concept of sustainability when the Presidential Council on Sustainable Development was established in 1993 (Hallmark, 2006). Environmental sustainability is now recognized as one of eight major United Nations Millennium Development (MGD) Goals, with ambitious targets set for 2015 (UN, 2007). The resurgence of international interest in sustainability has been termed a "sustainability revolution," characterized by a "new value system, consciousness, and worldview" (Edwards, 2005: 5).

As with other concepts presented herein, "sustainability" also has varying definitions, which some argue is to be expected, or even desirable, given that "such a dynamic concept must evolve and be refined as our experience and understanding develop" (Moldan and Dahl, 2007). An oft-quoted definition of sustainable development was offered in the 1987 Brundtland report as "development that meets the needs of the present without compromising the ability of future generations to meet our own needs" (UN, 1987).

We must manage the Earth's natural resources in ways that protect the potential growth and opportunity for future generations.

(U.S. National Security Strategy, 1991)

It is important to recognize the need for a suitable balance between (1) the long-term viability of the *environment* and use of natural resources; (2) the importance of *economic development* and valuation of natural infrastructure; and

(3) the concern for community and related *equity* in the distribution of natural resources (Edwards, 2005). The importance of balancing these "three Es" has been popularized by the increased attention being placed on sustainability by the private sector, and subsequent routine corporate reporting of performance based on a "triple bottom line" of economic, social, and environmental equity (Nattrass and Altomare, 2002).

4.2. THE ARMY AND SUSTAINABILITY—A NEW PARADIGM

The U.S. Army recognized the importance of sustainability, and has become a leader within the Department of Defense in developing and promoting its vision of this concept based on its underlying tenets. Sustainability is important to the Army, given its stewardship of over twelve million acres (4.86 million hectares) of land and 172 installations worldwide. The Army faces increasing challenges from local encroachment and regional development adjacent to its installations, which has the potential to adversely influence current and future military training. Sustainability is viewed from both a mission-oriented and systems-based perspective, equally applicable to both the institutional and operational Army, and not just the sole responsibility of the environmental community.

Sustainability can have two different, and not mutually exclusive, meanings (Davis, 2006). The first, more traditional, military definition of sustainability focuses on the capability to "maintain the necessary level and duration of operational activity [and combat power] to achieve [its] military objectives" (DoD 2007; NATO, 2007). The second is more akin to the recognized capability (and Brundtland definition) that "meets current as well as future mission requirements worldwide, safeguards human health, improves quality of life, and enhances the natural environment" (U.S. Army, 2004).

The *Army Strategy for the Environment* (2004) established a long-range vision to enable the Army to meet its missions today and into the future. This strategy incorporates the "triple bottom line" of sustainability with adaptation, by substituting "mission" for the "economic" (or corporate profit) dimension. It identifies six major goals to guide the Army as it makes the transition from a compliance-based environmental program to one focused on the importance of sustainability to mission, requiring involvement of the entire organization. The Army is in the final stages of developing a Strategic Plan for Army Sustainability that will assign specific objectives and accountability and help shape a framework for Army support to international cooperation on sustainability initiatives.

This is a new paradigm for the Army that will require placing increased emphasis on environmentally friendly materials, more efficient systems and facilities, increased use of renewable sources of energy, and closer working relationships with local and regional stakeholders. The concept is already proving its worth. Much of the early work on sustainability was initiated from the bottom up, with several major Army installations developing *Installation Sustainability Plans* (ISPs) that set twenty-five-year goals with their local communities, dealing with important common development issues, such as encroachment adjacent to military training areas.

The Army is also addressing larger scale (regional) sustainability issues and solutions that span beyond current Army boundaries, through a new Strategic Sustainability Assessment (SSA) initiative. The goal of the SSA is to ensure that the Army can continue to realistically train and provide a high quality of life for its soldiers and their families. The 2007 *Army Posture Statement* highlights the importance of sustainability to its mission (U.S. Army, 2007). The White House recently has recognized the Army for its leadership and innovation in implementing sustainability, not only within the DoD, but also within the entire Federal government, by bestowing the *Closing the Circle Award* on the Army for its groundbreaking work on the *Army Strategy for the Environment* and subsequent sustainability initiatives and accomplishments.

4.3. ENVIRONMENTAL SUSTAINABILITY AND STABILITY

The environment supports the most basic of human physiological needs, thus providing a foundation to support both national and regional stability. However, as discussed earlier, both the DoD and presidential directives are silent on this relationship. It is believed that countries that are less environmentally sustainable will be more prone to instability. A basic correlation analysis was used to assess this relationship between environment and stability using two commonly used and recognized indices. The first index provides a measure of a country's environmental sustainability, and was compared with an index representative of state failure.

Correlation is one of the most widely used analytic procedures to measure the degree of relationship, or association, between two sets of variables. Correlation does not necessarily imply causation. It is possible that two variables of interest may be affected by one or more other (unknown) variables. For the purpose of this analysis both variables were treated as normally distributed. A Pearson correlation coefficient (r) was used to quantify the linear relationship between these indices and significance tested at the 5 percent level for each r value. An r value of $+1.00$ or -1.00 represents a perfect positive or negative correlation, respectively. Such perfect correlations are rare. A value of 0.00 suggests no relationship.

A myriad of indicators, indices, and indicator sets of sustainability have been developed by various researchers, international working groups, and government institutions following the adoption of Agenda 21 at the Rio Earth Summit in 1992. Several frameworks have been proposed to help guide their development, to include the Organization for Economic Co-Operation and Development's Pressure-State-Response (PSR) model and the European Union's Driving Forces—Pressure-State-Impact-Response (DPSIR) framework (Hearne, 2001). The majority of indicators and indices are quantitative, with indices taking the form of a single, dimensionless number. It has been suggested that it is in the interest of the international community that more strategic leadership be provided in the future development of indicators of sustainability (Moldan and Dahl, 2007).

The Environmental Sustainability Index (ESI) was selected for use in this analysis, given its widespread acceptance in benchmarking the ability of nations to protect the environment over time. The higher the ESI score, the better the environmental stewardship and the ability, or capacity, of a nation to maintain favorable environmental conditions. The index was first introduced at an annual meeting of the World Economic Forum in 2001, ranking 122 countries as to their progress toward environmental sustainability (GLT, 2001). A deliberate choice was made to isolate the environmental dimension of sustainability for purposes of the correlation.

Importantly, the ESI does not track sustainability in the broad context of the "triple bottom line" of economic, social equity, and environmental dimensions that was discussed earlier. Rather, this aggregated index more narrowly captures the "environmental dimension of sustainability" as measured by twenty-one core indicators (Moldan and Dahl, 2007). The most recent ESI (2005) ranks 146 countries, integrating seventy-six data sets into these twenty-one indicators, which allows for comparison across five broad categories of environmental sustainability: (1) Environmental Systems; (2) Reducing Environmental Stresses; (3) Reducing Human Vulnerability; (4) Social and Institutional Capacity; and (5) Global Stewardship (ESI, 2005). Reportedly, these indicators and variables build upon the PSR model. The ESI score is determined by the equally weighted average of the twenty-one core indicators.

It was recognized by the collaborating research teams that developed the ESI that it is very much a multidimensional index, composed of a larger number of indicators, which is considered appropriate because sustainability also is seen as a multidimensional concept (ESI, 2005). Further, "for a complete measure of sustainability, the ESI needs to be coupled with equivalent economic and social indices to give an integrated set of measures of the efforts of countries to move towards full sustainability" (ESI, 2005: 12). In addition to the environmental, economic, and social dimensions of sustainability, a fourth "institutional" dimension was adopted by the UN Commission on Sustainable Development system of sustainability indicators, given the importance of institutions and enabling frameworks to take action and effect change (Moldan and Dahl, 2007).

As with many other terms introduced in this paper, there is no single definition of what constitutes *stability* or for that matter, defines when a country or region is considered *unstable*. As a nation moves toward full sustainability, it is thought to become increasing stable, and, thus, stability and sustainability are believed to be closely interlinked, and to share a variety of indicators in common. In light of this, the ESI was compared with an index of state failure (or instability) that builds upon indicators more closely linked to the three dimensions of sustainability (social, economic, and institutional), which are not addressed in detail by the ESI.

The Fund for Peace has, for the last two years, published its annual Failed State Index (FSI), which employs an internationally recognized methodology using twelve indicators to rate 148 countries in an effort to better identify and anticipate failing states (Fund for Peace, 2006). These twelve indicators are grouped into three categories: social, economic, and political.

Figure 3 illustrates the relationship between the Environmental Sustainability Index (ESI) with the Failed Stated Index (FSI). The number of countries (n) where data were available for both variables for the ESI-FSI comparison was 138.

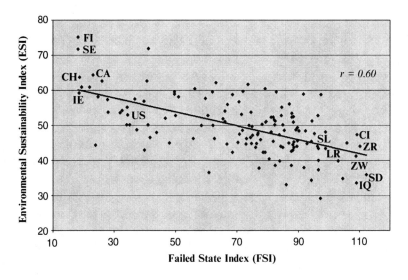

Figure 3. Relationship Between Indices of State Failure and Environmental Sustainability.

A marked degree of correlation was observed for the ESI-FSI comparison (n = 138, r = 0.60, statistically significant at the 5 percent level). The resulting trend line suggests that those countries that are more environmentally sustainable (i.e., those that have a higher ESI score) tend to be more stable (have a lower FSI score); but as the conditions that support a more favorable environment worsen, so does a country's ability to cope, resulting in increased propensity for state failure. Those countries most at risk are: the Democratic Republic of Congo (ZR); Sierra Leone (SL); Sudan (SD); Liberia (LR); Iraq (IQ); and Zimbabwe (SW). This is consistent with those countries found to be the world's least secure as listed in the *Human Security Report 2005* (Human Security Centre, 2005).

A more rigorous analysis is needed to further explore the relationship between sustainability and stability and the underlying variables and indicators that are used in developing the ESI and FSI aggregated indices. The ESI research team admits that "serious and persistent data gaps plague the ESI" (ESI, 2005: 2), but it is important to note that this is also the case with many such related indicators and indices (Moldan and Dahl, 2007).

The ESI is not without its critics. Some suggest that it contradicts another widely used and respected index of environmental sustainability, the Ecological Footprint (EF), "with the result that the most sustainable nations under the ESI are the least sustainable under the EF" (Ewers and Smith, 2007). Ewers and Smith suggest that opposite conclusions can be drawn depending on which index is used—a result of how the ESI data are aggregated in terms of the capacity to become environmental sustainable, and not necessarily being environmentally sustainable. Given these ambiguities, they believe that the EF is a more suitable index because it "converts human activities to a single, physical unit that has finite availability in the real world: land area."

The ESI collaboration research team also has provided a comparison of the ESI with other key sustainability indicators (ESI, 2005). They explain the inverse correlation with the EF as resulting from the ESI covering a wider range of sustainability issues than the EF. They acknowledge that high consumption of resources is not sustainable over the long term, but stress that countries with small footprints are not necessarily sustainable either, while more-prosperous nations with larger footprints generally have greater capacities to deal with environmental threats and stressors. They also note that the EF is included as one of the seventy-six variables that compose the ESI, and that some relationship was to be expected between the two indices. They conclude that EF primarily is a measure of environmental pressure and there is no "overt effort to balance pressure measures with systems, impact or capacity measures," explaining the higher positive correlation found with the ESI Waste and Consumption indicator as part of a broader analysis performed between 2005 ESI components and the EF index.

A similar comparison was made by the ESI research team (ESI, 2005) between the ESI and another widely recognized index of environmental sustainability, the Environmental Vulnerability Index (EVI). The EVI utilizes fifty indicators, characterized by three components or sub-indices, to measure vulnerability and resilience of environmental systems to threats by anthropogenic and natural hazards (EVI, 2007). The EVI was designed to be used with indices of social and economic vulnerability. The ESI research team found a weak relationship between the ESI and EVI and could not detect any significant trend. They did not find this surprising, because each index has a very different conceptual foundation. The ESI researchers view a high EVI as an impediment to sustainable development, and conclude that while the EVI addresses the susceptibility of a country to natural hazards, it does not address its capacity to deal with such threats as effectively as does the ESI.

4.4. SUSTAINABLE SECURITY

It has been argued that the field of security should be broadened to encompass the more comprehensive concepts of sustainability and sustainable development (Khagram et al., 2003). UN Secretary-General Kofi Annan reinforced this in a 2005 statement, asserting that "we will not enjoy development without security, we will not enjoy security without development, and we will not enjoy either without respect for human rights" (Toepfer, 2005: xv). A new term—*sustainable security*—is gaining wider use in the literature, but requires further definition.

Sustainable security is suggested to be less anthropocentric, in which the environment is valued in its own right, and not merely viewed as a set of risks such as those in Table 1. It also "facilitates critical integrations of state, human, and environmental security, and parallels the three linked pillars of society, economy, and nature central to the field of sustainable development" (Khagram et al., 2003) and, similarly, the three dimensions associated with the triple bottom line of sustainability. Khagram et al. (2003: 302) further suggest that human security be recast into the broader context of sustainable security in an effort to focus the "complex interaction between States, human beings, and nature." Sustainable security would focus on those threats having an impact on the generational quality of life and equity (see Table 2).

Sustainable security, as a more encompassing concept, would also provide an alternative response to address instability. As envisioned, it may offer a better framework from which policy makers can blend hard- and soft-power responses. Because it emphasizes long-term resolution of the root causes of insecurity, it would demand greater political commitment and resources to promote such an approach (Abbott et al., 2006).

5. Peacetime Military Engagement—Pathway to Stability

Early identification of instability is important in structuring appropriate interventions—whether economic, political, or environmental—before conditions worsen and the capacity of a nation-state to respond is threatened or weakened. Intervention does not have to be capital-intensive, as evinced by support to nongovernment and international government organizations, which generally are better pre-positioned in a country to provide a base to support such initiatives. Delayed intervention characteristically is messy, unpopular, and costly. The use of external military forces may be necessary, but such forces often are more difficult to extract once emplaced. In determining when to intervene, it is important to question whether it is best to pay less now or pay more later.

The United States remains heavily engaged in peacetime defense cooperation around the world. A major focus of this engagement has been on promoting professionalism within other militaries. This has been accomplished through various security cooperation programs, including International Military Education and Training (IMET), where foreign military students attend U.S. service schools, and education in civil-military relations provided by regional centers, such as the George C. Marshall European Center for Security Studies, the African Center for Strategic Studies, and the Asia-Pacific Center for Strategic Studies. The value of such engagement in promoting stability is difficult to quantify; however, fewer failures in "anocratic" regimes over the last fifteen years is believed to be due to a "lessening in political activism within militaries, which have been far less likely to intervene in politics or support forceful repression of public challenges to ruling elites" (Marshall and Goldstone, 2007). General Charles Wald (Retired), the former deputy commander of the U.S. European Command (EUCOM), also has credited the unwillingness and restraint of the Georgian military in responding to earlier internal public unrest, in part, to senior Georgian military participation in IMET-sponsored civil-military education (ECSP, 2007).

5.1. REGIONAL DEFENSE ENVIRONMENTAL COOPERATION

The U.S. EUCOM has maintained an active environmental engagement program. Successful initiatives include support to the Baltic Sea region Defense Environmental Cooperation (BALDEC) organization, also known as the Riga Initiative (Pocock and Smits, 2005); an annual Partnership for Peace Environmental Conference, which focuses on relevant issues facing the regions' militaries; and support to various NATO environmental initiatives, work groups, and pilot studies. The U.S. Central Command (CENTCOM) also has been active in defense environmental cooperation, including hosting symposia on responding to regional environmental challenges, with an emphasis on early warning and disaster response.

The Army and U.S. Southern Command (SOUTHCOM) currently are supporting a U.S. congressionally funded initiative known as the Western Hemisphere Information Exchange (WHIX) Program, which promotes DoD security cooperation between U.S. and Latin American and Caribbean militaries. WHIX facilitates information exchanges focused on technology demonstration and validation in the areas of environment, sustainability, energy security, and health and safety. Several WHIX projects likely will have utility in future military deployments and stability operations.

Defense cooperation with Africa is extensive; however, early environmental cooperation has been spotty and produced modest results. In the 1990s, the U.S. Congress provided funding to encourage U.S. and African militaries to work closely to curtail international fisheries poaching in West African littoral states, and to establish a number of land-based biodiversity and conservation projects (Butts, 1993). Funding to maintain these initiatives was discontinued, with the result that many languished, and thus have made little long-term difference (Henk, 2006).

EUCOM has provided support to the Environmental Security Working Group of the bilateral South Africa and U.S. Defense Committee (DEFCOM), and related workshops and initiatives. Participation has been resourced through DoD's Defense Environmental International Cooperation (DEIC) program. DEIC has also been a primary means for resourcing other regional environmental cooperative efforts elsewhere in southern Africa.

More recently, EUCOM and the Pacific Command (PACOM) have supported maritime security initiatives focused on the Gulf of Guinea to help manage regional natural resources. Illegal fishing has had an adverse economic impact on coastal communities. Such initiatives not only provide stability in the fishing sector, ensuring food security for current and future generations, but also help to combat other regional threats from terrorism, piracy, and various forms of trafficking (e.g., arms, drugs, persons).

5.2. ENVIRONMENTAL PEACE-BUILDING—THE MEANS

The case for added military involvement in international environmental cooperation is compelling. The core competencies of most militaries provide a comparative advantage in many areas, including rapid response and robust planning capabilities; unequaled strategic lift and ground transportation assets; advanced communications, surveillance, and mapping systems; force protection assets; and innovative research. Others may argue that the military has too centralized a command structure, too little exposure to other cultures, and a footprint that is overly oriented toward larger conflicts.

Defense environmental cooperation can utilize and leverage these core competencies to engage with other militaries and potentially hostile parties on nonthreatening but important issues to affected communities, and at a lower cost and with more political inclusiveness than might otherwise be possible. U.S. defense environmental cooperation has had some successes, but is constrained by resources—the *means*—and command emphasis.

Environmental issues of mutual interest and concern can serve as a valuable means to help bridge past animosities and other impasses in an effort to build peace and regional stability. Eliciting support for such cooperation has been difficult, given the long-term nature of environmental issues and time lag for their full effects to be felt. Concern over emerging environmental threats is increasing, as evinced by the attention now being given to climate change and the impact it may have on societies, economies, and critical resources, and recognition that it can exacerbate conditions and serve as a multiplier of instability and conflict for weak and fragile states.

Unfortunately, there has been insufficient systematic analysis to allow for any real determination as to how environmental cooperation might enhance conflict prevention and environmental peacemaking (Carius, 2007). Water, for instance, has served throughout history as an example of how hostile parties have resolved conflict over a shared and critical resource in a peaceful manner, even when other issues remained strained. Cooperation in nature conservation (peace parks) is more problematic, and only successful when economically and politically integrated, as with the Trans-Frontier Conservation Areas initiatives, encouraged by the Southern African Development Community (SADC), one of the most important regional organizations in southern Africa.

A good example of this effect is EUCOM support to the Environmental Security Working Group of DEFCOM. Notably, this environmental connection has been preserved, and reportedly has even flourished, at a time when relations between U.S. and South African governments have been strained (Henk, 2007). The U.S. military is continuing to assess how it might better participate in regional environmental cooperation activities. The aforementioned BALDEC initiative has been singled out as a "model for U.S. support to other regions of the world" (Pocock and Smits, 2005).

The African continent has continued to grow in strategic and economic importance. This is made evident by the resurgence of interest in expanding U.S. military engagement in Africa and the president's decision to stand up a new command effective October 1, 2007 to focus on African strategic issues. Prior to this, DoD divided Africa among three unified commands, i.e., EUCOM, CENTCOM, and PACOM, resulting in African issues not receiving adequate priority, resourcing, nor senior leadership advocacy.

The new Africa Command (AFRICOM) was established on a par with other unified [geographic] commands in an effort to address these concerns, but will have a heavier interagency focus to better identify and address instability (e.g., fragile and failing states) on the continent, and to promote liberty, peace, and economic prosperity through a relationship based on partnership, not paternalism. Discussions are ongoing as to how best to organize AFRICOM and where to locate a permanent headquarters.

The new command's focus is envisioned to emphasize the use of soft power, or non-kinetic instruments of national power, primarily oriented toward regional engagement. It would be capable of limited operations and, thus, be more dependent upon external support for large military missions. There is concern that, increasingly, traditional security types may press for an organization more focused on hard power, or kinetic, responses to threats. General Wald has expressed concern that those "hard headed over soft power" may unduly influence the command structure (ECSP, 2007).

This would be unfortunate, as the new Africa Command provides a recognized and unique opportunity for promoting defense environmental cooperation as a relatively nonthreatening means of engagement. The Kavango-Zambezi (KAZA) Trans-Frontier Conservation project in southern Africa is thought to present one such opportunity through the development of focused military-to-military engagements and partnerships on environmental issues (Henk, 2007). Accordingly, it is seen as a potential peace multiplier. Understandably, this enthusiasm may not be shared by some stakeholders out of concern for the securitizing, or militarization, of environmental agendas, and an unfamiliarity and skepticism of military involvement, regardless of how well intended. Success will, as always, be contingent upon forthright dialogue and support for consistent resources.

Our natural security must be seen as part of our national security ... decisions today regarding the environment and natural resources can affect our security for generations.
(U.S. National Security Strategy, 2000)

5.3. ALIGNMENT OF STRATEGIC OBJECTIVES—ENDS AND WAYS

The U.S. National Security Strategy (NSS) provides top-level guidance that is operationalized as specific regional objectives within the Theater Security Cooperation Plan (TSCP) of each geographic combatant command. The Army further operationalizes this guidance in The Army Plan (TAP) and the Army International Activities Plan (AIAP). The sequential alignment of guidance from the NSS down to the AIAP is illustrated in Figure 4.

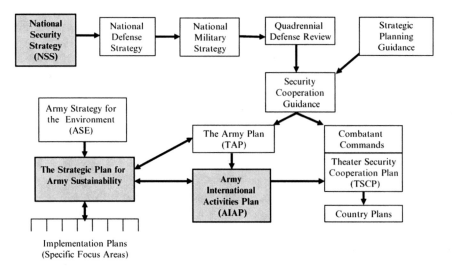

Figure 4. Linking National Security Strategy to Peacetime Military Engagement.

Importantly, the objectives and critical tasks being developed for inclusion in the Strategic Plan for Army Sustainability also are being fully aligned with the TAP. They will also be developed to promote U.S. interests and regional stability by providing training assistance, partnerships, and pilot programs on initiatives with other nations' militaries.

The AIAP provides goals and objectives and identifies countries of emphasis for Army security cooperation activities in implementing DoD Security Cooperation Guidance and Army strategic objectives outlined in the TAP. The AIAP provides a comprehensive listing of programs—the *ways*—to meet the various strategic objectives of the TAP and higher-level guidance documents—the *ends*. Excepting WHIX and DEIC, there are no programs that focus on environmental and sustainability initiatives. Rather, defense cooperation initiatives must be submitted for consideration by other applicable AIAP programs. Defense environmental initiatives, however, do not generally compete well, given other program priorities and constraints, suggesting the need for more specified programs and dedicated funding.

A recently completed RAND study of the AIAP assessed the value of the Army's noncombat (peacetime) interaction with other militaries (Marquis et al., 2006). This assessment identified two types of indicators of defense security cooperation. *Output indicators* reflect the immediate product of AIAP activities that tend to produce a near-term result, and can be counted more easily. *Outcome indicators* are more qualitative in nature, and derive from a long-term socialization process that involves building trust and changing foreign perceptions, and often are the by-products of prior AIAP outputs. Not surprisingly, the quantitative nature of output measures often is favored over more difficult-to-measure outcomes.

Sustainability provides another important resource for international defense cooperation that extends beyond just environmental compliance by focusing on the underlying roots of instability. U.S. Army experience with sustainability is extensive, and technical exchanges can provide an excellent and inexpensive means to engage with other militaries and regional stakeholders on issues of common concern. The draft Strategic Plan for Army Sustainability promotes security and stability in international activities as one of its main objectives, which would be accomplished by providing assistance and support through international military-to-military partnerships and pilot programs.

The success of environmental and sustainability efforts are dependent upon a socialization process that is best measured by outcome, rather than output, because positively affecting national and regional stability remains a lengthy proposition. Unfortunately, the long-term nature of environmental stressors contrasted with short-term planning, programming, and budgeting horizons, often only five to seven years, continues to make it difficult to attract support for these nontraditional types of engagement initiatives.

The importance of the environment to national security must be articulated better in top-level national and military policy guidance if support is to be provided through combatant command and respective Service security cooperation programs (e.g., the AIAP). Excerpts have been offered throughout this paper to highlight the relative importance given environmental threats and vulnerabilities in past national security strategies. The global terrorist threat will remain a reality for the foreseeable future. The threats posed by global environmental stressors also are very real. Not surprisingly, many have recommended that the next update to the U.S. National Security Strategy must address these environmental threats and vulnerabilities, and the important linkage between environment and security recognized in past strategies (Spencer, 2004; Damonte, 2006; Henk, 2006).

6. Implications and Opportunities

Environmental threats and vulnerabilities are important to long-term U.S. national security interests. It is in the very nature of these stressors that their effects are not fully appreciated for many years, at which point intervention is considerably more difficult and costly. A simple correlation analysis suggests that countries that are less environmentally sustainable are more prone to instability. Defense environmental cooperation is an effective engagement tool that can be expanded easily, and in a relatively nonthreatening manner, to promote regional environmental stewardship and stability. The DoD can provide needed international leadership in this area.

Earlier terrorist attacks refocused national security priorities on a very real and immediate asymmetric threat, but at the expense of more poorly understood environmental threats. This is of particular concern because environmental stressors have the potential to exacerbate underlying, and often unsustainable, conditions that favor terrorism and fuel instability. Prior administrations have recognized the risks posed by environmental threats in their respective national security strategies. The need for a more coherent U.S. National Security Strategy that addresses environmental threats and vulnerabilities is widely recognized, and recommended in the next update. Only then will sufficient attention and resources be directed by U.S. government agencies and the military to applicable initiatives and programs to address these nontraditional threats.

The U.S. Army is a recognized leader in sustainability, with a unique military perspective on implementing its basic principles and in adapting its triple bottom line. Environmental and sustainability issues provide an excellent and inexpensive means for the Army to engage with other nations, militaries, and regional stakeholders in a relatively nonthreatening way. The U.S. Africa Command presents a unique opportunity—a blank slate of sorts—to focus defense cooperation on environmental and sustainability initiatives that can help promote stability on a continent that is of increasing strategic and economic importance.

References

Abbott, C., Rogers, P., and Sloboda, J., 2006, *Global Responses to Global Threats: Sustainable Security for the 21st Century*, Oxford Research Group Briefing Paper.
Association of the United States Army (AUSA), 2006, *The U.S. Army's Role in Stability Operations,* Torchbearer National Security Report.
Botkin, D., and Keller, E., 1995, *Environmental Science—Earth as a Living Planet*, Wiley, New York.
Butts, K. H., 1993, *Environmental Security: What is DoD's Role?* Strategic Studies Institute Paper, U.S. Army War College, May 1993.
Butts, K. H., 2007, Environmental security and regional stability: The power of sustainability, *Army Sustainability Lecture Series*, Arlington, VA, February 9.
Carius, A., 2007, Environmental peacemaking: Conditions for success, *Environmental Change and Security Program Report*, Forthcoming Issue 12.
CNA Corporation, 2007, *National Security and the Threat of Climate Change*, CNA Report.
Damonte, E. M., 2006, *National Security Strategy: What About the Environment?* U.S. Army War College Strategy Research Project, 15 March; http://www.strategicstudiesinstitute.army.mil/ pdffiles/ksil322.pdf, accessed April 22, 2007.
Davis, T., 2006, Sustainability front stretches beyond environmental theater, *Environmental Update*, Spring.

Defense Science Board (DSB), 2004, *Transition to and from Hostilities*, DSB Summer Study.
Department of Defense (DoD), 2007, *Dictionary of Military and Associated Terms*; http://www.dtic.mil/doctrine/jel/ doddict/data/s/05262.html.
Department of Defense Directive (DoDD), 2005, *DoDD 3000.05, Military Support for Stability, Transition, and Reconstruction (SSTR) Operations*, 28 November.
Dobbins, J., McGinn, J. G., Crane, K., Jones, S. G., Lal, R., Rathmell, A., Swanger, R. M., and Timilsina, A., 2003, *America's Role in Nation-Building: From Germany to Iraq*. Santa Monica, California, RAND Corporation, MR-1753-RC.
Edwards, A. R., 2005, *The Sustainability Revolution—Portrait of a Paradigm Shift*. New Society Publishers, Gabriola Island, Canada.
Environmental Change and Security Program (ECSP), 2007, National security and the threat of climate change, a panel presentation, Woodrow Wilson International Center for Scholars, May 14.
Environmental Sustainability Index (ESI), 2005, Yale University and Columbia University, *2005 Environmental Sustainability Index—Benchmarking National Environmental Stewardship*; www.yale.edu/esi, accessed April 20, 2007.
Environmental Vulnerability Index (EVI), 2007, EVI Indicators; http://www.vulnerabilityindex.net/EVI_Indicators.htm, accessed March 31, 2007.
European Union (EU), 2004, *A Human Security Report for Europe: The Barcelona Report on Europe's Security Capabilities*, Barcelona.
Ewers, R. M., and Smith R. J., 2007, Choice of index determines the relationship between corruption and environmental sustainability, *Ecology and Society* **12**(1).
Faris, S., 2007, The real roots of Darfur, *Atlantic Monthly* **299**(3).
Fund for Peace, 2006, *Failed State Index*; http://www.fundforpeace.org/rg/programs/ fsindex.php, accessed January 15, 2007.
Gleditsch, N. P., 1998, Armed conflict and the environment: A critique of the literature, *Journal of Peace Research* **35**(3):381–400; http://www.un.org/milleniumgoals.
Gleick, P., 2006, *The World's Water, 2006–2007—The Biennial Report on Freshwater Resources*, Island, Washington, DC.
Glenn, J., Gordon, T., and Perelet, R., 1998, *Defining Environmental Security: Implications for the U.S. Army*, Army Environmental Policy Institute Report (AEPI-IFP-1298).
Global Environmental Change and Human Security (GECHS) Project, 2000, *The Index of Human Insecurity*, AVISO, GECHS Information Bulletin 6, January.
Global Leaders for Tomorrow (GLT), 2001, *2001 Environmental Sustainability Index*, Environmental Task Force, World Economic Forum, Report to Annual Meeting, Davos, Switzerland; http://www.ciesin.colombia.edu/indicators/ESI/downloads.html.
Hallmark, M. E., 2006, *Sustainability: Cultural Considerations*, U.S. Army War College Civilian Research Project, Internal Academic Research Paper.
Haugue, W., and Ellingsen, T., 1998, Beyond environmental scarcity: Causal pathways to conflict, *Journal of Peace Research* **35**(3):299–317.
Hearne, S. R., 2001, *Environmental Indicators: Regional Stability and Theater Engagement Planning*, Army Environmental Policy Institute Report No. AEPI-IFP-1001A, October.
Henk, D., 2006, The environment, the U.S. military, and southern Africa, *Parameters* (Summer): 98–117.
Henk, D., 2007, *The Kavango-Zambezi (KAZA) Trans-frontier Conservation Area Project—Is There A U.S. Military Interest?* Working INSS Paper sponsored by the Army Environmental Policy Institute (AEPI).
Homer-Dixon, T. F., 1999, *Environment, Scarcity, and Violence*, Princeton University Press, Princeton, NJ.
Human Security Centre, 2005, *Human Security Report 2005*, Oxford University Press, New York.
Joint Publication (JP), 2006, *JP 3-0 Joint Operations*, September 17.
Khagram, S., Clark, W. C., and Raad, D. F., 2003, From the environment and human security to sustainable security and development, *Journal of Human Development* **4**(2):289–313.

King, C., 2000, *Understanding International Environmental Security: A Strategic Military Perspective*, Army Environmental Policy Institute Report (AEPI-IFP-1100A).

King, C., 2006, Concepts of Strategic Environmental Security, Paper presented to the 15th OSCE Economic and Environmental Forum, Vienna, Austria, January 21–22.

Liotta, P. H., 2003, *The Uncertain Certainty: Human Security, Environmental Change, and the Future Euro-Mediterranean*, Lexington Books, New York.

Liotta, P. H., 2005, Through the looking glass: Creeping vulnerabilities and the reordering of security, *Security Dialogue* **36**(1):49–70.

Liotta, P. H. and Owen T., 2006a, Why human security, *Whitehead Journal of Diplomacy and International Relations* (Winter/Spring): 1–18.

Liotta, P. H. and Owen T., 2006b, Sense and symbolism: Europe takes on human security, *Parameters* (Autumn): 85–102.

Manwaring, M. G., ed., 2002, *Environmental Security and Global Stability*, Preface—Interview with General Anthony L. Zinni, USMC (Ret.), Lexington Books, New York.

Marquis, J. P., Darilek, R. E., Castillo, J. J., Turston, C. Q., Wong, A., Huger, C., Mejia, A., Moroney, J. D., Nichiporuk, B., and Steele, B., 2006, *Assessing the Value of U.S. Army International Activities*, RAND Arroyo Center publication.

Marshall, M. G., and Goldstone, J., 2007, Global report on conflict, governance and state fragility, *Foreign Policy Bulletin* **17**(S01):3–21.

Matthew, R. A., and Fraser, L., n.d., *Global Environmental Change and Human Security: Conceptual and Theoretical Issues*.

Merriam-Webster, 2003, *Collegiate Dictionary*, Eleventh Edition.

Miguel, E. S., Satyanath, S., and Sergenti, E., 2004, Economic shocks and civil conflict: An instrumental variables approach, *Journal of Political Economy* **112**(4):725–753.

Miguel, E. S., Satyanath, S., and Fisman, R., 2007, *Climate Change, Conflict, and Foreign Aid*, Presentation at University of California Washington Center, May 10.

Moldan, B., and Dahl, A. L., 2007, Challenges to sustainability indicators, in: *Sustainability Indicators: A Scientific Assessment*, Island, Washington, DC.

Mueller, J., 2004, A false sense of insecurity? *Regulation* (Fall): 42–46.

National Intelligence Council (NIC), 2000, *Global Trends 2015: A Dialogue About the Future with Nongovernment Experts*. NIC Paper 2000–2002, December.

National Intelligence Council (NIC), 2004, *Mapping the Global Future*, NIC 2020 Project Report, December.

National Security Presidential Directive (NSPD) 44, 2005, *Management of Interagency Efforts Concerning Reconstruction and Stabilization*, December.

National Security Strategy (NSS), 1991, *National Security Strategy of the United States*, August.

National Security Strategy (NSS), 1995, *A National Security Strategy of Engagement and Enlargement*, February.

National Security Strategy (NSS), 1997, *A National Security Strategy for a New Century*, May.

National Security Strategy (NSS), 1998, A *National Security Strategy for a New Century*, October.

National Security Strategy (NSS), 2000, *A National Security Strategy for a Global Age*, December.

Nattrass, B., and Altomare, A., 2002, *Dancing with the Tiger: Learning Sustainability Step by Natural Step*, New Society Publishers, Gabriola Island, Canada.

North Atlantic Treaty Organization (NATO), 1999, *Environment and Security in an International Context*, Committee on the Challenges of a Modern Society Report 232.

North Atlantic Treaty Organization (NATO), 2007, *NATO Glossary of Terms and Definitions*, NATO Standardization Agency.

Organization for Economic Co-Operation and Development (OECD), 2001, *OECD Environmental Outlook*, Paris, OECD Environmental Directorate.

Owen, T., 2003, Measuring human security: Overcoming the paradox, *Human Security Bulletin* (October) **2**(3):1–25.

Owen, T., 2004, Challenges and opportunities for defining and measuring human security, *Disarmament Forum* 2 (June).

Pearce, F., 2006, *Water—When the Rivers Run Dry—The Defining Crisis of the Twenty-First Century*, Beacon, Boston, MA.

Pocock, J. B., and Smits, K. M., 2005, *Developing Alliances Through Regional Defense Environmental Cooperation: Building on the Successes in the Baltic Sea Region*, Working INSS Paper sponsored by the Army Environmental Policy Institute (AEPI).

Reed, R., 2004, *Dehydrated National Security: Water Scarcity, the Emerging Threat of the 21st Century*, Thesis, School of Advanced Air and Space Studies, June.

Renner, M., 2006, Chapter I-C, Introduction to the concepts of environmental security and environmental conflict, *Inventory of Environmental and Security Policies and Practices—An Overview of Strategies and Initiatives of Selected Governments, International Organisations and Inter-Governmental Organisations,* Institute for Environmental Security Policies and Practices, 1st ed., R. A. Kingham, ed., Institute for Environmental Security, The Hague, The Netherlands, October, 2006.

Richmond, O., and Franks, J., 2005, Human security and the war on terror, in: *Human and Environmental Security: An Agenda for Change*, F. Dodds and T. Pippard, eds. Earthscan, London.

Solana, J., 2007, Managing Global Insecurity, EU High Representative for the Common Foreign and Security Policy, Brookings Institute Speech, Washington, DC., March 21.

Spencer, T., 2004, Pathways to Environmental Security, Chairman's Summary of the Hague Conference on Environment, Security, and Sustainable Development. June 1.

Speth, J. G., and Haas, P. M., 2006, *Global Environmental Governance*, Island, Washington, DC.

Stern, 2007, *Review on the Economics of Climate Change;* http:// www.hm-treasury.gov.uk/ independent_reviews/stern_review_economics_climate_cha, accessed April 5, 2007.

Toepfer, K., 2005, Foreword, *Human and Environmental Security—An Agenda for Change*, Dodds, F., and Pippard, T., eds., Earthscan, London, 2005.

United Kingdom (UK), 2005, *Investing in Prevention: An International Strategy to Manage Risks of Instability and Improve Crisis Response*, Prime Minister's Strategy Unit, February.

United Nations (UN), 1987, *Our Common Future*, World Commission on Environment and Development.

United Nations (UN), 2007, *Millennium Development Goals*; www.un.org/millenniumgoals/ pdf/mdg2007.pdf, accessed January 15, 2007.

United Nations Development Program (UNDP), 1994, New Dimensions of Human Security, in: *Human Development Report.*

United States Army (U.S. Army), 2004, *Army Strategy for the Environment*, October, 1.

United States Army (U.S. Army), 2007, *Army Posture Statement—Army Sustainability— Strategy—Appendix K*; http://www.army.mil/aps/07/addendum/k.html, accessed April 6, 2007.

Wolf, A. T., Kramer, A., Carius, A., and Dabelko G., 2006, *Navigating Peace: Water Can Be a Pathway to Peace, Not War*, Woodrow Wilson Center Environmental Change and Security Program brochure no. 1, July.

World Bank, 2004, *Worldwide Governance Indicators: 1996–2005*; http://info.worldbank.org/ governance/kkz2005/q&a.htm, accessed April 1, 2007.

World Wildlife Fund (WWF), 2006, *Living Planet Report 2006*; http://www.panda.org/newsfacts/ publications/living_planet_report/index.cfm, accessed January 21, 2007.

Yu, C., Quinn, J. T., Dufournaud, C. M., Harrington, J. J., Rogers, P. P., and Lohani, B. N., 1998, Effective dimensionality of environmental indicators: a principal component analysis with bootstrap confidence intervals, *Journal of Environmental Management* **53**(1):101–11.

Appendix. Core Set of Indicators for Assessing Stability

Indicators by major dimension	Institutions and researchers using indicators*						
	EEA	OECD	CSD	SDI	ESI	SFP	H&E
Environment							
NO_x emissions in urban areas (kg per capita)	X	X	X	X	X		
Urban exposure to particulate matter (micrograms/M3)	X		X	X	X		
CO_2 emissions (metric tons per capita)	X	X	X	X	X		
Change in forest area (% of total land area)	X	X	X	X	X	X	X
Arable and permanent crop land (% of total land area)	X	?	X	X		X	?
Withdrawal of ground and surface water (% total available)	X	X	X	X	X	X	?
Organic matter emissions in water bodies as kilogram BOD/capita	X	X	X	X	?		
Wetland loss through drainage in hectares	X						
Protected areas (% of total land—IUCN Categories)	X	X	X	X	X		
Socioeconomic							
Population growth rate as %		X	X	X	X		
Life expectancy at birth (years)			X	X		X	
Urban population (% of total population)			?	X			
Access to safe drinking water (% of population)			X	X	X		
Under-5 mortality (reported deaths to 1,000 live births)			X	X	X		
Daily per capita supply of calories			?	?	X	?	
Energy intensity (GDP output per kilogram (US $)-oil equivalent)	?	X	X	X	X	X	
Trade openness (imports + exports as % of GDP)			?	?		X	
GDP per capita (for 1998 in 1995 US $)		X	X	X		X	X
Income inequality (Gini coefficient)			X	X		?	?
Institutional							
Level of Democracy (Polity IV Dataset: range −10–10)						X	X
Civil liberties (Freedom House: range 1–7)					X	X	

Corruption score (Freedom House: range 1–7)

Research and development (R&D) as percentage of GDP

R&D scientists and engineers per million population

Number of internet hosts per 10,000

Main telephone lines per 1,000 population

Cellular mobile phone subscribers per 1,000 population

Defense expenditures (% of GDP)

Notes: X denotes that the indicator is being used by the listed institution; ~ denotes a similar indicator being used.

* Acronyms:

EEA: European Environment Agency—Towards environmental pressure indicators for the EU (pressure indicators)

OECD: Organization for Economic Co-Operation and Development—OECD core set of environmental indicators

CSD: UN Commission for Sustainable Development—CSD core set of sustainable development indicators

SDI: Sustainable Development Index—Indicators used in the dashboard of sustainability

ESI: Environmental Sustainability Index—Indicators used by World economic task force in developing the ESI

SFP: State Failure Project—Independent variables used in multiple linear regressions

H&E: Hauge and Ellingsen—Independent variables used in multivariate analysis

(*Source:* Hearne, 2001)

THE HUMAN SECURITY DILEMMA

Lost Opportunities, Appropriated Concepts, or Actual Change?

RYERSON CHRISTIE*

York University Centre for International and Security Studies

Abstract: Human security discourse and practice have been embraced by a number of states as a means to address our current security dilemmas. At the same time, human security literature is advanced as an alternative vision of world politics that moves past the state-centric analysis of traditional security studies. What is unclear is whether these two aspects of human security, its practice and intent, are compatible. There are two related questions that must then be addressed. The first is whether human security practice has actually benefited peoples, which will be approached by exploring the types of policies that have been put in place in the name of human security. This then brings up the second, and perhaps more profound, question: has the adoption of human security frameworks had the effect of changing the way that we "do" security policy, or has it merely widened the range of issues to which traditional security practices are applied? This paper will draw examples predominately from the current international involvement in Afghanistan.

Keywords: Human security; Canada; critical security studies; peacebuilding; securitization

1. Introduction—Has Human Security Been Adopted, Adapted, or Co-Opted?

I have been an advocate for human security for over a decade, and have actively promoted the concept's adoption by both academics and policy makers. However, Canada's and NATO's evolving mission in Afghanistan

---

*Address correspondence to Ryerson Christie, York University Centre for International and Security Studies, 375 York Lanes, 4700 Keele St. Toronto, ON, Canada, M3J 1P3

P.H. Liotta et al. (eds.), Environmental Change and Human Security, 253–269.
© *2008 Springer Science + Business Media B.V.*

has led me to question the faith I have placed in this idea. Its ability to change the way that we perceive security, and to in turn affect policy decisions, needs to be carefully assessed. It is not immediately clear whether the discourse of human security has been able to change the ways in which we examine the current operations in Afghanistan, or whether it has altered our engagement with the country. Nor is it clear, if human security has affected our interaction with Afghanistan, whether this has been in a way that is faithful with the intent of the literature, or if it has had unfortunate and unexpected effects. To put this question a different way, I want to know if the concept of human security has been adopted by governments and policy makers, adapted to the particular needs of a given situation (but still holding on to some of the core ideals), or if it has been co-opted. This question has profound implications for the ways that we think and write about human security, as well as how we can (or indeed whether we should) translate this into concrete policy recommendations.

Human security has been presented not just as an alternative to the dominant state-centric forms of security practices, but in part as a remedy to their worst abuses. By demanding that we consider the needs of peoples, rather than just the needs of states, human security is intended to control the behavior of states, and to reframe the context within which states set their foreign and domestic policies. It also provides a means by which the general public as well as academic and policy communities are able to judge the impact of policy decisions. The concern is that the language of human security is not always used in this manner. The use of the concept of human security can also be used to justify and legitimize forms of state interventionism that might otherwise have been deemed illegitimate, and it can risk securitizing a broad range of issues that might be best dealt with in different contexts.

This paper will explore the question by paying particular attention to the Canadian context, but reference will be made to the broader NATO situation. It is being written in a time when there is an ongoing debate within Canada about whether there is an appropriate balance being struck between the military and development components of the mission, and of the place of the Department of Foreign Affairs and International Trade (DFAIT) within the overall operation.

I will argue that the recent history of human security is not in fact simply one or another of these. In fact, what I find is that there is a combination of adoption, adaptation, and co-option. Dangerously, however, we are increasingly moving toward co-option. This shows that the re/defining of "security" is inseparable from the broader power relations of society, and that the attempt to dislocate the language of security is profoundly difficult at best. This may not mean that we should abandon the concept, but perhaps that having advanced it we must now guard against its use in ways that contravene the intent of human security advocates.

2. The Intent of Human Security

The intent of human security has been well discussed in academic literature. Since 1994, with the advent of the UN Human Development Report, human security has been seen as a way to discuss, promote, and frame policies that place the needs of peoples ahead of those of states.

> Although human security was not precisely defined in the report, it was said to have two main components: safety (of individuals and groups) from such chronic threats as hunger, disease, and repression; and protection from sudden and hurtful disruptions in patterns of daily life. (Busumtwi-Sam, 2002: 258)

The move, which if applied and taken to heart, is far from rhetorical; it argues that the state system cannot be assumed to guarantee the safety and dignity of peoples. In fact, the internal state conflicts that have absorbed the attention of the international community from the late 1980s onward are often defined by situations where the government is one of the most significant threats to peoples. According to Newman, "[i]n its broadest sense, "human security" seeks to place the individual—or people collectively—as the referent of security, rather than, although not necessarily in opposition to, institutions such as territory and state sovereignty." (Newman, 2001: 239) This does not mean that there has been a complete break with the state. Most of the current work, in fact, argues that the state remains the most effective guarantor of peoples' human security needs (Axworthy, 2001; MacLean, 2000). Much attention has therefore been paid to the reconfiguration of states to make them able and willing to protect their citizens.

Human security literature has evolved to the point where there are intriguing debates taking place among its advocates. These can be broken down into two main themes. The first is a discussion over policy focus. This has translated into a sense of a division between Japanese and Canadian orientations where there has been some debate about whether the lack of local economic/social development or physical insecurity should be the priority. In fact, we find that this division is lessening as most policy makers are now arguing that both sets of issues need to be addressed. What this debate does not take into account is how, once moved into the policy realm, human security is affected by bureaucratic struggles and mediated by broader discourses of security.

The second debate, which has taken place more marginally, has to do with whether human security ought to be moved into concrete policies, or if it is best retained as an ethical framework. This is an argument that has been located, for the most part, in the pages of *Security Dialogue* and has tended to have the critical security studies advocates in one camp, and those of a more liberal bent in the other. I have, in the past, sided with Kyle Grayson (2003) on this point, agreeing that the power of the concept resides in its capacity to critique the mainstream, and its effectiveness in challenging dominant security practices.[1] However, at the same time there is another aspect of this argument that needs to be explored in greater depth, which has been noted by Newman (2001).

> The underlying argument is that behavior, interests, and relationships are socially constructed, and can therefore change. Values and ideas can have an impact upon international relations; norms, systems, and relationships can change as an aggregation of agent-oriented processes. (Newman, 2001: 247)

What this suggests is that the adoption of human security has the possibility to not only broaden the issues that we include under the rubric of security, but to also change society's notions of both identity and international relations. This is in many ways a radical assertion, but one that speaks of the possibility of change rather than simply doing business in a slightly different way. To appreciate this we need to accept that language/narratives have the potential of being transformative, even while they predominately operate in conservative ways. In this way the promotion of human security is not only a means to address some problems in the practice of international politics, but holds out the possibility of altering the whole field of security.

3. Securitization Literature

While this topic could be addressed through a variety of means, refining the Copenhagen School's (CoS) securitization approach provides a way to both ask and address issues at the heart of how human security has been taken up and used by the policy community. I am going to follow securitization as a means of study rather than as a government policy. Traditionally the securitization literature has focused on the speech act of "security," examining how the utterance of the term in connection with any given issue serves to alter the way the state can respond to it. "It is by labelling something a security issue that it becomes one" (Wæver, 2004: 13). Now this does not mean that the state is always able to secure an issue. It needs to make a case to the population that the thing, the people, or whatever represents a real threat to the society. If the people agree, or at least permit the act of securitization, then the state is able to take action that it would not otherwise be able to, and to a limited extent is actually expected to do so. This then points to a

---

[1] Ken Booth has put forth the following definition of security that he feels is consistent with a commitment to critical studies. "Security in world politics is an instrumental value that enables people(s) some opportunity to choose how to live. It is a means by which individuals and collectivities can invent and reinvent different ideas about being human" (Booth, 2005: 23).

perceived difference between threats and vulnerabilities to individuals and communities, and security issues that come to represent an existential threat to the group itself.

> [The] first step towards a successful securitization is called a securitizing move. A securitizing move is in theory an option open to any unit because only once an actor has convinced an audience (inter-unit relations) of its legitimate need to go beyond otherwise binding rules and regulations (emergency mode) can we identify a case of securitization. (Taureck, 2006: 55)

Following from Taureck (2006), I believe that we must see the process of securitization as a political act; it involves public debate and struggle (although clearly this occurs within a field of unequal power relations where the agent of security—traditionally the state, militaries, etc.—have an unequal voice). However, once a thing is effectively securitized the normal functioning of politics comes to an end. Precisely because an issue or thing is labeled as a threat the agent of security is expected to take action to alleviate the concern. This then works to alter the terms of debate around a subject, both cutting off many forms of public debate and simultaneously making possible extra-juridical state responses.

The alternative to securitization is the pursuit of desecuritization, which is effectively the removal of a topic from the security narrative. This literature is somewhat more controversial than the securitization literature that it derives from. It is unclear, for example, how one might go about desecuritizing an issue, as discussing it within the context of security most likely will not serve such an end. In fact, attempting to do so would in all likelihood reproduce the broader narratives of security; the desecuritization would most likely revolve around whether the thing has ceased being a threat. This simply reproduces the notion that it was a threat in the first place. Making the slightly different argument that it was never a threat also reproduces the notion that there are such things as threats. The other aspect of this that needs to be noted is that desecuritization does nothing to challenge the terms of the debate about the validity of trying to secure the "state" itself. "[D]esecuritization still leaves the previously securitized issue or actor at the mercy of the undemocratic and exceptional power of this Sovereign" (Behnke, 2006: 63). Behnke interestingly argues that subjects cannot be desecuritized, that once this discursive move is made (and accepted), the subject cannot then be moved out. I will not dwell on this topic, other than to suggest that if human security has been co-opted then alternative strategies will need to be discussed to remedy this situation.

In attempting to relate human security and securitization literature we find that the Copenhagen School was not attempting to address this particular movement. Human security represents a particular challenge to securitization in a number of ways, but, as I will argue, these are surmountable. The most obvious issue is that securitization is largely premised on the idea of a state-centric system where the main focus of peoples is the continued existence of their own particular state/society. Human security is deliberate in its attempt to unsettle this. The power of securitization makes sense within the context of a society that is seeking to secure itself. If the government of the day is able to demonstrate that, say, environmental change represents a clear and present danger to the continued existence of the people, that is to securitize the issue, then the government will be allowed all sorts of extraordinary measures to address this concern. The people are willing to permit this precisely because they see themselves at risk. Human security rejects this logic, arguing that we must move our calculations away from our own self-interest on to the "Other." The fact that it deliberately seeks to undermine the traditional strategic logic requires that we consider its implications for securitization itself. The Copenhagen School presumes the centrality of the state to its framework, which is precisely what is being rejected by a re-centering away from the state onto substate actors (such as individuals). If human security is attempting to place attention on the needs of peoples, rather than the state, is this as likely to lead to the population's acceptance of extraordinary measures by the state?

What securitization literature does for us in this context is to provide a framework to assess whether discourses of human security have the same potential to suspend politics, and to ask how and why human security practices and discourse might come to be adapted and co-opted by the state. I am moving beyond the strict focus of the Copenhagen School in some other respects. To begin with I want to look at the broader discursive production of security, not just at the "speech act." Discourse is obviously much broader than the public statements and claims of governments. Following the securitization literature we can expect to find an initial stage of political debate about the adoption of human security and subsequent debates about whether or not particular issues also represent dangers. Once accepted we need to ask if this has led to policy decisions that exceed normal government activities.

However, after this (I will argue that the adoption and adaptation of human security is not commensurate with its securitization) there is another aspect of human security and securitization. This is what happens after the language is adopted by both the state and society. Does it alter the way that securitization of issues occurs, and the subsequent state policies to resolve future threats? Put another way, does the adoption of human security profoundly alter the broader discourse of security, or does it get subsumed under the traditional security language? This requires that we adjudicate whether the language has been able to change the behavior and calculations of states, and if so whether this has occurred in a way that protects peoples, or if it has instead expanded the issues that are securitized.

Not everyone sees an inherent conflict here; indeed many advocates see traditional and human security as being compatible. The Commission on Human Security, for example, stated that state security and human security are complementary (Commission on Human Security, 2002: 4). Unfortunately this belies the ways in which human security was seen as a necessary change to offset the problems of state-centric security. Arguing that the two are compatible and complementary does not provide a means of escaping from the trap of "security for whom," and crucially what to do when security needs are in conflict.

4. Adoption

The first point to be addressed is whether human security as a concept has been adopted by international policy makers. There is little doubt that it has picked up a degree of credence among academia and the broader society. Conferences such as the one for which this paper was written demonstrate the manner in which it has acquired legitimacy within the field of international relations. Additionally, its longevity also illustrates how much traction it has acquired; it has been over a decade since it was first found in the major academic journals. But aside from the rarefied conversations within the ivory tower, how has it been integrated into the policy community? Here the record is remarkable for the extent to which it became a core focus of a number of middle-power states.

Some have seen the adoption of this language by middle powers as strictly serving a traditional state-centric self-interest (Busumtwi-Sam, 2002; Tomlinson, 2002), the idea being that in pursuing such policy goals such states were able to extend their political reach, and to continue playing the "soft-power" card that allowed them to exert an influence in the international system that was beyond their strict hard-power capacity. Others have seen it as being in keeping with a pattern of behavior by middle powers that is beyond the logic of self-interested behavior.

[E]mbracing the human security discourse has followed a well-established trajectory for Canada and Norway, two of the leading states in the so-called human security coalition. Both countries use the discourse of human security to distinguish themselves as "progressive middle powers" and to protect identities that are in opposition to the United States for Canada and to the European Union (EU) for Norway. (Busumtwi-Sam, 2002: 269–270)

Without an in-depth analysis of why the policies were brought in, we cannot know what the impetus was. If this description of the way the concept has been integrated into the policy community is accurate, then it points to broader processes of policy adoption and adaptation. It has been adopted due in part to personal leadership of individuals within government, such as Lloyd Axworthy as the Minister of Foreign Affairs and International

Trade, and others within the bureaucracy, and in the broader academic and NGO community. As with any actual policy decision, the reasons would have been complicated, overlapping, and most likely occasionally contradictory. However, the reasoning for the adoption of the concept by policy makers is a result of a political debate, a struggle between various stakeholders both within and outside of government.

In Canada the adoption of human security was whole-hearted and resulted in numerous different policy initiatives. These include, but are by no means limited to

- The establishment of desk officers responsible for promoting and advocating human security initiatives.
- The creation of a human security program with a mandate for advancing and funding programs both inside and outside of government for the exploration and promotion of the ideal.
- The funding of the Canadian Consortium for Human Security, which fostered a generation of junior scholars with a focus on the study and application of the concept.
- In conjunction with Norway, the signing of the Lysøen Declaration on 11 May, 1998 in Bergen, Norway.

In many respects the Lysøen Declaration, and the creation of the Human Security Network (HSN) that grew out of it, can be seen as the high-water mark of the human security movement. At that point the Ministers of foreign affairs of the two states agreed on a broad agenda of action that was premised and deliberately grounded in the conceptual framework of a re-centered security. This resulted in a commitment to move on issues such as landmines; the establishment of an international criminal court; human rights; international humanitarian law; women and children in armed conflict; small arms proliferation; child soldiers; child labor; and Northern and Arctic cooperation. The later HSN encompassed another dozen governments: Austria, Canada, Chile, Costa Rica, Greece, Ireland, Jordan, Mali, the Netherlands, Norway, Slovenia, Switzerland, Thailand, and South Africa (as an observer). Besides these agreements, Canada and other middle powers have embraced the Responsibility to Protect, which was promoted as an instrumental mechanism to protect the rights and dignity of peoples. According to the Government of Canada, "The 'responsibility to protect' was a key component of Canada's human security agenda. The human security approach to foreign policy puts people—their rights, their safety and their lives—first."[2] What these various initiatives illustrate is that Canada

---

[2] http://geo.international.gc.ca/cip-pic/library/canadaandresponsibilitytoprotect-en.asp.

and other states have taken up the language of human security and have largely agreed with the academic literature's call to begin focusing on the security needs of people rather than states.

While the language of human security has been adopted, there has been a debate about whether this has profoundly changed the way that the North reacts to perceived human insecurity in the South. Some, such as Richard Bowes (2001) argue that the North is not willing to place lives at risk for the human security agenda. He uses a discussion of the Bosnian crisis as a way to illustrate this point. This however assumes that the only way to respond to human insecurity is with the use of military force. Given the range of issues that come into focus through the adoption of human security, military force (and thus the potential loss of life) should be a relatively small component. Regardless, discourses of human security were adopted by policy makers and academics in the mid-1990s and it clearly began to help shape policy initiatives.

5. Adaptation

What needs to then be examined is the extent and manner to which the commitment to human security has been translated into policy. Unlike the actual adoption of human security, this aspect has taken place in a much more opaque environment. This is not to say that there have been deliberate attempts to obscure this process, just that it is inherently bureaucratic. This is distinct from the securitization literature's assertion that once an issue is successfully made a security concern it would be removed from the realm of politics. In fact, the initial embracing of human security has not resulted in the use of extra-juridical measures. The issues that have come into focus as a result of the application of human security coincide closely with those advanced in the academic literature. The majority of measures that have been advocated are those that we would recognize as development measures, typically those that are associated with the international development agencies and reconstruction.

Much of the early critique of the human security literature focused on its inability to effectively inform policy making. The concept was argued to be too vague, too empty of meaning to provide guidance for bureaucrats and academics.

For policymakers, the challenge is to move beyond all-encompassing exhortations and to focus on specific solutions to specific political issues. This is a difficult task not only because of the broad sweep and definitional elasticity of most formulations of human security but also—and perhaps even more problematically—because the proponents of human security are typically reluctant to prioritize the jumble of goals and principles that make up the concept. (Paris, 2001: 92)

Roland Paris has missed the mark in his criticism in this regard. Human security has proven to be quite malleable to the policy community. In fact, the very characteristics of elasticity and nonspecificity have been assets, allowing policy makers to fit a range of programs within its framework. What it has done is to tell the policy makers where to look (at people inside of the state) and what to look for in broad terms (things that threaten, risk, or impoverish people). In turn what the policy makers take from this is that these things, previously seen as more general "development" or "quality of life" issues, are matters of security. In many regards this was an easy sell; the international community was engaged in numerous peace operations throughout the 1990s and had begun to pay a great deal of attention to the causes of internal conflict and civil strife. As a result, explaining that issues as diverse as organized crime, lack of potable water, and gender violence contributed to insecurity (rather than simply being domestic safety concerns) made sense against a background of attempts to understand why states were falling apart.

For these reasons governments have proven remarkably adept at extracting concrete policies from a commitment to human security. Once policy makers began to pay attention to what endangered the lives of individual people and what eroded their dignity, determining what was a security issue was not a particularly difficult affair. In contrast, deciding what to leave out of this framework was more difficult.

The Canadian government has established a policy framework that focuses on five priority areas for government action:

1. Protection of civilians

2. Peace support operations

3. Conflict prevention

4. Governance and accountability and

5. Public safety

While each of these in turn has subcomponents, they provide a picture of the direction that the government has proceeded in. We can see from this, for example, that Canada has prioritized the establishment of physical security over social development. This is not to say that the latter has been ignored or perceived as irrelevant, just that physical security is seen as essential before development issues can be addressed. The focus of human security initiatives has been on a range of issues that in the past have been linked to the language of economic and social development. Engaging in the promotion of good governance, the attention paid to public safety is not new, but their linkage with the security concerns of governments is.

Other observations of the embracing of human security by governments are based on more personal observations. The term comes up regularly in talks with government representatives, and members of DFAIT, the Department of National Defence (DND) and the Canadian Forces (CF), and the Canadian International Development Agency (CIDA) all use the language of people-centered security. As the concept has entered into common discourse it must necessarily be shaping (as well as being shaped by) broader conceptualizations of international relations and security.

In Canada there is a debate under way about whether human security has run its course (was it relevant in the 1990s but no longer so?), become irrelevant in an era of the Global War on Terror (GWOT), or if it has matured to the point where specific policies no longer need to be justified through its use. These discussions have been exacerbated by a change of government. The human security initiative has been closely linked to the Liberal government, and as such, the terminology has been pushed aside by the current Conservative government under Prime Minister Harper. However, the language has not disappeared completely, nor does the loss of the specific term necessarily indicate an abandonment of the ideals of human security. After all, if human security has served to bring into focus the importance of people-centered policies, even if the terminology itself is not used, if we can continue to identify programs and policy decisions as deriving from it, then arguably it continues to be an influential discourse.

Certainly there have been some clear moves to separate the current government from past policy choices. The current DFAIT Plans and Priorities for 2007–08 makes no reference to the term at all. The most public change has been the cessation of funding to the Canadian Consortium for Human Security (CCHS) in March 2007. This initiative, funded through DFAIT, fostered a body of academic and policy expertise in Canada that directly contributed to Canada's position of leadership in this area. At the time this paper was written (May 2007), it had not yet been announced whether the old Human Security Program would receive any future funding. The end of the program will undoubtedly impact Canada's future leadership role in this area. The second significant change has been in the reduction of the human security program within DFAIT from a broad area of focus of the department to being just one of many global issues.[3] Furthermore, when recent human security projects have received funding they have been much more narrowly focused on specific issue areas (such as on the human security in cities initiative).[4]

---

[3] http://geo.international.gc.ca/cip-pic/about/global_issue-en.asp.

[4] This is not to denigrate the quality of such initiatives. The human security in cities project was an excellent example of the quality of work that can be produced around a narrowly focused topic. The quick studies were able to broaden the issues that are typically discussed within the halls of a ministry of foreign affairs and illustrated how the developed North needs to pay attention to micro-level concerns.

What does signal a turning away from human security is a return of the language of state-centric security, and of a security orientation that places the concerns of Canada at the fore. Canada's "First Strategic Goal" is now "A safer, more secure and prosperous Canada within a strengthened North American partnership." This represents a shift back to prior forms of security calculations where the needs of Canada and Canadians must outweigh the security needs of others. The security of others becomes relevant to us when and where it potentially risks us. However, this last point does not represent a complete rejection of human security discourse (though obviously it is contradictory with its initial intent), but rather is a culmination of the re/securitization of the human security. To put this another way, it represents the co-option of the narrative.

We should not see this as a simple matter of policy prescription, rather the introduction of policies under its banner serves to reproduce and reinforce the human security discourse and validate its application. At the same time its use also fixes the particular categories in meaning, justifying our focus on substate security. However, this does not mean that the adoption and adaptation of human security has happened in a unidirectional manner. At the same time as human security practice has taken hold, the concept itself has slowly been altered and changed.

6. Co-option of Human Security

From the outset some human security advocates have been wary of the manner in which the concept might be integrated into policy making. In particular, those of a critical bent, including myself, have been reticent to see the notion fixed in meaning, and have been skeptical that it could retain its emancipatory value while attached to a state-level institution. Scholars and some policy makers were raising danger flags over the potential risks inherent in the adaptation of human security to policy making. Specifically they were concerned that attaching issues of education, health, and other social development issues to the security needs of states would not benefit the targeted peoples in the long run.

The securitization literature provides a means to address this issue, directing us to look at the ways in which crises, identified as security threats, can result in a significant departure from normal modes of politics. If the strength of human security has been to bring to the global community's attention the needs and insecurity of peoples in the developing world, it has been an effective tool in part because it has politicized these issues. It has led to debates about how international policies have to consider local community needs, and how the security needs of people should not be trumped by the security needs of states. But the securitization thesis warns that the moment that an issue is securitized these debates stop. Accordingly, the adoption and to a lesser degree the adaptation of the concept "human security" was not a securitizing move. It was a challenge of the state-centric assumptions of Ole Wæver's approach.

Even though the concerns of human security and its terminology have taken root in society, as we found on 12 September 2001, it had not supplanted or significantly altered the state-centric security calculation. The commencement of the GWOT demanded that the West fight and defeat terrorism. If ever there was a textbook example of successful securitization, this would be it. Western liberal states mobilized all of their coercive powers in this effort, suspending civil liberties at home and engaging in combat abroad. If the initial adaptation of human security is seen by some as problematic, what was to begin happening after 11 September 2001 could be read as a rejection of human security. This, however, would be a profound misreading of what has been occurring. The issues that came to the fore under the adoption and adaptation of human security have remained through the Global War on Terror. While the term "human security" has perhaps receded in use, we have nevertheless witnessed the more thorough securitization of the concerns raised under its name.

From early on in Canada's role in Afghanistan, first as part of Operation Enduring Freedom (begun on 7 October 2001), and as of August 2003 under NATO control (authorized by UN Resolution 1510.1), the government has paid close attention to the condition of people on the ground. The needs of local communities were seen to be crucial to the completion of the mission and to the winning of the hearts and minds of the Afghani peoples. Nowhere is this clearer than in the creation of small military-led teams with the goal of improving the lives of locals in areas where the NATO armed forces were operating. The initial form of such activities, proposed by the United States in November 2002, were the Joint Regional Teams, which evolved into the Provincial Reconstruction Teams (PRTs). These teams are envisioned as a means of providing security to local communities, of bettering their lives through small quick-impact development projects, and to enable the central government reach beyond Kabul. There are a number of different PRT models in effect, identified with the American, German, and British forms.[5] They differ on the degree to which they integrate noncombat elements into

---

[5] There are three primary PRT models: the American, British, and German. For a discussion of the various models and their respective strengths and weaknesses refer to: Pearson Peacekeeping Centre Backgrounder (http://operationsdepaix.org/en/ci/ci_afghanistan _ background.asp); McNerney, 2005–06; Rubin, 2006.

the teams, as well as on the extent to which they focus on self-protection. The other area where the various models vary is the extent to which the PRTs fall within the Command and Control of the military component of the various states, or fall under the leadership of diplomatic or development Ministries. However, what each emphasizes is that local security-building is essential to the overall combat mission, and that this requires efforts to alleviate local suffering, as well as development needs.

Canada's PRT has been operating since August 2005, and is an extension of the Canadian Battle Group's combat mission. This dictates that the role of the Canadian PRT be designed, primarily, to complement the overall military mission, rather than meeting diplomatic or development priorities. This also means that the members are primarily soldiers who, while some have limited training in Civil Military Cooperation (CIMIC), are not development experts. Consequently it falls under the control of the military, not DFAIT or CIDA.

We cannot examine the emergence of the PRT model in isolation from the human security literature; after all, both are premised on the importance of the safety and well-being of individuals within a zone of conflict for the eventual resolution of the civil strife. This is particularly evident when we look at the issues that the PRTs are addressing. Here we see a clear parallel to the issues advanced by human security policies. The PRTs are interested in providing first and foremost a situation where people are able to live "free from fear," which echoes the Canadian government's human security literature. Next, the PRTs are engaged in a host of programs that are aimed at improving the lives of peoples, such as fighting crime, providing opportunities for education, and promoting gender rights. We also discover that the PRTs and the broader government mission are attempting to strengthen the local security forces. This has included measures such as police training, and the provision of trucks and other equipment to police forces. All of these have been done in a context where the language of human security has taken hold.

The PRTs have become the subject of a great deal of controversy, with a number of International NGOs, including CARE Canada and Save the Children, questioning the appropriateness of a military-led humanitarian program. On the one hand such initiatives are argued to be ineffective, as they are not led by individuals with professional training or extensive experience in humanitarian relief or development. Beyond this, the more profound critique is that the PRTs are contributing to an environment where the humanitarian and development missions are becoming militarized.

However, while the language of human security, and many of its policies, can be traced to the development of the PRTs and the overall mission in Afghanistan, there have been some crucial alterations. The adoption and adaptation has led to the integration of considerations of local level insecurity into the overall strategic calculation. The securitization of the terrorist threat, and its link to Afghanistan, has resulted in the change of human security discourse. What we find is that the issues themselves are now being actively considered by traditional security actors, and that they have begun to see addressing them as a core component of their overall combat mission. But the reason this is being pursued no longer conforms to the intent of the Lysøen Declaration. Consider the following statement by the Canadian government.

Canada is making important diplomatic, defence and development contributions to the stabilization and reconstruction of Afghanistan. Our objectives are threefold: to defend our national interests, ensure Canadian leadership in world affairs, and help Afghanistan rebuild into a free, democratic and peaceful country.[6]

Here we find the defense of Canadian interests taking the forefront; this is at odds with an ethos that places individuals at the heart of ethical considerations.

7. Conclusions

In the months following the attacks on the World Trade Center and the Pentagon a number of scholars, including P. H. Liotta (2002) and Kyle Grayson, cautioned of the dangers of the integration of human security discourse into military policy. "I would argue that the key to effective security policy in the post–11 September world is that human security should be treated as an ethos rather than as an agenda to be slotted into existing security paradigms" (Grayson, 2003: 340). Liotta, for example, argued that state and human security concerns were liable to become blurred, which was one of his main reasons for the advocacy of vulnerability rather than security issues. Even earlier, in 2001, Hugo Slim cautioned that there were dangers inherent in violent humanitarianism, and that such activities were likely done more for self-serving reasons than out of some commitment to a security-first human security platform.

Promoting the humanitarian idea can project credibility and legitimacy. The very fact that all power tends to want to adopt humanitarian discourse indicates the very real strategic significance of the idea and its language. (Slim, 2001: 337)

Hugo Slim's warning was certainly prescient, but not for the reasons he described. Rather it would appear that the discourse of human security, having gained traction within governments, helped inform what states should be

---

[6] geo.international.gc.ca/cip-pic/current_discussions/afghan-en.asp.

R. CHRISTIE

concerned with, and what sorts of activities would be necessary to respond
to them. The extent to which humanitarianism and development have
become militarized (and securitized) is evident in the establishment of the
PRTs under the control of militaries as a part of the overall combat mission.
Canada's experiences show a situation where human security discourse has
been in some ways incorporated into foreign and defense policies. Yet the
manner in which this has been done has not been in keeping with the intent
of human security.

Human security seems to have worked well in a time of stability at home,
but once human security itself becomes securitized, there being a perceived
risk to the state, it actually becomes detrimental. The promotion of the language of human security has served to broaden the conception of security
issues; it has led to the consideration of micro-level concerns and has made a
clear link between development issues and instability. While this has provided
a greater sense of urgency to issues affecting human dignity, it has also had
the unanticipated (at least by some) result of securitizing human security.

How we exit this situation is not clear. The notion of desecuritization
raised by Ole Wæver has been roundly rejected in much of the literature
as impractical and ultimately ineffective. Now that the Pandora's Box of
the militarization of aid has been opened, shutting it is likely to prove difficult. The other option is to continue pushing the ethical component of the
framework, demanding that the lives and dignity of people trump that of
state security, either our own or that of governments that nominally claim
jurisdiction over people elsewhere.

References

Axworthy, L., 2001, Human security and global governance: Putting people first, *Glob. Gov.*
7(1):19–23.
Behnke, A., 2006, No way out: Desecuritization, emancipation and the eternal return of the
political—a reply to Aradau, *J. Int. Rel. Dev.* 9(1):62–69.
Booth, K. (ed.), 2005, *Critical Security Studies and World Politics*, Lynne Rienner, Boulder, CO.
Bowes, R., 2001, Sacrifice and the categorical imperative of human security, *Int. J.* **56**(4):649–664.
Busumtwi-Sam, J., 2002, Development and human security, *Int. J.* **53**(2):253–272.
Commission on Human Security, 2002, Report of the Third Meeting of the Commission on
Human Security, Haga Castle, Stockholm, 9–10 June.
Grayson, K., 2003, Securitization and the boomerang debate: A rejoinder to Liotta and
Smith-Windsor, *Secur. Dialogue* **34**(3):337–343.
Liotta, P. H., 2002, Boomerang effect: The convergence of national and human security,
Secur. Dialogue **33**(4):473–488.
MacLean, G., 2000, Instituting and projecting human security: A Canadian perspective, *Aust. J.
Int. Aff.* **54**(3):269–276.

McNerney, M. J., 2005–06, Stabilization and reconstruction in Afghanistan: Are PRTs a model or a muddle? *Parameters* (Winter): 32–46.

Newman, E., 2001, Human security and constructivism, *Int. Stud. Persp.* 2(3):239–251.

Paris, R., 2001, Human security: Paradigm shift or hot air? *Int. Security* **26**(2):87–102.

Rubin, B. R., 2006, Peace building and state-building in Afghanistan: Constructing sovereignty for whose security? *Third World Q.* **27**(1):175–185.

Slim, H., 2001, Violence and humanitarianism: Moral paradox and the protection of civilians, *Secur. Dialogue* **32**(3):325–339.

Taureck, R., 2006, Securitization theory and securitization studies, *J. Int. Rel. Dev.* **9**(1):53–61.

Tomlinson, B., 2002, Defending humane internationalism: The role of Canadian NGOs in a security-conscious era, *Int. J.* **57**(2):273–282.

Wæver, O., 2004, Aberystwyth, Paris, Copenhagen New Schools in Security Theory and the Origins between Core and Periphery, Paper presented at the International Studies Association (ISA) Annual Conference, Montreal, March.

SECURING HUMANS AND/OR ENVIRONMENT IN THE POST-CONFLICT BALKANS

BILJANA VANKOVSKA[1]* AND TONI MILESKI[2]

[1,2] *Skopje University, P.O. Box 567, Skopje 1000, Macedonia*

Abstract: The growing debate over the significance (and even primacy) of human and environmental (in)security is typical merely for global powers and in more developed countries. The most vulnerable countries are mostly objects and not equal participants in the security debate. This paper discusses theoretical and political tendencies through the prism of the (de)securitization process. It aims to enrich the discussion with a "bottom-up" perspective. Its focus is on the discrepancies between the dominant debate in the area of "high politics" (or even bio-politics) and how these insecurities are viewed from the perspective of the (European) periphery, specifically Macedonia and the post-conflict Balkans.

Keywords: Human security; environmental security; securitization; the Balkans; Macedonia

1. The Link Between Human and Environmental Security: Dialogue Between the Core and the Periphery

At the end of the Cold War, it seemed as if the redefinition of the classical security concept expanded dramatically and proportionally with the "peace dividend" and new hopes invested in the coming era. The global and national actors were allegedly willing to pay attention to numerous social issues that had been neglected or barely mentioned in the previous period dominated by the traditional security concept. Apparently, the State had lost its privileged position of a central referent object to security. But it did not necessarily mean that security lost its significance. On the contrary, the peace

---

* Address correspondence to Biljana Vankovska, Skopje University, Faculty of Philosophy, P.O. Box 567, Skopje 1000, Macedonia; e-mail: bvankovska@mt.net.mk

P.H. Liotta et al. (eds.), Environmental Change and Human Security, 271–297.

dividend faded away swiftly, while the security debate remained as newsworthy as ever. It even got a new academic and policy-oriented boost, especially with the introduction of a range of new referent objects, such as "societal groups" (societal security), "human being" (human security) and "environment" (environmental security). From today's perspective the world is indeed "richer" for numerous security concepts that may explain or justify some development; yet in general it has not helped build a safer and more secure world. Securitization of values does not necessarily mean better protection; rather it gives better ways to mask political and prevention failures and to build a new global agenda by framing more issues as "urgent security priorities." In other words, more security is not always better.

The human security concept is still in the center of a vigorous theoretical debate, which boils down to a conclusion that "too much disagreement exists" and that it would be much more productive to shed more light directly on security threats and strategies for coping with them (Florini and Simmons, 1998). It appears to be a form of emergency politics that focuses on consequences rather than on causes of human insecurity. Acceptance of such a pragmatic approach still meets serious problems on more analytical and strategic policy making. For instance, environmental threats are part of a long list of security threats that endanger individual humans (thus constituting a legitimate part of human security), while at the same time some authors argue that the environment should be a referent object in its own right. There have also been opinions that link environmental security more closely to national security (as the state is seen as a legitimate "uncertainty reducer"), thus defining new military missions in case of emergencies caused by environmental threats (Mathews, 1989; Ullman, 1983).

It is generally accepted that the areas of national, human, and environmental security concepts clearly overlap with each other. One could also claim that the identified threats may be, more or less, the same ones—the only difference refers to the lens through which an issue is seen as a security problem and when, why, and how it is being securitized. A quick overview of some of the discussion related to defining security supports our view:

- The link between human and environmental security can be illustrated by the following definitions: (1) "Human security can no longer be understood in purely military terms. Rather, it must encompass economic development, social justice, *environmental protection*, democratization, disarmament, and respect for human rights and the rule of law" (Annan, 2006). (2) "Threats to human security are varied—political and military, but also social, economic and *environmental*" (Ogata, 1999). (3) "According to both 'critical' and 'human' security approaches, security is about attaining the social, political, *environmental* and economic conditions conducive to a life in freedom and dignity for the individual" (Hammerstad, 2000: 395). (4) "One alternative is to focus on *human security*, recognizing the inter-*linkages of environment and society*, and acknowledging that our perceptions of our environment and the way we interact with our environment are historically, socially, and politically constructed" (Lonergan et al., 2000).

- The link between national and environmental security is usually explained in the context of widening the traditional security agenda. In these views, destruction of the environment is just one additional threat to national security. A practical expression of this logic is the U.S. National Security Strategy, which included environmental threats in 1991. According to some (realist) authors, "the emergence of the concept of environmental security became inevitable once national security became associated with 'quality of life' within a sociopolitical context marked over recent years by the introduction of environmental questions into overall national and international concerns" (Frederick, 1999: 94). Also in the words of Al Gore, environmental deterioration has become an "issue of national security because it threatens not only the quality of life but life itself" (1990: 60).

- The link between national, environmental, and human security is pointed out in the following statement: "Human security in its broadest sense embraces far more than the absence of violent conflict. It encompasses human rights, good governance, access to education and health care, and ensuring that each individual has opportunities and choices to fulfill his or her own potential. ... Freedom from want, freedom from fear and the freedom of the future generations to inherit a healthy natural environment— these are the *interrelated building blocks of human, and therefore national security*" (Commission on Human Security, 2003; emphasis added).

Bearing in mind the vagueness of the definitions, one could paraphrase Harold Lasswell: there are no experts on security! There are only experts on aspects of the problem, or even ones who can master construction of an issue into a security problem. Obviously, many things have changed in the IR theory and practice except that security remains the focal point of analysis. One major difference is in identifying the referent object of security: sometimes it is State, sometimes human beings; and other times it is environment/nature/life. Which of these referent objects is going to get our full attention depends on the will and ability of the securitizing agents. Nevertheless, there appears to be agreement that it is all about *survival* and protection of *core values*, although these are defined and prioritized differently.

The ideal situation would be for the so-called holistic approach to security to include of all referent objects in a simultaneous and harmonious way. However, this is unlikely because there are competing and conflicting claims among securitizing agents and affected groups, as they each define security threats and priorities differently. For instance, some authors discover a philosophical conflict between eco-centric and humanist thinkers regarding the nature of the relationship between humanity and nonhuman nature (Humphrey, 2002). In this line of thinking, there is the arrogant assumption that only we humans are of utmost importance and that nature and other beings exist for our use.

Unfortunately, there are even more arrogant presumptions, such as some human lives have more worth than others, and environmental threats count only when they endanger lives of developed countries' populations. It is not surprising at all that the concept of environmental security gained relevance quite late (i.e., when the possible consequences became more imminent to those who were most responsible for the environmental deterioration, scarcity of recourses, and climate change).

The new security agenda of the twenty-first century is far from being a result of a humane and more responsive international regime. There is no consensus over the possible explanation of its complexity and rich spectrum, but here we shall focus on two interesting academic approaches. In this analysis we shall focus on two popular concepts and use them in the context of a "core-periphery" dialogue framework. The concepts of securitization and of the world risk society may shed some more light on this issue, and particularly why some security issues are seen as more important than others and who is responsible to deal with them, be it on a national or global level.

According to the Copenhagen School, security is both a social construct and a speech act (Wæver, 1995; Buzan et al., 1998). In the view of Ole Wæver, security is not an objective, given necessity, but an intersubjective construction. In other words, almost any issue can be constructed as an issue of security (i.e., an issue of survival). Securitization is nothing but an illocutionary speech act through which securitizing agents may shift an issue from the political to the security realm. By doing so, they acquire and legitimize extraordinary powers, action means, and methods of dealing with the "security issue." Securitizing agents are supposed to be in a position to raise issues and attention (e.g., politicians, ethnic or religious leaders, scholars, environmentalists, or journalists) as well as to acquire support for their appeals, while the audience is made of decision makers including citizens (i.e., the general public). The process is seen as a part of power politics, which may be carried out in all political systems as well as in the international arena.[1] Securitization legitimates action otherwise deemed non-legitimate; it makes morally unacceptable policies morally acceptable, because they are seen as ensuring the existence of something that should survive. From this perspective, it does not matter if the security policy and the action taken are truly necessary, apt, or effective—what matters are the extraordinary powers of the actors and general support for their actions (including past actions as well). Needless to say, securitization is handled differently depending on the political system (democracy vs. non-democracy) and societal and cultural factors (which call for specific language skills and forms).

Most importantly, the concept of securitization offers an explanation as to why and how even in a democratic society the public may be successfully manipulated by securitizing agents (which are, by default, authoritative personalities who have audience and legitimacy). Hence, some authors quite correctly define human security as a politically powerful concept that mobilizes numerous actors and make them focus their efforts on saving human lives. Human security is an effective rallying cry with unifying power to get together various groups and actors and to turn the world's attention to the most pressing issues (Paris, 2001). The vagueness of the concept may be seen as its strong point in terms of unifying endeavor; however, human security may mean almost anything—thus meaning nothing (i.e., phenomenon of "conceptual stretching"). Consequently, anything can be undertaken for the sake of human security, even when the responses to human insecurity generate new security dilemmas and other vicious circles. For instance, military humanitarianism is called for in order to save some human lives while at the same time endangering other innocent human lives, harming the environment by use of depleted uranium (DU) and cluster bombs, and finally strengthening the spiral of violence in the original conflict (e.g., in the Kosovo war). The very emergence of human insecurity may be a consequence of the failure of "normal politics" both at national and international levels. This failure brings in the securitization of human lives, but at the end of the day the only true remedy is desecuritization—moving the security issue back into the realm of "normal politics."

---

[1] As typical examples of current securitization, one may point out "humanitarian intervention" and the "war against terrorism." The former was used during the NATO campaign against the former Yugoslavia in 1999 when the action was justified by the claim that the military action was necessary to save human lives in the province of Kosovo. Securitization of terrorist threats has a dual effect: internally, there is a visible tendency of reduction of liberal freedoms for the sake of homeland security, and internationally, getting public support for the war in Afghanistan and Iraq.

When it comes to the environment as a referent object of security (environmental security) it is equally "elastic" and ambiguous. It also can be downplayed or strongly highlighted, depending on the situation. According to this theoretical approach, environment and nature as understood by humans is a social construct; thus it is easy to portray it in terms of politicization and/or securitization. Lonergan argues that our perceptions of the environment and the way we use the environment are historically, socially, and politically constructed (2000). Consequently, environmental security will mean different things to different populations at different times (Matthew, 1999).

When supported by arguments of biopower, it appears that the purpose of power is to manage humans at the population level to ensure some sort of balance with their resources. In this context, the environment is defined as the amount of physical resources upon which a now manageable population depends (Rutherford, 1999). Warner argues that the securitization of the environment helps legitimize state intervention in environmental affairs, and thus legitimizes the state. This accusation casts the growing "environmental security" discourse as a tool of the state to justify intervention into environmental and political affairs and thus exert political influence (Warner, 2000). The same may apply to human security as well: security discourse helps the so-called international community justify its interventionist agenda and hide its geostrategic interests. One could point out that security indeed becomes a political desideratum rather than an analytical category. It provides narratives of danger as the stimulus to collective action (Dalby, 2002).

Beck's concept of risk society undoubtedly attracts scholars' attention and raises pragmatic concerns. At the beginning of the new millennium it seems that Beck's words ring even more accurately: "Dangers are being produced by industry, externalised by economics, individualized by the legal system, legitimised by the sciences, and made to appear harmless by politics" (Beck, 1998: 16). According to this view, nowadays issues of wealth distribution as the fundamental basis of political struggle are replaced by risks to human health and well-being. So the central point is how such risks may be minimized and more justly distributed. He "blames" science and technology not because they create these risks but rather because their power and authority are unchecked. Decision makers are getting more and more incompetent in problem solving and have no other choice but to trust others' advice on highly technical matters.

In liberal societies the problem of responsibility for the state of affairs such as resource exploitation, pollution, and technical innovations should be seen through the prism of growing privatization and commercialization (including some of the public services). Thus, when accidents and problems arise, government officials have to legitimize and take responsibility for decisions that are often taken elsewhere or taken under the expert advice. But in reverse, many issues become "political" especially when they involve risks for the population, environment, and society. At the end of the day, there is a vicious circle or unavoidable synergy between political and nonpolitical actors (e.g., scientists, businessmen), who meddle in one another's realms. However, each of these actors has different interests and demonstrates no willingness to take full responsibility for the results. This is a situation that Beck calls "a modern state of organized irresponsibility," which he compares to the "nobody's rule" that Hannah Arendt names the most tyrannical of all forms of power because under it nobody can be held responsible (Beck, 1998). Obviously, Beck's risk society is to be found mostly in developed and industrialized societies, but even there the science and technology should not bear the major burden for the creation of the growing risks and uncertainties. The departure to the "industrial society" has not been completed yet, while the capitalist political economy should also be taken into account, as free market economies also lead to the proliferation of risks, especially in the environmental sphere (Benton, 1997).

These two concepts applied to environmental threats are clearly Western oriented. Securitization has been greatly used in non-Western and underdeveloped societies but almost never in relation to the environmental sector. Poor and failing states even when faced with serious environment-related conflicts rarely identify them as such. Socioeconomic divisions, lack of social justice, and unequal access to national wealth and resources are more easily translated into ethnic intolerance and grievances rather than something that originates in the bad management or scarcity of the environmental sphere. At a glance, it looks as if the "wars of the third kind" occur because belligerent and poor people of different ethnic and religious origins simply hate each other. The turmoil that takes place in these societies is a good platform for the concerned "international community," which is more ready to "securitize" vulnerable groups of individuals (human security) than to raise justified alarm for the economic and environmental deterioration that is going to be paid precisely by those who are "referent objects" of their humanitarian interventions and aid assistance.

At one time it seemed as if the national security concept had lost its primacy, particularly because of the dominant world's picture of the "West and the Rest." At least, the picture was such until 11 September 2001. The Western democracies apparently got together in a stable and harmonious security community zone of peace, while the "Rest" was less concerned with national security but rather with its own internal dismay. The statistics showed a decrease of the number of interstate wars and an increase of intrastate conflicts (Wallensteen and Sollenberg, 1999; Scherrer, 2002; Wimmer, 2004). The developed part of the world has, consequently, become more and more concerned with failed states and the spread of insecurity from the Rest toward the West, while the logical remedy was seen in finding solutions "in situ." Dealing with problems "on the spot," however, has taken some legally and ethically dubious forms. For the sake of human rights and human security, various concepts got prominence, such as "humanitarian intervention" and "responsibility to protect," or the softer variant of "developmental security."

At a glance, these concepts look amazingly altruistic, both from the point of view of some states' foreign policies and from the global perspective. They were supposed to be proof of the allegedly growing humanism within the international system. However, some authors rightly point out the limited scope of these concepts or even merely neologisms. What gives some issues a "security aura" is the urgency, gravity, and intensity of the threats so that they refer to survival of the referent object(s). For instance, human security is contained in two main aspects. First, it does not cover all necessary and profound aspects of human living (Alkire, 2003). Second, it refers only to *some* human beings who are seen as more endangered vis-à-vis some other human beings. The task of prioritizing among human beings is basically a value judgment, albeit a difficult one. It always depends on so-called securitizing agents and rarely on the individuals concerned. The most prominent agents, however, are international agencies and NGOs and rarely domestic public institutions. In other words, human security is most often to be found as one's foreign and security policy or international donors' agenda rather than as an inherent part of one's domestic policy agenda (Vankovska, 2007). The "vital core" values and basic human needs are usually defined by others, and quite often in a situation when a respective national state appears to be too weak to deliver or even as a primary intimidator and tyrant of its own citizens.

One could even talk about self-appointed securitizing agents who de facto possess extraordinary powers that go far beyond "ordinary (international) politics." Despite the efforts to portray their actions as apolitical, unbiased, and a result of their "enlightened self-interest," down-to-earth strategic and political interests of the human security providers have been too obvious. One could mention double standards of the so-called international community in the cases of the former Yugoslavia and Rwanda, both in terms of the decisions (not) to intervene and in terms of the level of developmental assistance. Some authors rightly stress that human security could be seen as the essence of bio-politics, that is, the core of a new form of global governance that differentiates between "homeland" and "borderland" populations and between so-called developmental and humanitarian life (Duffield, 2004; Duffield and Waddell, 2006). Politics is taken out of the conflict by portraying the intervention, military or otherwise, of Western powers as being above politics (Chandler, 2006). However, there is growing strategic unevenness of aid dispensation as well as variations in levels of response to humanitarian crises, depending on powerful actors' political will (Macrae et al., 2002).

In contrast to the idea that human security represents a fusion between development and security, today's situation is defined in "either/or" terms. Additional problems arise when the security needs of "homeland populations" tend to outweigh the developmental needs of "borderland populations." In the artificial clash between the two versions of human security, "freedom from fear" is winning. The defining characteristic of "humanitarian life" is extreme vulnerability; thus, freedom from fear and violence means protection of those who are most vulnerable (Suhrke, 1999). The Western governments assume that once people are liberated from fear (physical vulnerability) they will focus more on "freedom from want" and will devote their societal efforts to poverty eradication and development. Their "emergency argument" reads: as a result of physical force and oppression lives may be put at risk, while "freedom from want" is less severe and immediate. Hence the threshold of an acceptable/tolerable death-toll from direct and/or structural violence (to use Johan Galtung's terms) is never decided by those who are dying but by those who have the power to intervene and save lives. The bio-politics of the rich inevitably leads to politicization of aid. By accepting the official version of apolitical humanitarian aid, and despite their best intentions, humanitarians "maintain a secret solidarity with the very powers they ought to fight" and join the game in which humanitarianism reproduces the isolation of bare life (Agamben, 1998).

Global prejudices usually depict a black-and-white image of allegedly primitive and warmongering people who hate each other and particularly hate the West. The truth is, however, more complex. Global, especially financial, institutions remain blind and deaf to global injustices and the growing gap between the developed North and the poor South. They actually embody structural violence, that is, they bear some amount of responsibility for the deaths of millions without a single bullet being fired. Media reports indicate that vulnerable people tend to become even more vulnerable and can even be "sentenced to death" by the business interests of multinational corporations and the national interests of powerful states.

The most recent debates over environmental threats—more precisely, climate change—resulted in a new notion of aggression in the form of an explicit allegation of the South to the North. As Ugandan President Museveni put it at the African Union summit held in February 2007, "we have a message here to tell these countries, that you are causing aggression to us by causing global warming" (Revkin, 2007). UN experts gathered at the Intergovernmental Panel on Climate Change (IPCC) agree on the fact that the first and most affected victims of the environmentally deteriorated world will be precisely the poorest populations, that is, those who least contribute to the unfortunate situation (IPCC, 2007). According to Catherine Pearce, Friends of the Earth International's climate campaigner, "climate change is no longer just an environmental issue; it is a looming humanitarian catastrophe" (McCarthy and Castle, 2007). Many believe that the situation may be mitigated relatively easily only if there is good (governments') political will. But they usually forget one small detail: "Like the sinking of the Titanic, catastrophes are not democratic" (Henry I. Miller, a fellow with the Hoover Institution).

2. A Look at Human Insecurity in the Post-Conflict Balkans

While international academic and political circles were debating the new security agenda, the states and nations of the former Yugoslavia were "busy" providing new empirical material for the debate. Actually, many of the new concepts were supported or even formulated thanks to developments in this and, of course, some other parts of the world. To mention just a few: the theory of "new wars" was promoted; the Copenhagen School could practically test its sectoral understanding of security precisely on the example of Yugoslavia's dissolution; more recently, David Chandler examined the EU state-building strategy and ownership issue. On the side of the international-policy community, there were also many innovations, such as "preventive deployment" in Macedonia; "humanitarian intervention" in Serbia; and "democracy promotion through imposition of power-sharing models" in Bosnia, Macedonia, and Kosovo.

The human security concept was most often applied in the dimension of providing "freedom from fear" and "responsibility to protect," although some international agencies have shyly and modestly been engaged in the "freedom from want" strategy.

Actually, the bloody conflicts in the Balkan region served well the creation of moral and political ground for securitization and legitimization of human security in its most radical variant—through military and police actions of the international actors, let alone the dubious tool of economic sanctions that deepened human insecurity and environmental neglect, while the responsible elites (i.e., the ones that should have been "punished") were unharmed. As soon as the mass violence was contained (although not totally eliminated), the internationals' interest decreased and fatigue set in. The post-conflict societies are now expected to play according to the rules of the liberal democracy game—or more precisely, to adapt to the capitalist political economy.

There are fewer grants and aid, but the road to loans and credit is being opened for the new governments. During the war turmoil the "freedom from fear" was celebrated, especially when huge masses of people appeared to be vulnerable. It was a good case for securitization and interventionism. Now, when there is wide-spread agreement that the region has been stabilized and democratic governments have been elected and EU/NATO strategies advocated, the interest in human security is being very limited, especially in terms of "freedom from want." Interestingly, even one of the newest security initiatives in the EU—the so-called Human Security Doctrine for Europe—merely invests its hopes in the formation of Human Security Rapid Forces even when referring to the Western Balkans (Vankovska, 2007). In other words, there still is a militarized approach to human security even though most of the respective countries are on their way to acquiring EU membership.

The post-conflict Balkan landscape displays an ambiguous picture. At first sight, it seems that since 2001 the region has entered a more peaceful stage of stabilization and reconstruction. New state entities have been born and new borders drawn; elections have been held regularly, and, at least nominally, there are democratically elected governments with the main strategic goal to acquire NATO/EU membership. International actors may have good reasons to be satisfied with the achieved level of peace and stability; but a reality check shows a somewhat different picture. With few exceptions, the new state entities are territorially determined, so that the responsibility for human rights protection and human (in)security is supposed to rest with the national governments.

The relationship between citizens and the government is not straightforward. There is the so-called ownership issue. According to Kofi Annan, "national ownership is the core principle of peacebuilding, and the restoration of national capacity to build peace must therefore be at the heart of the international efforts" (Annan, 2006). Yet there is a paradox: the longer the international missions last, the more likely the dependency syndrome is. On the other hand, worsening global security elsewhere (Iraq, Afghanistan) calls for more troops and more interventions. The time constraint is evident in the tension between the desire to withdraw international military troops as soon as possible in order to deploy them elsewhere and the desire to leave in place stable national structures capable of providing security and upholding law and order. This means strengthening the police, the army, the judiciary— i.e., the institutions that had been involved in repression and violence. Instead of establishing democratic and peaceful structures, the internationals usually rely on security sector reform that is resource intensive, and time and money is something that fragile societies do not have at hand. Security sector reform accompanied by the state's strategic goal named NATO membership means quicker, more superficial, and more expensive reforms of these sectors at the expense of social needs (e.g., health care, education, rehabilitation).

The Balkan region indeed enjoys a negative peace, which is the absence of direct violence; yet there are few signs of a burgeoning positive peace, which includes the presence of social justice and development. The sources of insecurity differ slightly from state to state and they are not only to be ascribed to authoritarian regimes, inter-ethnic clashes, and incompatibility of cultures. The sense of insecurity more intensely derives from nonconventional threats to security like organized crime, terrorism, drug trafficking and abuse, environmental degradation and pollution, and social and economic insecurity, as well as a chronic inability to handle intra-societal conflicts in a nonviolent way. Regardless of the methods used, numerous surveys are almost unanimous in depicting human insecurity.[2] Very few, however, point out that the imposed constitutional arrangements are hardly affordable for impoverished societies. The weaknesses of power-sharing models are evident in societies with no mature ethnic elites, where politics is carried out through blackmail and "partitocrazia." Politics becomes a nontransparent and corrupt sphere, and it is hard to tell where the state ends and where the party rule begins (Karasimeonov, 2005).

The governing elites in the so-called Western Balkans[3] have never cared much about the concept of human security. This is understandable because for so long they were a theatre for the concept's application. Not even developed countries or international actors view the human security concept through a lens of internal policy. Thus, as in other parts of the world, the essence of human security is far from being an inherent part of countries' national policies. Furthermore, in the collectivistic ideology where the nation-state is worth more than anything including especially human life and well-being, which are expected to be sacrificed for the sake of "the state of our own," the human security idea and priority of individual citizens resemble mission: impossible.

Accepting this as a paradigm for action, as some regional experts suggest, will require changing attitudes toward the state (Krastev et al., 1999). Accordingly, the state has to be seen not as an end in itself but as a provider of human security, which is a somewhat strange proposition. If the basic postulate of the human security concept is that human beings matter more than states, then logically it would be a very long and painful process to persuade these societies to embrace such an approach. For years the local elites

---

[2] In this context, one can mention IDEA surveys, the UN Development Programme's Human Development and Early Warning Reports, and World Bank analyses and reports.

[3] There is no geographic or historic justification of the notion of the so-called Western Balkans. It is purely political term invented in Brussels. It is supposed to be a new official title for the countries of the former Yugoslavia (plus Albania, minus Slovenia), but the content is that of Europe's grey zone of instability.

were fighting bloody wars because of collectivist ideologies and for the sake of state-building, causing everything else to be sacrificed.[4] An independent nation-state was pictured as worth individual sacrifice, while the newly established state was supposed to take care of its "own people" (ethnos) against the "Other" (everyone who does not belong to "us"). The state-building policy was based on ethno-nationalism, which had power to determine who is "ours" versus "theirs," that is, to decide on life and death of the Other (who, surely, was deprived of all characteristics of a fellow human being). This gruesome and, unfortunately, true description is something that hangs over the region's image in the West. However, one can pose the equation the other way around: how much different is the biopower of the Western states from the Balkan overt and unpolished ethno-nationalism and care for the human lives in a very selective manner? It often seems that the difference is only in a way politics is justified before the audience.

Total GDP in the region is smaller than that of Greece alone. The comparisons with the EU average are much worse (Gligorov, 2002). An increase in poverty and the extinction of the middle class (which is conditio sine qua non for democratic stability) in the Balkans were caused by war, destruction, a large decrease in GDP, and economic deprivation during the 1990s, as well as by the transition process (Matković, 2005). Truly, the data on poverty and other social indicators are not very reliable or complete in the whole region, mostly because of the administrative chaos, politically motivated statistical manipulations, and the existence of a grey and illegal economy that stays secretive. However, experts agree that large portions of the population live below the poverty line (for instance, in Bosnia, 25 percent; Macedonia, 30 percent; and in Kosovo, 50 percent). Many people are extremely poor, meaning not even basic food needs can be met. The very high rates of unemployment testify to the abysmal employment record (Gligorov et al., 2003). The countries in the region also "compete" with each other when it comes to organized crime and corruption.

As a rule, the political establishments and citizens live in parallel worlds, each one preoccupied with their own perceptions and priorities. In contrast to the rather euphoric political claims about the countries' readiness to join NATO or the EU, the citizens are caught in a vicious circle of insecurity that originates mainly in social and economic spheres, which creates a rather schizophrenic situation. In usual circumstances, the perception of the national problem is based on the perception of the problem the individual

---

[4] Estimates of casualties are not exact, but it is believed that 250,000 people were killed and three million more uprooted, either displaced within their own countries or driven away across borders as refugees.

is facing. Unemployment, poverty, and corruption are almost unanimously seen as the most pressing problems everywhere in the region. Environmental issues are not even noted as present.

Being unable to deliver any substantial change in the quality of life, the national elites and policies are either populist and nationalistic, floating on the high expectations of the exhausted and aged voters, or try to "transcend" reality by offering promises about incoming entrance into the "Promised Land" of NATO and EU. Thus the assumption that the issues that people think are important will figure prominently on the political agenda often proves wrong. Quite often, the policy agenda in the region has very little or nothing at all to do with the issues that people care about the most. The policies that the governments in the region claim to implement are mostly concerned with NATO/EU integration, economic stability, and systemic reforms, of which the latter is mostly a cover for wasteful redistribution and for corruption. Obviously, there is no room for "green policies" or "green political forces," as they would not prove useful in the political competition.

The policy framework of the international community has been based on three main steps: (1) an introduction of a constitutional *provisorium* (which usually lacks both legitimacy and functionality, as the Bosnian case proves very well); (2) adoption of the politics of ambiguity; and (3) acceptance of aid and other financial incentives aimed at building more stable political and economic structures through "carrot and stick" methods (Gligorov, 2006). In sum, there has been a benevolent international dictatorship that believed that the new structures and free market's "invisible hand" would certainly lead toward the emergence of democracy and a developed economy. Yet there is no such thing as a "free lunch"; policy objectives and funds coming from outside have been conditional on the adoption of certain institutional arrangements and policies (Gligorov, 2006).

As soon as the international actors undertake an action in a failing state, the main concern is to shift responsibility and accountability onto the recipient state itself. It is usually just repackaging of external coercion in the warm and fuzzy language of empowerment, partnership, and capacity-building. The result is a vicious circle for the states involved: the more intervention, the more the target state is held to be responsible and accountable for the consequences of those practices (Chandler, 2006). The basic contradictions of the international guidance lie in the following:

- Constitutional provisorium does not work as the "international founding fathers" envisioned. Given the absence of some basic preconditions, the expensive political arrangements can hardly function in a democratic manner. Constitutionalization of the power-sharing models so far has proved deficient, which has led toward further weakening of the state. In the words of a distinguished expert in ex-Yugoslavia, Susan Woodward, all international arrangements in the former Yugoslavia are "friendly to the radical ethnic elites," that is, to those who were responsible for starting the wars in the first place (Woodward, 2006). According to Robert Badinter, "thinking within ethnic borders cannot be qualitative and an effective solution. It paralyzes the administration and initiates other problems with a package of advantages and privileges, instead of creating common people and one joint nation. This over-ethnical approach risks to worsen what is already bad in the country" (quoted in Siljanovska-Davkova, 2004: 203). Furthermore, consensus democracy ("community of communities") is expensive due to the replication of specific institutions on various levels and the need for new unproductive (administrative) employment.

- Liberal economic advice and mentoring of the international financial institutions appeared to be controversial medicine for the burning human security problems and have been consequently keeping the potential for internal conflict high. Most of the regional economic experts as well as the crude statistics agree that international medicines put much more emphasis on stabilization than on prompting development and economic growth. On the other hand, the need to speed up economic and structural reforms in order to catch up with the developed world economy accents painful social reforms.[5] The price is usually paid by the most vulnerable groups while less pressing problems are being pushed aside. Inadequate and slow economic reforms resulted in an endless and unsuccessful transition from socialism to capitalism, which additionally contributed to passive acceptance of the fact that "we were trapped in a cage of problems specific to the capitalist periphery" (Nikolovska, 2004: 13). However,

---

[5] The dilemma of catching up with global economic race and internal political and social problems was nothing new for the former Yugoslavia; in fact, was noted by some authors as one of the main causes for its turmoil: "Burdened by a territorial, ethnically-defined 'division of labor' established following the violence of World War II, by the 1980s, Yugoslavia faced the choice of economic and political reform or peripheralization. The competition for Western markets between the Yugo and the Hyundai symbolized the dilemma and the dead end facing the country: What export opportunities were there for a cheap, shoddily-made car, when a cheap, well-made one was also on the market? Even at the time, as late-night TV talk show jokes suggested, the answer to the dilemma was not so obscure. In retrospect, it is crystal clear: economic reform. But attempts to reform the economy of Yugoslavia, to make it better able to compete in world markets, foundered again and again on the shoals of the republican distribution of resources, which had been defined in terms of the 'nations' of Yugoslavia (Slovene, Serb, Croat, Bosnian, Montenegrin, Macedonian). In the end, to push an unfortunate metaphor farther, the crew fell to fighting amongst themselves and the ship of state went straight to the bottom." (Lipschutz and Crawford, 1996)

newly established states have been almost immediately thrown into an international economic competitiveness bind while heavily dependent on international aid and loans. Economically speaking, there has been a paradoxical macroeconomic framework that rests on pillars of simultaneous liberalization and stabilization. According to the new Orthodoxy praised by the international centers such as the International Monetary Fund, the World Bank, and the EU "shock therapy" of market liberalization was supposed to lead the way toward quick development and democratization. States have yielded to the capital, while its government plays the role of a procurer.

The new democratically elected governments, heavily dependent on their international mentors, are more ready to comply with the external demands than to listen to their citizens' demands. Having been unable to deliver and satisfy citizens' needs, the governments lack internal legitimacy. Therefore, the external legitimacy in the form of international support is of utmost importance. A Bulgarian scholar put it nicely: "Our Balkan governments make love to their peoples but are loyal to the international community." Instead of focusing on the internal societal problems, they uncritically undertake costly reform endeavors such as sending troops to Iraq and intensifying defense reforms, which are justified by the strategic national goals, namely, NATO and EU membership.

The local elites' logic is as follows: Let us first invest into defense reforms (mostly military professionalization and modernization), then we should join international peace missions in order to prove our military capabilities; once we are admitted into NATO the country's image will be radically improved and foreign investors will rush to launch economic projects into the country. And *then* we shall be able to meet human security priorities such as job creation, better social and health care, and education. The democratic dilemma "more weapons or more butter" has been resolved and here it reads: "let us invest first in more weapons in order to eventually get more butter." During his visit to Macedonia in May 2006, the NATO Secretary General publicly commended the government for the decision to increase the military budget allocation.[6]

---

[6] In an interview for a Macedonian daily, on the eve of his visit to Skopje, Jaap de Hoop Scheffer was asked how pleased he was with the military reforms in Macedonia. He answered: "Much is achieved indeed. Macedonia is now a security provider, through your participation in ISAF and the Belgium contingent 'Althea' in Bosnia. That's all a result of serious military reforms. I see positive changes in the defence expenditures too. The percent of the GDP allocated for defence is much higher than the one in many of our member-states. I wish they followed the Macedonian example where 2.3% of GDP is allocated for the military sphere." (*Dnevnik*, 12 May 2006, www.dnevnik.com.mk).

3. A Look at the Environmental Security in the Balkans

The most neglected part of the academic and policy debate over Balkan insecurities is in the environmental sector, whether environmental security is seen as a crucial part of the human security or the national security concept. The explanation for this state of affairs may follow a quite cynical line of argumentation: ex-YU conflicts and the long post-conflict recovery apparently had a favorable impact on environmental security.[7] Truly, the industrial production and transportation throughout the region declined dramatically during the war years, while the means and results of mass killings were quite often ecologically clean.[8] Even in the Macedonian "oasis of peace"[9] the factories built during the socialist era closed down, while agriculture, especially early vegetable and fruit production, could have competed for ecological excellence had there been an international market for it.

The direct negative ecological effects of military actions are connected to the destruction of chemical and other facilities, use of land mines, DU and cluster bombs, and the refugee crisis during NATO air campaigns in the region of Bosnia, Serbia, and Kosovo. But, quite interestingly, these issues have never been truly raised because to do so is seen as unpopular and politically incorrect now when the affected countries want to join NATO and EU.[10] Thus, the whole region somehow missed the global debate over

---

[7] Interestingly, even UNEP has come to such a conclusion: "Economic decline and the UN sanctions against the FRY have in general led to reduced pollution of air and water" (UNEP and UNCHS, 1999).

[8] "It's surprising that war had very positive ecological effects. In the industrial town of Zenica in Bosnia the air has never been as clear as nowadays—its gigantic metal industry had to cease its production because of the war; river Neretva in Herzegovina for decades has not been so clear as now since the aluminum production facility in Mostar has been destroyed. And it is well known that the corpses on the river bottom are biologically degradable. The same conclusion applies to the rest of the continent, because the closed roads through the former Yugoslav territory enforced the use of ships as main transport means, which ecologically disburdened whole parts of Austria." (Puhovski, 2001).

[9] Macedonia was the only republic of the former Yugoslavia that gained independence in a peaceful way in 1991. For a decade it had been known as an "oasis of peace." Nevertheless in 2001 an internal conflict occurred.

[10] For instance, the remnants of cluster bombs used during the 1999 NATO intervention are the cause of death of civilians in Serbia on a daily basis. According to Norwegian experts, the cleansing operation would cost over 20 million EUR, while the Serbian government has allocated only half a million EUR for the operation. One of the authors of the documentary film on cluster bombs "Yellow Killers" stresses the responsibility of the Serbian state that does not care for the victims. While blaming the state's unwillingness to provide at least prosthetics for the victims, the local experts do not even try to raise the issue of an international ban of use of cluster bombs. (See: Preventing "Yellow Killers," *B92-News,* 17 May 2007, http://www.b92.net/info/vesti/index.php?yyyy=2007&mm=05&dd =19&nav_category=12&nav_id=247335, accessed 19 May 2007).

the significance of environmental protection because the focus was on some other referent objects of security. Now as it awakens from its conflict-related nightmare, the region tries to catch up with everything it missed in the previous years. The environment is hardly relevant because of its own merit but mostly because it is something that the international community and most especially the EU, care about. For instance, for most of these post-conflict and developing countries climate change and the issue of their geographical vulnerability to extreme climate events still appear marginal to mainstream development imperatives such as food security, poverty relief, and energy access. The framework of "burden sharing" in which climate policy has been framed has often been seen as an undeserved obligation or even as a non-issue because the countries are totally powerless to contribute in any way to the global powers' policy.[11]

Today's assessments of the state of the environment clearly display a complex picture of the Balkan region. The problems are immense and originate from different periods and causes; thus the remedy is expected to be long-lasting, expensive, and complex. Experts agree that today's environmental degradation is not a result only of war-related events. Yet it is almost impossible to determine the state of the environment prior to the wars/conflicts. Socialist regimes were not totally ignorant of the significance of the environment and had enacted a certain amount of environmental legislation. The problem was in the lack not of a normative framework but in implementing the law, the lack of transparency in environmental affairs and, in general, the lack of data for the related issues. The gain of independence by the former Yugoslav republics did not abolish these problems but added new war-related problems.

Even though Macedonia was an exception to the bloody pattern, at least for ten years it was still dealing with burning human and environmental problems. For instance, the town of Veles has always been an overt and blatant public health catastrophe and none of the ruling governments implemented appropriate measures to alleviate the catastrophic situation. A smelter for lead and zinc built barely 100 feet away from the nearest residence buildings, in defiance of expert advice, and the lack of financial means has brought horror and suffering into the lives of the town's 60,000 inhabitants. According to

---

[11] Such statements often come from many MPs or other officials, whenever NGO activists, experts, or intellectuals raise the issue publicly. It is interesting that these very personalities strongly believe that Macedonia's contribution to the global war on terrorism (in Iraq, for instance) is huge and very significant, while at the same time they claim Macedonia's powerlessness in regard to climate and other environmental dangers. The explanation is simple: Macedonia's military participation is minor but brings positive points with the United States, which is considered a crucial actor to bring the country into NATO.

a local pediatrician, Veles offers scenes comparable to those from horror movies: "Babies are being born with entire organs missing. The deformities are frightening" (Jovanov, 2003). The World Health Organization added the town to their list of critically dangerous places in 2001 and the situation has not improved since then. Actually, there are plans to reopen the smelter facility by a foreign investor who would allegedly provide adequate filters and employ some of the old workers. The dilemma is archetypal: Veles needs economic growth and employment and the smelter was one of the main employers for more than thirty years, but the environment and human health conditions call for the opposite solution (i.e., keeping the smelter closed). Similar problems exist in the neighboring region, Tikves, where the FENI (ferronickel) industry is endangering the environment and human health for the sake of the large profit that the owners get from the high prices of nickel in the world markets (Danov, 2007).

In the rest of the region, the violent conflicts in the 1990s caused immense consequences for humans, the environment, and flora and fauna. There are data on migration not only of masses of people but also of animals. There are many peculiar stories and anecdotes from this critical period. For instance, Sarajevo's war herbicide had also strange effects on the environment: during the siege of the city the inhabitants had to cut down all available trees during the winter months. A local cartoonist published a drawing of a frustrated would-be suicide standing with a noose around his neck in a field of stumps. More serious problems are related to the intentional targeting of industrial and chemical plants by warring parties in Croatia and Bosnia, but also during the 1999 NATO campaign (Pancevo was the most alarming case),[12] generation of waste and other problems created by refugees,[13] and the massive number of landmines and cluster bombs. During

---

[12] The Serbian town of Pancevo featured a petrochemical plant, a fertilizer plant, a vinyl chloride monomer plant, and an oil refinery. According to UNEP's Balkan Task Force Report of October 1999, Pancevo was bombed at least nine times during the NATO campaign.

[13] For instance, during the refugee crisis in 1999 Macedonia hosted over 360,000 Kosovo refugees, which equals an increase of its total population of 14.77 percent (for comparison, imagine that the United States would have to accept over 37 million people overnight). At the very climax of the situation, the UNHCR advised other countries not to host the refugees because in Macedonia they would be closer to their homes, which dramatically worsened the situation both for the refugees and for the host country. Five refugee camps were deployed in the first and second protected zones of the main natural water resource basin Rasce, 15 km away from the capital, Skopje. It was a serious breach of the Macedonian ecological legislation, but the security and political situation demanded the disregard of these environmental criteria for the sake of "human security." Actually, everybody was seriously endangered.

the 2001 conflict, Macedonia offered one more peculiar example in which a natural resource (water) was used as a "weapon" against the enemy. The Albanian rebels seized two dams (Lipkovo and Glaznja) and thus caused a humanitarian crisis for over 130,000 inhabitants of the Kumanovo region for more than thirty days.[14]

The post-conflict period imposed a series of problems of different time, scope, and significance. However, it seems that most of the Balkan countries are mostly concerned with meeting EU standards in the environmental sphere, which are a part of the European *acquis*. But in order to do so they have to first politicize the environmental issues. Currently environmental affairs are on the bottom of the political agenda, which can be illustrated by the fact that during governments' coalition building the domain of environmental protection is the least attractive for the political elites. This ministry is usually given to the minor parties, while the minister can enjoy his mandate in a quite comfortable shadow. The political life is usually overburdened by scandal on a daily basis, so few clearly see if there are serious or scandalous developments in this sector. Even worse, people turn a blind eye to environmental degradation because it is seen as the "normal state of affairs." Many issues can be successfully securitized in the Balkans, but environment is not one of them. The state officials do not show particular interest in raising public awareness about environment-related problems because they perceive them as a nuisance and a sort of luxury.

Within a usually extremely expansive administration, which has to be severely cut down in accordance to IMF and WB requirements, the so-called environmental administration is quite weak. Furthermore, there is total chaos in terms of allocation and overlapping competences of various ministries as well as a lack of interministerial coordination and cooperation. The process of institution building is ongoing and often ends up with strange solutions, mergers, or separations. One of the most interesting examples comes from Croatia, where the delegation of nature preservation powers lies with the Ministry of Culture (Juelich, 2005). According to the official evaluations, the Macedonian case stands the best, at least in terms of legislative and administrative solutions in the field of environmental protection. However, there is a chronic lack of funds for implementation of the environmental policy. In countries with a so-called culture of impunity, breaches of environmental norms do not even ring the public alarm. The normative framework is more or less in place because it calls for nothing but a bit of a good political will.

---

[14]Today this incident represents one of the four cases of possible war crimes in which Albanian rebels were directly involved and which are expected to be processed in the Macedonian court system.

In today's world it is as important to have democratic legitimacy as being "green," even if it is not genuine.

The Montenegrin case is very illustrative in this regard. While still a federal unit of Federal Republic of Yugoslavia, Montenegro had self-declared the first ecological state in the world even prior to its getting international recognition and state independence.[15] The beginning was marked by an extremely ambitious approach toward the implementation of foreseen objectives, but quite soon implementation slowed because of war in which Montenegrin forces took part as aggressors in Croatia, particularly in the Dubrovnik front; economic crisis; and international community sanctions. As some observers conclude, natural resources remained relatively preserved because of insufficient economic development, while the problems related to a few major industrial polluters could not possibly be tackled because of the lack of financial resources, particularly for those measures relating to introduction of "clean technologies."[16] Additional decrease of environmental quality took place during the 1999 NATO air campaign, especially due to strikes on the seacoast by DU ammunition.[17] In any case, despite the declared political will and "unique approach" toward environmental protection, Montenegro faces similar problems to any other country in the region. Most importantly, identification of the environment as a value to be protected has very much to do with national identity and state-building rather with extraordinary awareness of the need for environmental protection.

Most of the environmental projects are financed by international donors. Or more exactly, the environmental policy agenda has been merely set by the internationals, be they individual states and donors, NGOs, or international organizations. The countries' environmental agenda tends to be donor-driven,

---

[15] Article 1 of the Constitution defines Montenegro as a "democratic, social and ecological state." In September 1991 the republican Parliament also adopted a Declaration that reads: "We, deputies of the Parliament of Republic of Montenegro, see that the protection of identity of our people and land on which we live and work, because of destructing the nature, has become our opportune and most important job. By respecting our due to the nature, which gives us the strength of health, freedom and culture, we turn ourselves to the protection of hers, in the name of our own survival and the future of our successors. Not feeling any difference between us so strong, as the changes, which our natural environment is exposed to, we subject our national, religious, political and other feelings and trusts to the plan to turn Montenegro into an ecological state. We announce by this act of the Declaration, that Montenegro begins to make state relationships with nature." (http://www.montenet.org/econ/ecostate.htm, accessed 1 May 2007)

[16] www.un.org/jsummit/html/prep_process/national_reports/montenegro1_natl_assess2008. pdf, accessed 10 May 2007.

[17] Depleted uranium decontamination at the place known as Arza represents the first project of this type ever conducted in the world.

rather than one that is led by the recipient and, by definition, more coherent. According to UNEP, however, environment does not appear to be given the high priority that it warrants on donors' agendas—even among those donors for whom environment is generally a central component of international cooperation (UNEP, 2001). The lack of strategic coherence among the environmental projects supported by foreign donors is partly a consequence of insufficient leadership from the national authorities in setting a clear environmental agenda for donors. Weak state institutions usually accept anything that donors offer, no matter whether the grant and project is consistent to the current situation in the respective field or not. Some of the projects have even been accepted at the expense—and against the will—of citizens.[18] Some NGOs raise complaints that some of the donors care more for the interests of the multinational companies than for the local population's needs.

The great international assistance does not, however, pardon the very "international community" for its wrongs during the conflict period. Some of the grants were, however, directed toward elimination of the ecological degradation and harm made during the interventions. During the process the assessment showed the real picture of the environmental wealth and risks in particular regions. Surely, it appeared that many of today's problems are historical; they are consequences of the prewar period and the initial industrialization and the subsequent neglect of the environment so typical for the failing ex-Yugoslav federation. In the early 1950s, for instance, socialist Yugoslavia achieved significant economic growth due to the traditional industrialization model, based on intensive use of energy and raw materials, which pressured natural resources and the environment. The result was decreased forest area, deterioration of water quality in rivers and lakes, and increased air pollution in urban areas. There was also a significant wave of migration from the rural toward urban areas, which created problems typical for rapid overpopulation. Later on, the overall decentralization and delegation of power from the federal to republican and local (municipality) level brought additional decentralization of irresponsibility in many sectors, including the environmental one. Municipalities as well as the federal republics and provinces behaved as "states within states," and corrupted national

---

[18] Such examples are given in an edition prepared by a Macedonian NGO "Eko-svest" (Eco-Awareness), which clearly illustrates that national authorities are much more prone to listen to international donors and to care more about economic effects than ecological harm. On the other hand, in these examples EBRD has also proved to be more in favor of economic logic of the projects than to the environmental needs of some smaller communities. It seems that economic drive is still stronger than human and environmental security concerns (Colakovic and Jankovska, 2004).

elites cared more about their income than about mutual cooperation and concerted care about the Yugoslav natural wealth and resources.

To make things worse, right after the cease-fire in the late 1990s, the newly established elites again faced the need to recover from the war-devastated economy at the expense of the environment and exploitation of the cheap labor force. The environment as well as human well-being could hardly be securitized, or more precisely the related problems were pushed aside and totally desecuritized. In countries in which people die because of the lack of elementary health or social care, there is almost a national consensus that the environment should not be politicized—even less securitized.

From the perspective of public opinion there is no need to invest into the environment because the problem "does not even exist." The majority of citizens are merely focused on day-to-day survival, such as providing food and elementary health care. When advised to buy, for instance, energy-efficient light bulbs or solar collectors, the usual feedback from the citizens is negative.[19] The civil society organizations are weak, voiceless, and divided along ethnic lines rather than united over common problems. Most of all they are dependent on grant donors and follow their agendas rather than the society's basic needs. Many of these NGOs operate under difficult conditions, including chronic underfunding. As a consequence, they tend to be driven by the priorities of external donors, and are often in competition with each other for limited funding. This can result in poor collaboration and information exchange, duplication of effort, and other inefficiencies. For the elites, problems do exist but only in the context of EU accession and for which they are not ready to invest unless there is foreign assistance. Most of the countries, and even the ones that are closer to EU membership such as Croatia and Macedonia due to their candidate status, are barely aware of the width and depth of the environmental *acquis communautaire*. According to the Croatian Minister of Foreign Affairs and European Integration, the negotiation

---

[19] According to an owner of such a shop in Macedonia, "only the 'elite cream' buys these light bulbs. Truly they are 10 times more expensive than ordinary ones but they save five times more electric energy and also last five times longer. But people do not think of saving energy, they think of saving money. Obviously they don't grasp that saving energy is the same as saving money" (*Ceased Development*, 2005: 25). Another owner of a Macedonian company for trade with solar collectors testifies about the huge problem his company meets in a country that is situated in a very favorable sunny region. He says: "Only wealthy people are my customers. The middle class cannot even dream about buying a solar collector, which would save them a lot and would pay back a within two to three year period. People rather care about existence-related needs, not about saving. The Government does not provide these families with some financial assistance that would help then provide collectors" (Ibid.: 32).

process in the context of the "environmental chapter" is expected to be the toughest.[20] It is certainly not because the Croatian government objects to the improvement of environmental protection and preservation of national resources; the reasons are found in the financial burden that is unavoidably involved in the process.

4. Instead of Conclusion: Need for More Securitization or Desecuritization?

Listening to the heated debate, even the most knowledgeable security experts often wonder what it is all about. Do we discuss "objective" or "subjective" security—or both? Are human security and environmental security the most recent securitized issues? Are they complementary, or is, as some authors say, the environmental debate being intentionally animated in order to distort the calls for international social justice and democratic control of globalization? Debates remain hollow unless followed by organized action and positive results. Science has a very formidable task, but unfortunately there are growing complaints about the lack of credibility or even politicization in academic circles, which can produce results or justification depending who makes the order.

The experts and scholars in underdeveloped countries are not an exception. However, their actual powers to (de)securitize humans or the environment is very limited due to the fact that the state budget does not support research and inventions. Recognition of hazards or proposing solutions is often a very academic issue with no real impact even on a local/national level. When it comes to the grand debate, the people and societies concerned are outsiders and passive observers of the developments, and often of their own fate. Self-help and local initiative are almost unknown phenomena, while the general feeling of powerlessness paralyzes the subjects and anomy is what prevails. At the same time, the governing elites are the ones who can afford full information, international communication, and summits.

The Balkans indeed face huge human and environmental problems, but the dilemma often reads: humans or environment—who has priority when the resources are very limited, ambitions big, and general patience low? Or more correctly, where should one start with the survival policy? Instead of a clear idea and decision, the usual response is—NATO/EU integration. It is a specific case of virtual reality creation and "transcending" the most pressing problems.

---

[20] Interview with Minister Kolinda Grabar-Kitarovic for the Croatian TV NOVA, broadcast on 26 April 2007, on the occasion of the adoption of EU Parliament Declaration.

References

Agamben, G., 1998, *Homo Sacer: Sovereign Power and Bare Life*, Stanford University Press, Stanford, CA.

Alkire, S., 2003, A conceptual framework for human security, Centre for Research on Inequality, Human Security and Ethnicity (CRISE), Working Paper 2; http://www.crise.ox.ac.uk/pubs/workingpaper2.pdf, accessed 26 April 2007.

Annan, K., 2006, Remarks on the launch of the Peacebuilding Fund in New York, 11 October; www.un.org/News/Press/docs/2006/sgsm10677.doc.htm.

Beck, U., 1996, When experiments go wrong, *Independent*, 26 March.

Beck, U., 1998, Politics of risk society, in: *The Politics of Risk Society*, J. Franklin, ed., Polity, Cambridge.

Benton, T., 1997, Beyond left and right? Ecological politics, capitalism and modernity, in: *Greening the Millennium? The New Politics of the Environment*, M. Jacobs, ed., Blackwell, Oxford.

Buzan, B., De Wilde, J., and Wæver O., 1998, *Security: A New Framework for Analysis*, Lynne Rienner, Boulder, CO.

Ceased Development, 2005 (in Macedonian), Eko-svest, Skopje, October.

Chandler, D., 2006, *Empire in Denial: The Politics of State-Building*, Pluto, London.

Colakovic, A., and Jankovska, M., 2004, Assessment of the impact on the environment: How far have we gone? (in Macedonian), Eko-svest, Skopje, December.

Commission on Human Security, 2003, Communications Development, New York.

Dalby, S., 2002, *Environmental Security*, University of Minnesota Press, Minneapolis, MN.

Danov, A., 2007, FENI cares more for the profit than for public health, *Utrinski vesnik*, 26 April; http://www.utrinskivesnik.com.mk/?ItemID=A61BB69AA8F2D545B85D7A1C 5 B84C835, accessed 1 May 2007.

Duffield, M., 2004, Carry on killing: Global governance, humanitarianism and terror, DIIS Working Paper 23.

Duffield, M., and Waddell, N., 2006, Securing humans in a dangerous world, *Int. Polit.* **43**(1):1–23.

Florini, A., and Simmons, P. J., 1998, The new security thinking: A review of the North American literature, in: *Project on World Security*, Rockefeller Brothers Fund, New York.

Frederick, M., 1999, A realist's conceptual definition of environmental security, in: *Contested Grounds: Security and Conflict in the New Environmental Politics,* D. H. Deudney and R. A. Matthew, eds., Albany State University Press, New York.

Gligorov, V., 2002, South eastern Europe: New means for regional analysis (Policy Brief), *IDEA*, March.

Gligorov, V., 2006, Aiding Balkans, *Global Development Network Southeast Europe (GDN-SEE)*; http://www.wiiw.ac.at/balkan/files/GLIGOROV3.pdf, accessed 5 May 2007.

Gligorov, V., Holzner, M., and Landesmann, M., 2003, Prospects for further (South-)Eastern EU enlargement: From divergence to convergence? WIIW Working paper 296.

Gore, A., 1990, SEI: A strategic environmental initiative, *SAIS Rev.* **10**(1);60.

Hammerstad, A., 2000, Whose security? UNHCR, refugee protection and state security after the Cold War, *Secur. Dialogue* **31**(4):395.

Humphrey, M., 2002, *Preservation Versus the People: Nature, Humanity, and Political Philosophy*, Oxford University Press, Oxford.

IPCC, 2007, *Climate Change 2007: Impacts, Adaptation and Vulnerability, Summary for Policymakers*; http://www.ipcc.ch/SPM13apr07.pdf.

Jovanov, M., 2003, Poisoned town pleads for help, *The Institute for War & Peace Reporting*, London, 29 April; http://www.worldpress.org/Europe/1078.cfm, accessed 9 May 2007.

Juelich, R., 2005, *Progress in Environmental Law Drafting in South Eastern Europe*, Regional Environmental Center, Szentendre, December; http://www.rec.org/REC/Programs/EnvironmentalLaw/PDF/Progress_in_Environmental_Law_Drafting_in_SEE.pdf.

Karasimeonov, G., ed., 2005, *Organizational Structures and Internal Party Democracy in South Eastern Europe*, FES, Sofia.

Krastev, I., et al., 1999, Human security in South-East Europe, *UNDP Human Security Report*, 12.

Lipschutz, R., and Crawford, B., 1996, Economic globalization and the "new" ethnic strife: What is to be done? IGCC Policy paper 25; http://igcc.ucsd.edu/pdf/policypapers/ pp25.pdf, accessed 9 May 2007.

Lonergan, S., 2000, Human security, environmental security, and sustainable development, in: *Environment and Security*, M. Lowi and B. Shaw, eds., St. Martin's Press, New York, 70.

Lonergan, S., Gustavson, K., and Carter, B., 2000, The index of human insecurity, *AVISO Bulletin* 6, January; http://www.gechs.org/aviso/06/index.html, accessed 25 April 2007.

Macrae, J., Collinson, S., Buchanan-Smith, M., Reindorp, N., Schmidt, A., Mowjee, T., and Harmer, A., 2002, Uncertain power: The changing role of official donors in humanitarian actions, HPG Report 12, Humanitarian Policy Group, Overseas Development Institute, London.

Mathews, J. T., 1989, Redefining security, *Foreign Aff.* **46**(2):162–177.

Matković, G., 2005, Overview of poverty and social exclusion in the Western Balkans, presentation at the UNDP Western Balkans Forum on Social Inclusion and the Millennium Development Goals, Tirana, 23–24 June; intra.rbec.undp.org/mdg_forum/ overview.htm.

Matthew, R., 1999, Introduction: Mapping contested grounds in: *Contested Grounds: Security and Conflict in the New Environmental Politics*, D. Deudney and R. Matthew, eds., State University of New York Press, Albany, NY.

McCarthy, M., and Castle, S., 2007, How the worst effects of climate change will be felt by the poorest, *Independent*, 7 April; http://news.independent.co.uk/environment/climate _change/article2430118.ece, accessed 2 May 2007.

Nikolovska, N., ed., 2004, *Macedonia on Globalization*, Global Scholarly Publications, New York.

Ogata, S., 1999, Human security: A refugee perspective, Keynote speech at the Ministerial Meeting on Human Security Issues of the "Lysoen Process" Group of Governments, Bergen, Norway, 19 May; www.unhcr.ch/refworld/unhcr/hcspeech/990519.htm, accessed 28 March 2007.

Paris, R., 2001, Human security: Paradigm shift or hot air? *Int. Security* **26**(2):87–102.

Puhovski, Z., 2001, War: Continuation of morality by other means, *Rec.* **62**(June).

Revkin, A. C., 2007, Poor nations to bear brunt as world warms, *New York Times*, 1 April; www.nytimes.com/2007/04/01/science/earth/01climate.html?ex=1178510400&en=b9bc91b 7c10d2300&ei=5070.

Rutherford, P., 1999, The entry of life into history, in: *Discourses of the Environment*, Eric Darier, ed., Blackwell, Malden, MA.

Scherrer, C., 2002, *Structural Prevention of Ethnic Violence*, Palgrave Macmillan, New York.

Siljanovska-Davkova, G., 2004, Democracy in multiethnic and multicultural society— Between Demos and Ethnos, in: *Macedonia on Globalization*, N. Nikolovska, ed., Global Scholarly Publications, New York.

Suhrke, A., 1999, Human security and the interests of states, *Secur. Dialogue* **30**(3)265–276.

Ullman, R. H., 1983, Redefining security, *Int. Security* **8**(1):129–153.

UNEP, 2001, *Strategic Environmental Policy Assessment—FYR of Macedonia: A Review of Environmental Priorities for International Cooperation.*

UNEP and UNCHS, 1999, *The Kosovo Conflict: Consequences for the Environment & Human Settlements*; http://postconflict.unep.ch/publications/finalreport.pdf, accessed 6 May 2007.

Vankovska, B., 2007, Human security doctrine for Europe: A view from below, *Int. Peacekeeping* **14**(2):264–281.

Wallensteen, P., and Sollenberg, M., 1999, Armed conflicts, 1989–98, *J. Peace Res.* **36**(5):593–606.

Warner, J., 2000, Global environmental security: An emerging "concept of control"? in: *Political Ecology: Science, Myth and Power,* P. Stott and S. Sullivan, eds., Arnold Publishers, London.

Wimmer, A., ed., 2004, *Facing Ethnic Conflicts: Toward a New Realism,* Rowman and Littlefield, Boulder, CO.

Wæver, O., 1995, Securitization and desecuritization, in: *On Security,* R. D. Lipschutz, ed., Columbia University Press, New York, pp. 46–48.

Woodward, Susan, 2006, Presentation at the International Conference "Regional Cooperation, Peace Enforcement and the Role of the Treaties in the Balkans," Istituto per l'Europa Centro-Orientale e Balcanica, Alma Mater Studiorum, University of Bologna, Forli, Italy, 20–21 January.

ENVIRONMENTAL JUSTICE AND HEALTH DISPARITIES IN APPALACHIA, OHIO

Local Cases with Global Implications

MICHELE MORRONE*

School of Health Sciences, Ohio University

Abstract: The *Appalachian Regional Development Act of 1965* defines Appalachia as a region of the eastern United States that historically has relied on mining and other extractive industries. A major purpose of this Act was to decrease economic inequities in more than 400 counties in thirteen states. These counties have contributed to national prosperity but have not reaped the benefits of this contribution. As part of Appalachia, southeastern Ohio posts the highest poverty levels in the state. Combined with this poverty are numerous environmental problems that expose area residents to environmental health hazards. There is a relationship between socioeconomic status (SES) and environmental conditions that leads to both positive and negative health outcomes. This paper discusses environmental health and health disparities in Appalachia, Ohio. The meanings of "environmental justice" and "health disparities" are explored. Several case studies are presented, including a case of perfluorooctanoic acid (C8) exposure, a case involving the siting of a medical waste incinerator, and a case in which an entire community was purchased by a major power company because of uncontrollable emissions. All of the cases set precedents that may lead to improvements in our understanding of the political, social, and environmental factors that lead to health disparities in poor rural populations worldwide.

Keywords: Environmental justice; Appalachia; health disparities; poverty

1. Background

As part of Appalachia, southeastern Ohio has the highest poverty levels in the state. Combined with this poverty are numerous environmental problems that expose area residents to environmental health hazards. Such disproportionate

---

*Address correspondence to Michele Morrone, School of Health Sciences, Ohio University, Athens, OH 45701

P.H. Liotta et al. (eds.), Environmental Change and Human Security, 299–323.

exposure to adverse environmental conditions creates an intersection between environmental justice and health disparities.

There is a relationship between socioeconomic status (SES) and environmental conditions that leads to both positive and negative health effects (Evans and Kantrowitz, 2002). SES often is measured using demographic variables such as income, education, occupation, and employment status (Farmer and Ferraro, 2005; Nordstrom et al., 2004). People who have low incomes and educational levels, or are unemployed, are categorized as having low SES. Low SES often leads to poor living conditions, including proximity to hazardous facilities and hazards directly related to inadequate housing (e.g., lead).

Health-disparities research has identified several factors that contribute to poor health among poor people. These include lack of access to health care, lifestyle choices such as smoking and poor nutrition, and lack of adequate insurance. However, it is possible that some of the disparities in health status can be explained by disparities in environmental exposures. Simply put, low SES can be an indicator of poor human and environmental health, while high SES is associated with good overall health.

A significant body of research has examined the SES–environmental health relationship (Prus, 2007). However, when examining relationships among SES, health, and the environment, research has tended to focus on urban areas (McMichael, 2000), even though populations in rural areas may suffer more from environmental exposures. Murray et al. (2005) divided America into eight distinct areas based on demographic characteristics such as education, income, population density, and homicide rates. They determined that the area they identified as "poor whites living in Appalachia and the Mississippi Valley" has higher mortality rates among the young and middle-aged than some third-world countries. They called for more research examining the factors that lead to these high rates.

Appalachia, Ohio, is a rural area with a history of environmental contamination and some of the most poverty-stricken population in the United States. This paper examines rural health disparities in the context of environmental justice.

2. Environmental Justice and Health Disparities

Environmental justice is a broad concept that traditionally has been defined as the proximity of minority and low-SES populations to hazardous facilities (Bullard, 1994; Bryant and Mohai, 1992). Examining environmental justice from an environmental-health perspective broadens the definition of environmental justice to include disproportionate exposures due to poor living conditions, lack of access to adequate health care, and minimal opportunities for community involvement in defining and solving environmental health problems. Disproportionate environmental exposures and lack of access can be major factors in health disparities between the rich and the poor.

A serious limitation of environmental-justice research is in identifying a significant causal relationship between environmental exposures and health outcomes. It is difficult to identify the cause of chronic health conditions, such as cancers, that are commonly linked to environmental exposure. Acute conditions, such as respiratory effects that may be obviously related to the environment, can be associated with numerous confounding factors as well. Because of the difficulty of scientifically associating health with the environment, most environmental-justice research focuses on quantifying unequal exposures, then drawing qualitative conclusions about health disparities.

2.1. UNEQUAL EXPOSURES

Where people live is related to demographic characteristics such as income, race, and education. Geography also is an important determinant of exposures to adverse environmental conditions. Taking a simplistic approach to place of residence, it is possible to characterize most of the population as urban, suburban, or rural, this classification can apply globally, not just in the United States. Populations living in each of these areas are exposed to different types of environmental problems. For example, urban-based environmental studies have looked at exposures to ambient air pollution, lead, hazardous substances, and indoor-air contaminants. In terms of air pollution, research indicates that residents of low-income and minority urban areas are exposed to high levels of air pollution, and that there is a link between these exposures and respiratory health effects (Houston et al., 2004; Lopez, 2002; Perlin et al., 1999, 2001). In addition, minority urban children are exposed to higher levels of heavy metals, air pollution, and noise pollution, and have worse housing conditions than white children (Hambrick-Dixon, 2002).

Suburban populations have a different set of environmental health issues that are more likely to be related to lifestyle than pollution. One of the most prominent environmental health issues being discussed in the United States today is the relationship between so-called urban sprawl and obesity (Lopez, 2004; Ewing et al., 2003, 2006). The Centers for Disease Control and Prevention (CDC) has made addressing obesity a priority, and the National Center of Environmental Health in CDC has focused on the role that the built environment plays in increasing levels of obesity in the United States (Jackson and Kochititzky, n.d.).

Poverty also is associated with obesity, but poverty levels in suburban areas are not as high as those in urban and rural areas. In Ohio, for example, the suburbs that surround the major cities of Cleveland, Columbus, and Cincinnati have the highest median household incomes and the lowest poverty rates in the state. In examining the rising percentages of African Americans moving to suburban areas, Clark (2007) argues that income is a better predictor of whether a family will move to a suburb than race.

Rural populations are perhaps the group that is most at risk in terms of environmental exposures. These populations are exposed to many of the same environmental issues that urban residents are, such as poor housing, air pollution, and heavy metals. Similar to suburban populations, lifestyles of the rural poor can lead to high levels of obesity and related health effects. However, low accessibility and availability of adequate healthcare magnifies the plight of the rural poor.

Combined with similar urban and suburban exposures are environmental issues that pose additional risks to rural populations. These are issues such as inadequately treated wastewater and drinking water, proximity to facilities that manage waste from urban and suburban areas, and proximity to large industrial complexes that produce goods used by populations outside of their regions. Exacerbating all of the environmental exposures of the rural poor is inadequate public-health infrastructure. The tragedy is that rural areas, where public-health protection is most needed, also are the areas where resources are the most limited.

2.2. UNEQUAL HEALTH OUTCOMES

Unequal health outcomes, also known as "health disparities," are defined as differences among groups of people in terms of their health status. This is different from health "inequity," which "refers to differences among people that are unjust or unfair" (Resnik and Roman, 2007: 231). It is possible for there to be differences in health status among populations that are not unfair. For example, smokers have higher levels of lung cancer than nonsmokers and, while this is an unequal health outcome, few would argue that it is unfair. On the other hand, there are health disparities that can be categorized as health inequities, especially when it comes to conditions that are related to environmental exposures beyond the control of the individual or population.

Carter-Pokras and Baquet (2002) identified eleven different definitions of "health disparity." They argue that the definitional issue has important policy implications because how disparities are measured can influence the allocation of public health resources. Ultimately they suggest that

A health disparity should be viewed as a chain of events signified by a difference in: (1) environment, (2) access to, utilization of, and quality of care, (3) health status, or (4) a particular health outcome that deserves scrutiny. (p. 427)

In the case of rural Appalachia, all of the factors identified by Carter-Pokras and Baquet contribute to health disparities between this population and those who live in non-Appalachian areas. In other words, rural, disadvantaged populations suffer from poor environmental conditions, lack of health-care access, poor health status, and some specific health outcomes that distinguish them from advantaged, non-rural populations. Furthermore, when it comes to environmental conditions, some health disparities in Appalachia are inequitable as well as unequal.

3. The Study Area: Appalachia, Ohio

The *Appalachian Regional Development Act of 1965* defines Appalachia as a region that historically has relied on mining and other extractive industries. A major purpose of this Act was to decrease economic inequities in more than 400 counties in thirteen states that encompass Appalachia. According to the Act, these counties have contributed to national prosperity through their extractive industries but have not reaped the benefits of this contribution. The amendments to the Act in 2002 state the following with regard to equity in Appalachia:

The Congress further finds and declares that while substantial progress has been made toward achieving the foregoing purposes, especially with respect to the provision of essential public facilities, much remains to be accomplished, especially with respect to the provision of essential health, education, and other public services. (Appalachian Regional Commission)

As Figure 1 shows, approximately one-third of Ohio officially is delineated by the federal government as part of Appalachia. This is the area of the state that historically has relied on coal mining, leading to a legacy of environmental and economic issues. Among the more prominent large-scale environmental problems in the area are acid mine drainage and illegal dumping (including hazardous materials), followed closely by poor-quality housing stock, poorly funded education, and aging infrastructure. Economic issues manifest themselves in high poverty and unemployment rates, which further contribute to the disparities noted.

Poverty is endemic in southeastern Ohio. As Figure 2 shows, the counties in Appalachia, Ohio, also are those that exhibit the highest poverty rates in the state. Four counties, Athens, Meigs, Lawrence, and Scioto, have poverty rates between 16 and 18 percent. With only a few exceptions, the population in the area delineated as Appalachia is more likely to live in poverty than those who do not live in this geographic region.

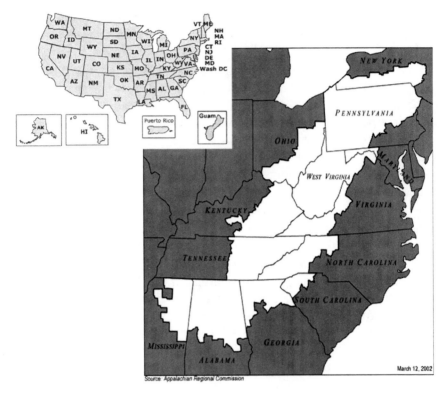

Figure 1. The Appalachian Region Showing the Southeastern Portion of Ohio.
Source: Appalachian Regional Commission.

The impact that poverty has in the region can also be found in examining its educational infrastructure. Of the 608 school districts in the state of
Ohio, 127 are located in the Appalachian counties. These mainly are rural
districts in which children are bused long distances from outlying areas to
attend school. In several counties, there is only one high school, and all of
the children who live in the county attend it. As Table 1 shows, the average
size of the school districts in Appalachia, Ohio, is almost twice as large as
the districts in non-Appalachian counties.

Table 1 indicates that a student in Appalachia is more likely to be poor
and live further away from school than one in a non-Appalachian county.
In Ohio, schools are funded largely on the basis of property values, a funding system deemed unconstitutional by the Ohio Supreme Court in 1997.

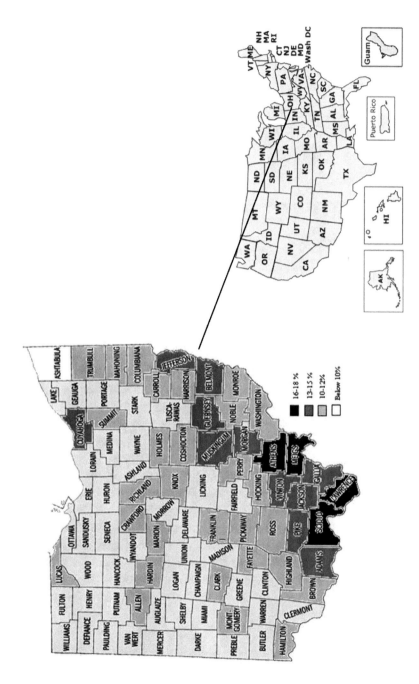

Figure 2. Poverty Rates in Ohio.

TABLE 1. Comparison of Appalachian and Non-Appalachian Public Schools on Selected Indicators, Including Results of One-Way Analysis of Variance (ANOVA) (N = 608)[1].

Indicator	Category	Mean (Std Dev)	F (Sig)
Average square miles in district	Non-Appalachian	56.14 (46.67)	80.75 (.000)
	Appalachian	111.52 (100.91)	
Number of students per square mile	Non-Appalachian	136.13 (193.89)	29.78 (.000)
	Appalachian	40.88 (62.72)	
Percent of students in poverty	Non-Appalachian	2.51 (3.22)	63.19 (.000)
	Appalachian	5.15 (3.72)	
Assessed property valuation per pupil (dollars)	Non-Appalachian	139,012.23 (66,812.69)	50.97 (.000)
	Appalachian	94,814.20 (38,942.15)	
Revenue raised by 1 mill (ability of district to generate property tax) (dollars)	Non-Appalachian	470,612.15 (776,745.69)	17.10 (.000)
	Appalachian	183,671.04 (169,475.84)	
Total property tax per pupil (dollars)	Non-Appalachian	4,500.08 (2,541.16)	75.09 (.000)
	Appalachian	2,484.56 (1,244.89)	

This decision, known as the "DeRolph decision," originated in Appalachian Perry County when a coalition filed a lawsuit against the state. The unconstitutionality of school funding in Ohio has been upheld by the Ohio Supreme Court numerous times since 1997; however, little has been done in the state to remedy this situation. Poor children still attend poor schools.

The inequity of school funding in the state can also be seen in Table 1, which identifies three indicators of the impact that property values has on schools. For the purpose of determining taxes, the communities in the United States use a millage system. Under this system, 1 "mill" equals $1.00 of property taxes for every $1,000 of assessed property valuation. The assessed value of a property is a percentage of its appraised value. For example, if a home is appraised at $100,000, the assessed value for a specific community might be $30,000 (or 30 percent of the appraised value). A 1-mill tax on this property would be $30. So, communities with lower property values have lower tax revenue as well.

---

[1] Results of the One-Way ANOVA should be interpreted with some caution due to the nature of the data. Although the groups (i.e. Non-Appalachian and Appalachian) are assumed to be independent, they are different in size, which could affect the means that are being compared.

The average assessed property valuation per pupil is almost $50,000 more in non-Appalachian school districts. This affects a district's ability to raise revenue with a tax levy, as the revenue from a 1-mill levy in an Appalachian district would equate to an average of about $184,000. In a non-Appalachian district, a 1-mill levy would equate to about $470,000, almost three times as much as the Appalachian district. Finally, the amount of property tax that the districts raise per pupil is $4,500 for non-Appalachian districts and about $2,484 for Appalachian districts. In comparing the means of these indicators, all of these differences between Appalachian and non-Appalachian school districts are statistically significant.

Inequities in school funding mean inequities in education. The result of this inequity may be directly related to health effects and environmental exposures because uneducated people feel less empowered to address these issues. In order to examine the relationship between environmental conditions and health outcomes, demographic characteristics of eight counties in the twenty-nine-county Appalachian region were examined and compared to statistics for the eighty-eight counties in the state. These counties were selected because of data that suggest significant environmental exposures. The population in these counties are compared by: (1) median household income; (2) percent below poverty line; (3) education level; and (4) unemployment rate (Table 2). When compared to state data, four characteristics, depicted in Table 1, highlight the inequalities that exist in the state.

With the one exception of Clermont County, all of the eight counties had median household incomes lower than the state value of $43,119. In addition, all of these counties except two have higher poverty rates than the state percentage of 10.6. None of the counties noted have a population with a higher than state percent of college-educated population, and only one county has a slightly lower unemployment rate. These data paint the typical

TABLE 2. Demographic Characteristics of Select Counties in Appalachia, Ohio, and the State of Ohio.

Appalachian county	Median household income (2003)	Percent below poverty line (2003)	Percent with college degree (2003)	Unemployment rate (2006)
Adams	30,421	14.9	7.2	7.6
Belmont	31,489	14.3	11.1	5.9
Clermont	51,863	6.9	20.8	5.1
Coshocton	35,191	10.3	9.8	6.9
Gallia	32,167	15.5	11.6	6.1
Jefferson	32,414	13.6	11.8	7.0
Lawrence	30,104	16.5	10.3	5.3
Washington	35,162	11.2	15.0	5.3
State	43,119	10.6	21.1	5.2

picture of an Appalachian county as an area of poor people with limited education and more likely to be unemployed than other areas in the state.

3.1. ENVIRONMENTAL JUSTICE ISSUES IN APPALACHIA, OHIO

In defining environmental injustice as a disproportionate exposure of a specific population to environmental hazards, there are several environmental issues that contribute to environmental inequities in Appalachia, Ohio. These include location of waste-disposal facilities, air-pollution releases, and the environmental-health legacy related to a long history of resource extraction, including coal mining.

3.1.1. Waste-Disposal Facilities

In the United States, waste is identified in several categories, including: (1) solid waste, (2) hazardous waste, (3) nuclear waste, (4) wastewater, (5) medical waste, and (6) construction and demolition debris. Waste is managed and regulated according to its categorization. For example, solid waste is largely managed by states and local governments, while nuclear waste is the responsibility of the federal government.

When it comes to solid and hazardous-waste facilities, states are responsible for permitting and inspecting these facilities to ensure compliance with federal laws and regulations. Solid waste can be thought of as generic trash and garbage generated by households and communities. The most common method of managing solid waste is by land disposal, also known as landfilling.

Hazardous waste is regulated more stringently than solid waste; however, it is still largely the state's responsibility to manage this waste. For a waste to be defined—and therefore regulated—as a hazardous waste, it must either demonstrate characteristics of toxicity, ignitability, corrosiveness, or reactivity; or it must be specifically listed in state regulations as a hazardous waste. It is more common for hazardous waste to be incinerated than landfilled.

Medical waste is generated by healthcare and research facilities, and includes blood, discarded organs and limbs, discarded equipment and clothing used in surgery, and cultures. Medical waste can be classified as infectious, radioactive, hazardous, or general waste, and it is managed and regulated based on its characteristics. Most infectious waste is incinerated, and often, hospitals and research facilities have on-site incinerators.

For those facilities that do not have on-site capabilities to incinerate medical waste, they must transport their waste, often at great expense. Currently there are only two commercial medical-waste incinerators in Ohio, suggesting there might be a profitable opportunity for a new facility in the state. When it comes to siting waste-disposal facilities, whether it is a landfill or an incinerator, two key factors come into play: (1) the availability of inexpensive land; and (2) a marginalized population that will not fight against the facility—and both factors exist in Appalachia, Ohio.

One example of the impact of waste disposal on poor communities occurred in the early 1990s when a consulting and engineering firm filed a permit to operate a medical-waste incinerator in Coolville, Ohio.[2] Coolville is in Athens County, one of the poorest counties in Appalachia, Ohio, and at that time, the town population was a little more than 600 people. Documented historical accounts of this proposal are sketchy, but ultimately, the community formed a group known as the Concerned Citizens of Coolville.

At the time, the debate about the proposal revolved around concerns that the residents had about possible health effects and the unknown environmental consequences of such a facility. Citizens became increasingly outraged when they perceived dishonesty on the part of the Ohio Environmental Protection Agency. In retrospect, this case was a true example of environmental injustice because Coolville was a likely target due to the characteristics of its population. As one resident put it in a letter to Senator John Glenn, "It seems like I am powerless to effectively oppose this medical waste incinerator."

Ultimately the incinerator was never built in Coolville for technical, economic, and political reasons. However, this case was a wake-up call to the residents of the Appalachian region that they might be the target of additional environmental hazards because of their population and land base.

3.1.2. Toxic Releases

Largely in response to the environmental catastrophes in Bhopal, India, and Chernobyl, Ukraine, in the 1980s, the U.S. Congress passed the Emergency Planning and Community Right to Know Act (EPCRA) in 1988. A major goal of EPCRA was to provide public access to amounts of pollution released by facilities in their neighborhoods. The public-access mechanism is known as the Toxic Release Inventory, or TRI.

The TRI is one indicator of environmental quality, albeit one with many limitations. The data compiled in the TRI are self-reported by facilities and often are estimated based on inputs rather than actual measures of releases.

---

[2] Most of the information about the Coolville medical waste incinerator is taken from archival files of Senator John Glenn, accessed at the Ohio State University. This case is presented in this section of the paper because there is not enough available information to provide a comprehensive case history; however, it is an important example of how poor communities are often the target for environmental contamination.

The releases are presented to the public in terms of pounds of release. For example, in 2005, one facility reported releasing 9,600,000 lb of hydrochloric acid into the air. This type of information is alarming to the public because they generally rely on the media to interpret the data and the media does so in a manner that may not provide appropriate context and explanation.

Additional limitations with the TRI data have to do with the amount of time it takes to make the data available to the public. For example, data for 2005 releases were not available until early 2007. Finally, looking for trends in releases over time should be done cautiously because new chemicals are added to the reporting requirements regularly. This means that it may appear that a facility never released any dioxin-containing compounds until 2000, when these were required to be reported in the TRI.

Even with these limitations, the TRI is one of the best public-information tools available when it comes to raising awareness of local environmental conditions. The data do provide one measure of potential pollutants in the ambient environment, and allow for geographic comparisons.

The 2005 TRI report for Ohio indicates that the Appalachian region is subject to some of the largest reported amounts of chemical releases in the state. Among the top fifty reported releases of chemicals into the air, twenty-eight of these occurred in the counties identified as Appalachia. As Table 3 shows, these twenty-eight releases occurred in the eight counties in Appalachia as discussed above. The eight counties also account for 42.9 percent of the total reported on-site releases in the state in 2005.

Some additional analyses of the TRI data with demographic information suggest that there is a relationship between poverty, health, and TRI releases. The correlation coefficients between health outcomes, demographic

TABLE 3. Data from 2005 Toxic Release Inventory from Counties Reporting the Highest Releases.

Appalachian county	Number of reported releases in top 50	Total on-site releases (lb) reported in 2005	Percent of total reported releases in state (220,202,286 lb)
Adams	4	16,899,114	7.7
Belmont	2	3,096,977	1.4
Clermont	4	8,345,877	3.8
Coshocton	2	11,799,525	5.4
Gallia	5	12,774,081	5.8
Jefferson	6	23,578,697	10.7
Lawrence	1	752,136	0.3
Washington	4	17,226,936	7.8
Totals	28	94,473,343	42.9

characteristics, and the TRI data are shown in Table 4.[3] These coefficients are based on the data available for all eighty-eight counties in Ohio; statistical significance of the correlation is in parentheses:

Although most of these correlations are statistically significant, the strongest relationship is between cancer (i.e., a health outcome) and poverty level. These data are exploratory only, because of inherent limitations and the fact that the correlations are only based on eighty-eight data points representing the eighty-eight counties in Ohio. Nevertheless, these results should generate interest in further examining the relationships among the environment, health, and demographics, including the need for additional research on causality.

3.1.3. Coal Mining

Coal mining began in Ohio in about 1800, and southeastern Ohio is the location of most of the mines in the state. Because coal in southeastern Ohio generally contains high amounts of sulfur, it is not as much in demand as it once was. Decreasing demand has led to numerous abandoned mines in the area. There are many environmental health concerns associated with abandoned mines, including landslides, subsidence, explosions due to gas buildup, flooding and acid-mine drainage, and safety issues associated with mine openings.

A serious problem in Ohio is that the records that indicate the location of these mines are poor. In 2005, The Ohio Department of Natural Resources

TABLE 4. Correlations Between Environmental Indicators, Health, and Socioeconomic Status in Ohio Counties (N = 88).

	Unemployment rate	Poverty level	Cancer death rates	New cancer cases
Unemployment rate	1.00			
Poverty level	.596** (.000)	1.00		
Cancer death rates	.306** (.000)	.622** (.000)	1.00	
New cancer cases	.219* (.041)	.364** (.000)	.539** (.000)	1.00
TRI releases	.122 (.265)	.245* (.024)	.228* (.036)	.355** (.001)

*Correlation is significant at the 0.05 level (2-tailed)
**Correlation is significant at the 0.01 level (2-tailed)

[3] Cancer data are from the 2005 Ohio Cancer Incidence Surveillance Study; poverty level data are from the U.S. Census for 2003; unemployment rates are from the U.S. Census for 2006; TRI data is from 2005.

estimated that only about 50 percent of an estimated 7,000 abandoned mines are on record. There are concerns that the unrecorded mines could pose a serious threat to environmental health, and citizens are being asked to help by contributing information about any known abandoned mines.

In 2003, the Office of Surface Mining in the U.S. Department of the Interior issued a white paper, "People Potentially at Risk from Priority 1 and 2 AML Hazards" (U.S. Department of Interior, 2003). Priority 1 and 2 hazards include clogged streams, dangerous highwalls, hazardous and explosive gases, polluted water, subsidence, and underground mine fires. These mines are considered the greatest threat to public health. According to this report, more than 169,000 Ohioans live within one mile of a priority 1 or 2 mine. While this pales in comparison to Pennsylvania (more than 1.6 million people) and West Virginia (more than 693,000 people), it is significant because all of the people who live in these areas are in the southeastern part of the state. Hence, there is no question that there are disproportionate exposures to the hazards associated with abandoned mines in Appalachia, Ohio.

3.2. HEALTH DISPARITIES IN APPALACHIA, OHIO

In general, people who live in poverty are less healthy than those who do not. The factors that contribute to health status include behavior and lifestyle, medical care, heredity, and environmental exposures.

3.2.1. Behaviors and Environmental Exposures

The Centers for Disease Control and Prevention (CDC) conducts the Behavioral Risk Factor Surveillance Survey (BRFSS) annually. The purpose of this survey is to gather baseline information about health behaviors and health status. Table 5 summarizes the results of the respondents to the 2002

TABLE 5. Comparison of Responses to Behavioral Risk Factor Surveillance Survey, 2002.

Health indicator	Appalachian respondents (%)	Ohio respondents (%)
Current cigarette smokers	33.3	26.6
No leisure time physical activity	28.7	25.4
Five or more servings of fruits and vegetables per day	18.2	20.5
Adult obesity	20.5	23
No health insurance	18.7	13.4
Have diabetes	7.8	7.7

BRFSS by comparing those who lived in Ohio Appalachian counties with the state as a whole.

The notable differences in reported indicators between Appalachian respondents and the state as a whole suggest that Appalachian respondents smoke more, exercise less, have a worse diet, and are less likely to have health insurance. On the other hand, Appalachian respondents note lower levels of obesity and similar rates of diabetes as whole state.

In terms of smoking, understanding why those who live in Appalachia smoke more than those who do not is complex. It can include a lack of awareness about the health effects from smoking, failure to recognize addiction, and smoking as a way to manage stress (Weaver et al., 2006).

Cancer is one outcome of unhealthy behaviors, especially related to tobacco use and diet. Table 6 summarizes cancer estimates from the same eight Appalachian counties discussed above in relation to the Toxic Release Inventory. As with the TRI data, this information must be interpreted with caution, because not all counties report complete data. Nevertheless, the data suggest that the cancer and death rates in the Appalachian counties are among the highest in the state. Smoking rates must be considered in interpreting the cancer-incidence data.

Of the eighty-eight counties in Ohio, four of the five highest rates of new cancer cases occur in the Appalachian counties with the highest TRI emissions. Jefferson County reports the highest rate of new cancers in the state, almost 100 more cases per 100,000 than the state as a whole. Interestingly, Jefferson County also contributes more than 10 percent of the releases reported in the TRI, and six of the top ten TRI releases in Ohio occurred in Jefferson County in 2005.

TABLE 6. New Cancer Cases and Death Rates, 1999–2003.

Appalachian county	New cancer cases (rate per 100,000, age-adjusted) (state rank)	Cancer death rate (rate per 100,000, age-adjusted)
Adams	473.2	260.8
Belmont	479.8	208.8
Clermont	510.1 (4)	219.1
Coshocton	457.3	208.7
Gallia	450.3	189.3
Jefferson	565.3 (1)	225.6
Lawrence	515.1 (3)	241.6
Washington	499.4	199.3
State	468.4	208.6

Source: Ohio Cancer Incidence Surveillance System (2006).

3.3. MEDICAL CARE

The cancer-incidence data can also be one indicator of medical care. The data related to cancer mortality suggest that a case of cancer in Appalachia is more likely to result in mortality than a case in a non-Appalachian county. The top five mortality rates for cancer in the state of Ohio are in five Appalachian counties: Adams (260.9), Lawrence (241.6), Guernsey (233.3), Hocking (232.9), and Ross (231.3).

Access to medical care most likely is one critical element when it comes to treating cancer. Those residents of rural counties in Appalachia, Ohio, must drive long distances for health care. In addition, the poverty and unemployment levels, which are the highest in the state, mean that this population is not likely to have adequate insurance, if they have any insurance at all.

4. Case Study: Perfluorooctanoic Acid (PFOA)

Residents of southeastern Ohio and western West Virginia who live along the Ohio River are subject to some of the most intense industrial and manufacturing pollution in the country. Within a few hours' drive along the river, there is a hazardous-waste incinerator, several coal-burning power plants, and numerous chemical facilities. One of these chemical facilities is currently under suspicion by local residents for polluting the drinking water supply in several Appalachian communities.

4.1. THE POLLUTANT

In 1999, the chemical perfluorooctyl sulfonate (PFOS) caught the attention of the United States Environmental Protection Agency (USEPA) because it was showing up in environmental and blood samples everywhere that samples were taken. PFOS was used by only one company in the United States, 3M.

There was very little information, and consequently great uncertainty, about the environmental and human health effects from exposure to PFOS. However, in 2000, 3M announced that it would phase out its use of PFOS. At the same time that 3M was announcing its phase-out, USEPA decided to investigate the implications of the presence of a related chemical, perfluorooctanoic acid, in the environment.

Perfluorooctanoic acid (PFOA) is an organic acid that contains a chain of eight carbon-fluorine bonds. The carbon-fluorine bond is extremely stable, which means that once it enters the environment, it is persistent. PFOA also is known as "C8" because of the eight carbon chains. C8 is used to make many plastics and in manufacturing non-stick coatings, such as Teflon.

DuPont developed Teflon in the 1950s, and has been using PFOA in the production process ever since.

PFOA has been found in blood samples in every geographic region of the United States. Six hundred and forty five samples taken from six American Red Cross blood banks showed the presence of PFOA in every region from which the samples were obtained (Olsen et al., 2003).

There is much uncertainty about whether there are health effects stemming from even the highest recorded levels of PFOA in blood (Butenhoff et al., 2004). However, because of the widespread existence of PFOA in blood samples, the USEPA issued a call for research to focus on the fate and transport of PFOA and possible health effects from exposure (Perfluorooctanoic Acid (PFOA), Fluorinated Telomers; Request for Comment, Solicitation of Interested Parties for Enforceable Consent Agreement Development, and Notice of Public Meeting, 2003).

From 2003 through 2006, USEPA coordinated a series of meetings of three groups: a plenary group of those who registered as "interested parties" and two technical workgroups charged with examining the scientific evidence related to PFOA.

4.2. ENVIRONMENTAL EXPOSURES

Since C8 is a processing aid, it becomes waste after the product is manufactured. In 1984, DuPont opened a landfill near Lubeck, West Virginia, and began disposing of C8-contaminated waste in this landfill. Due to the lack of scientific information about the health and environmental effects of C8, there were no permit limitations related to the amount of C8 that DuPont could put in the landfill. In addition to landfill disposal, DuPont released C8 into the air and water surrounding the West Virginia plant for many years.

The highest levels of PFOA found in water samples in the United States were found in the groundwater used by the Little Hocking Water Association (Lyons, 2007). The Little Hocking Water Association is a private nonprofit organization that provides drinking water in rural Washington County, Ohio, to more than 12,000 people. The source of the drinking water for this district is directly across the river from the DuPont plant.

Davis et al. (2007) conducted a study to determine how PFOA was getting into the groundwater across the river from the plant. They examined the likelihood of groundwater, surface-water, and air transport, and determined that the most likely scenario was air transport. The pollutant is released in air emissions, settles on the ground over the aquifer, and then leaches into the groundwater. The results of this research suggest one possible reason why PFOA has been found in blood samples across the United States, even in areas in which PFOA is not released locally. Earlier research by Barton et al. (2006) suggests that PFOA is found in the air in particle form rather than vapor form, which would enhance its ability to disperse from the facility and settle on the ground.

4.3. THE REMEDY

In 2005, DuPont settled a lawsuit with residents near the Washington Works plant by agreeing to pay more than $107 million to the plaintiffs, who were affected residents in the water district. In addition, DuPont is funding an independent health study that currently is under way.[4] This study is looking for exposure levels that could result in health effects. If the results indicate that exposures are not high enough to contribute to health effects, then DuPont will not pay an additional $235 million to the plaintiffs. The health study in the area affected by C8 may be one of the largest studies of this type ever conducted.

In the meantime, a local journalist documented the case in a book called *Stain Resistant, Nonstick, Waterproof, and Lethal* (Lyons, 2007). In this book, the author suggests a long history of deception on the part of DuPont, who may have been aware of the possible health effects from exposure to C8 more than forty years ago.

4.4. GLOBAL IMPLICATIONS

The situation in southeast Ohio in relation to C8 has global implications for several major reasons. First, DuPont uses C8 to manufacture products that are used worldwide. These include products that enhance architecture, transportation, and communications. However, the product that raised public awareness about C8 is Teflon. There is no doubt that Teflon has been a useful consumer product that has revolutionized cookware. DuPont argues that there are many more benefits to Teflon than risks, including the health benefits of using less fat when cooking. In addition, DuPont notes that Teflon extends the life of carpeting and textiles, and that it has fire-suppression uses. As DuPont searches for a suitable substitute for C8, the availability and price of Teflon may be affected.

In a deposition in November 2002, a DuPont company executive estimated that Teflon accounted for approximately $200 million in profit for

---

[4] Information about this study can be found at: http://www.c8healthproject.org.

the year 2000 (DUPONT: EPA Will Study Safety of Chemical Used to Make Teflon Products, 2003). Meanwhile, the people who live around the facility where Teflon is manufactured continue to live in poverty, exposed to the chemical used in the manufacturing process for this profitable product. The research currently under way to examine health effects from exposure also has global implications. The health study itself has the most significant global implication in that this is a rare case in which there will be data about the relationship between a specific environmental contaminant and health effects.

The current health study is focused on the potential health effects associated with C8. There are ten components to the health study that investigate health endpoints, such as heart disease, birth defects, and cancer. These studies involve analyzing the relationship between levels of C8 in blood and the health status of the participants. Most of the studies will be completed by 2011. Whatever the outcome, this case will provide data that address causality between exposure to a specific environmental contaminant and health outcomes in a low SES population. Lack of these kinds of data has seriously limited the validity of environmental justice research.

If the health study shows a significant relationship between negative health outcomes and exposure to C8, other multinational corporations should be concerned that their hazardous practices will be scrutinized more closely. In this case, it is also likely that the health study will be an excellent piece of evidence in lawsuits that are sure to be filed against DuPont. The C8 situation could then serve as a model for other communities fighting pollution from facilities in their regions.

On the other hand, if the health study does not identify a significant relationship, environmental activists should be concerned. Environmental activists generally agree that exposure to pollution causes adverse health effects. The health data may show that this is not the case and could lead to a new global strategy on the part of environmental activists.

5. Case Study: Electric Power Generation

Cheshire, Ohio, is located in Gallia County, and is home of one of the largest coal-burning power plants in the country. American Electric Power (AEP) operates the General J. M. Gavin Power Plant in Cheshire, Ohio. The Gavin power plant is a 2,600 MW facility which, according to the Energy Information Administration (EIA), is the 16th largest in the United States in terms of its capacity to generate electricity (EIA, 2005).

As noted above, Gallia County is an Appalachian County in the southeast corner of Ohio, bordering the Ohio River. Approximately 31,000 people

live in Gallia County, where the median household income in 2003 was
$32,167.

5.1. THE POLLUTANTS

Air pollution is the major concern resulting from operating the power plant.
Through the Toxic Release Inventory, the Gavin Power Plant has reported
releases of millions of pounds of chemicals into the environment. In terms
of quantity, sulfuric acid is the pollutant emitted from Gavin in the largest
quantities; in 2005, an estimated 1.8 million pounds of sulfuric acid was
released from the stack in Cheshire. The high quantities of sulfuric acid
are the result of burning high-sulfur coal, some of which is found in the
Appalachian coal-mining region. It is notable that these releases occur even
with state-of-the-art pollution-control equipment.

AEP reports on more than twenty pollutants in the TRI. In addition
to sulfuric acid, hydrochloric acid, vanadium, manganese, copper, barium,
and cobalt compounds are reported. Heavy-metal compounds such as lead,
arsenic, and mercury also are emitted into the environment.

5.2. ENVIRONMENTAL EXPOSURES

In order to comply with air-quality regulations for nitrogen dioxide, AEP
installed special pollution-control equipment at the Gavin Power Plant in the
spring of 2001. This equipment, known as the selective catalytic reduction (SCR)
system, introduces ammonia into the smokestack to reduce nitrogen oxides.
The result was unexpected in that a chemical reaction occurs in the smokestack
that created sulfuric-acid mist, causing localized air-pollution events.

Almost immediately following the installation of the pollution-control
equipment, residents of Cheshire began to experience "plume touchdowns"
in their community. This plume mainly contained sulfuric acid, although
nitrogen oxides and particulate matter also were constituents. Community
members began complaining of acute health effects, including eye irritation,
respiratory problems, and sore throats.

During the summer of 2001, AEP commenced an air-quality monitoring program to quantify the amount of sulfur compounds affecting the
local community. In addition, Ohio EPA gathered ambient air-quality data
from their monitoring station on top of the Cheshire Village Hall. The
data from AEP and Ohio EPA showed that sulfuric acid in the local ambient
air exceeded 100 ppb numerous times. One Ohio EPA sample, from July
2001, registered a level of 565 ppb for 5 min; then it went off the scale. This
reading is more than four times higher than the regulatory standard.

In February 2002, the Agency for Toxic Substances and Disease Registry completed a Health Consultation for Cheshire (ATSDR, 2002). Based on their analysis of the ambient air-monitoring data and published health-effects research, ATSDR concluded that the sulfur dioxide and sulfuric acid levels in Cheshire posed "a public health hazard to some residents." Specifically, ATSDR identified asthmatics and children as being most susceptible to the high levels of these pollutants. ATSDR recommended that AEP take steps to reduce the emissions of sulfur from the Gavin Power Plant.

5.3. THE REMEDY

In an unprecedented move, American Electric Power worked with attorneys for village residents, and offered to pay $15 million to purchase homes and attempt a buy-out of Cheshire in 2002. The total amount AEP ultimately paid was $20 million, with approximately $5 million going to the attorneys who negotiated the agreement.

Residents who accepted the offer to purchase their homes could choose to move from the village or could remain in the village in their homes. However, whether the residents moved or not, they agreed not to sue AEP for current or future health effects that might be linked with exposure to pollutants from the Gavin Power Plant.

While most of the residents did accept the utility's offer, several did not. Today, Cheshire is still a village with an elected government and a public school system.

5.4. GLOBAL IMPLICATIONS

The Cheshire case is fascinating from both legal and economic perspectives. According to Parchomovsky and Siegelman (2004), the buy-out in Cheshire contradicts expected results in such a situation. Specifically, they explain that even though the village residents would likely have won a nuisance lawsuit against AEP, perhaps leading to substantially more monetary damages than the buy-out offer, no one in the village took steps to sue AEP.

Economic theory anticipates that there would be some holdouts in the community that might refuse to sell in order to force AEP to pay them more than the standard price for their property. Parchomovsky and Siegelman note that there were no holdouts in the community. Either the residents took the offer and moved, or they took the money and stayed in their homes. There may be many reasons that there were no holdouts in the community, thus this situation might suggest that the right economic incentives can solve environmental problems by removing the exposed population.

Another interesting factor related to the Cheshire case is that there was almost no involvement from environmental activists outside of the community. This is the opposite of the C8 case in which national and state environmental activists are playing an important role. This suggests that minimizing involvement of non-local players, that is people who are not directly affected, may lead to enhance negotiation and an ultimate solution. The Cheshire residents never demanded a health study; they agreed to accept an economic solution.

One global implication of the Cheshire case is that it may be a model for other facilities that have attempted technological pollution controls with limited success to address public health concerns of surrounding populations. Essentially, this case is a model for polluting facilities to follow that could be financially desirable by minimizing litigation and damage claims. Since the Cheshire case, there have been reports of other scenarios being proposed similar to the AEP buy-out, leading some to suggest that the Cheshire case has set a "dangerous precedent" (Buckley et al., 2005).

Aside from the possible precedent that this case set, it is another example of how residents of a poverty-stricken area have endured a lower quality of life because of environmental conditions that are beyond their control. For many years, the residents of this small village in Appalachia, Ohio, endured environmental exposures so that others far removed from their village could have cheap electricity. This is a scenario that plays out all over the world, as natural resource-rich but poverty-stricken communities supply resources to fuel the global economy.

6. Conclusions

Residents in Appalachia, Ohio, similar to residents in many poor, rural areas, have been subject to numerous adverse environmental conditions that others in the state of Ohio have not. Some of these conditions, such as coal mining and electric power generation, are tied to the resources that are available in the region. Others, such as chemical manufacturing, are based on the long history of industry.

Regardless of the inequality of the environmental exposures, it is becoming evident that this population is experiencing health disparities. Some of these disparities are unequal and some are inequitable. The unequal health outcomes are tied to lifestyle choices, such as a lack of exercise and the propensity for low-SES populations to smoke. The underlying reasons for these lifestyle choices can be complicated and may be related to a lack of educational and health infrastructure.

Other health outcomes, especially those that are acute, such as the respiratory effects associated with the power plant in Cheshire, could be deemed inequitable. Populations who suffer acute health effects often are the targets for additional environmental exposures due to their lack of political power.

The cases in this paper present two different outcomes.[5] In the C8 case, the population has agreed to serve as human subjects for a large health study that could have global ramifications in the realm of public health. On the other hand, in the AEP case, the residents did not participate in a health study, but agreed to be compensated for their suffering and to move away from its source or remain near the source with their money.

A further distinction between the cases is that there appears to be a difference in how the local population reacts based on whether the environmental exposure is perceived as inequitable or unequal. In the C8 case, residents are outraged, in part, because they feel as if the exposure is unfair, since the contaminant is coming from a facility in another state.[6] In addition, there is no perceived benefit to the local community from production of the contaminant; rather the benefits are perceived to be to the national and international community because of the products derived using C8. The potential health effects are unknown, and most likely chronic, in the C8 case.

In the Cheshire case, the health effects are acute and more certain. In addition, the local community benefited somewhat, albeit only slightly, from the plant being located there in terms of economic development. So, while there were health disparities, they were more likely to be perceived as unequal rather than inequitable. Both of these cases set precedents that may lead to improvements in our understanding of the political, social, and environmental factors that lead to health disparities in rural populations.

As we examine health disparities and environmental exposures, these two cases can be instructive. In understanding whether health effects are perceived by those affected as fair or unfair may lead to different approaches to solving environmental problems. For example, if the population believes that they are suffering adverse health effects because of injustice, it may be more difficult to negotiate a solution. Furthermore, the involvement of professional environmental activists from outside the affected area might contribute to this perception of inequity.

The cases might also give hope to some facilities that emit pollution, in that there is precedence for an economic solution to their environmental problem. The implication of this is that once we head down this path of buying out communities who might be affected by pollution, where will it end? Hopefully we will never come to a point in which poor communities are vying for pollution as a way to solve their economic problems.

---

[5] As noted above, the Coolville Medical Waste Incinerator case is not discussed here due to incomplete information.

[6] Based on public comments made in an open forum at Ohio University on Tuesday, 8 May 2007.

322   M. MORRONE

References

Appalachian Regional Commission, Appalachian Regional Development Act of 1965, as amended February 2007; http://www.arc.gov/index.do?nodeId=1243, accessed 21 August 2007.

ATSDR, 2002, Health Consultation, Gavin Power Plant, Cheshire, Gallia County, Ohio; http://www.atsdr.cdc.gov/hac/pha/gavinpower/gpp_toc.html, accessed 6 May 2007.

Barton, C. A., Butler, L. A., Zarzecki, C. J., Flaherty, J., and Kaiser, M., 2006, Characterizing perfluorooctanoate in ambient air near the fence line of a manufacturing facility: Comparing modeled and monitored values, *J. Air Waste Manage.* **56**(1):48–55.

Bryant, B., and Mohai, P., eds., 1992, *Race and the Incidence of Environmental Hazards: A Time for Discourse*, Westview, Boulder, CO.

Buckley, G. L., Bain, N. R., and Swan, D. L., 2005, When the lights go out in Cheshire, *Geogr. Rev.* **95**:537–555.

Bullard, R., ed., 1994, *Unequal Protection: Environmental Justice and Communities of Color*, Sierra Club Books, San Francisco, CA.

Butenhoff, J. L., Gaylor, D. W., Moore, J. A., Olsen, G. W., Rodricks, J., Mandel, J. H, and Zobel, L. R., 2004, Characterization of risk for the general population exposure to perfluorooctanoate. *Regul. Toxicol. Pharm.* **39**(3):363–380.

Carter-Pokras, O., and Baquet, C., 2002, What is a "health disparity"? *Public Health Rep.* **117**(5):426–434.

Clark, W. A. V., 2007, Race, class, and place: Evaluating mobility outcomes for African Americans, *Urban Aff. Rev.* **42**(3):295–314.

Davis, K. L., Aucoin, M. D., Larsen, B. S., Kaiser, M. A., and Hartten, A. S., 2007, Transport of ammonium perfluorooctanoate in environmental media near a fluoropolymer facility, *Chemosphere* **67**(10):2011–2019.

DUPONT, 2003, EPA will study safety of chemical used to make Teflon products, Class Action Reporter; http://bankrupt.com/CAR_Public/ 030421.mbx, accessed 15 May 2007.

Energy Information Administration, 2005, *100 Largest Electric Plants*; http://www.eia.doe.gov/neic/rankings/plantsbycapacity.htm, accessed 18 April 2007.

Evans, G. W., and Kantrowitz, E., 2002, Socioeconomic status and health: The potential role of environmental risk exposure, *Annu. Rev. Publ. Health* **23**:303–331.

Ewing, R., Schmid, T., Killingsworth, R., Zlot, A., and Raudenbush, S., 2003, Relationship between urban sprawl and physical activity, obesity, and morbidity, *Am. J. Health Promot.* **18**(1):47–57.

Ewing, R., Brownson, R. S., and Berrigan, D., 2006, Relationship between urban sprawl and weight of United States youth, *Am. J. Prev. Med.* **31**(6):464–474.

Farmer, M. M., and Ferraro, K. F., 2005, Are racial disparities in health conditional on socioeconomic status? *Soc. Sci. Med.* **60**(11):191–204.

Hambrick-Dixon, P. J., 2002, The effects of exposure to physical environmental stressors on African American children: A review and research agenda, *J. Children Poverty* **8**(12):23–34.

Houston, D., Wu, Jun, Ong, P., and Winire, A, 2004, Structural disparities or urban traffic in southern California: Implications for vehicle-related air pollution exposure in minority and high-poverty neighborhoods, *J. Urban Aff.* **26**(5):562–592.

Jackson, R. J., and Kochtitzky, C., n.d., *Creating a Healthy Environment: The Impact of the Built Environment on Public Health*, Sprawl Watch Clearinghouse, Washington, DC, review article; http://www.sprawlwatch.org/health.pdf, accessed 10 May 2007.

Lopez, R., 2002, Segregation and black/white differences in exposure to air toxics in 1990, *Environ. Health Perspect.* **110**(suppl. 2):289–296.

Lopez, R., 2004, Urban sprawl and risk for being overweight or obese. *Am. J. Public Health* **94**(9):1574–1579.

Lyons, C., 2007, *Stain Resistant, Nonstick, Waterproof, and Lethal*, Praeger, Westport, CT.

McMichael, A. J., 2000, The urban environment and health in a world of increasing globalization: Issues for developing countries, *B. World Health Organ.* **78**(9):1117–1126.

Murray, C. J. L., Kulkarni, S., and Ezzati, M., 2005, Eight Americas: New perspectives on U.S. health disparities, *Am. J. Prev. Med.* **29**(suppl. 1):4–10.

Nordstrom, C. K., Diez Roux, A. V., Jackson, S. A, and Gardin, J. M., 2004, The association of personal and neighborhood socioeconomic indicators with subclinical cardiovascular disease in an elderly cohort: The cardiovascular health study, *Soc. Sci. Med.* **59**(10):2139–2147.

Ohio Cancer Incidence Surveillance System, 2006, *Cancer Incidence and Mortality among Ohio Residents, 1999–2003*; http://www.odh.ohio.gov/ odhPrograms/svio/ci_surv/ci_reports1. aspx, accessed 18 April 2007.

Olsen, G. W., Church, T. R., Miller, J. P., Burris, J. M., Hansen, K. J., Lundberg, J. K., et al., 2003, Perfluorooctanesulfonate and other fluorochemicals in the serum of American Red Cross adult blood donors, *Environ. Health Perspect.* **111**(16):1892–1901.

Parchomovsky, G., and Siegelman, P., 2004, Selling Mayberry: Communities and individuals in law and economics, *Calif. Law Rev.* **92**(1):75–146.

Perfluorooctanoic Acid (PFOA), Fluorinated telomers; Request for comment, solicitation of interested parties for enforceable consent agreement development, and notice of public meeting, 2003, 68 Fed, Reg. 18,626, 16 April.

Perlin, S., Sexton, K., and Wong, D., 1999, An examination of race and poverty for populations living near industrial sources of air pollution, *J. Expo. Anal. Environ. Epidemiol.* **9**(1):29–48.

Perlin, S., Wong, D., and Sexton, K., 2001, Residential proximity to industrial sources of air pollution: Interrelationships among race, poverty, and age, *J. Air Waste Manage.* **51**(3):406–421.

Prus, S. G., 2007, Age, *SES*, and health: A population level analysis of health inequalities over the lifecourse, *Sociol. Health Illness* **29**(2):275–296.

Resnik, D. B. and Roman, G., 2007, Health, justice, and the environment, *Bioethics* **21**(4):230–241.

U.S. Department of Interior, Office of Surface Mining, 2003, People potentially at risk from priority 1 and 2 AML hazards, white paper; http://www.osmre.gov/ pdf/wp041703.pdf, accessed 22 March 2007.

Weaver, M. E., Katz, M., Fickle, D., and Paskett, E. D., 2006, Risky behaviors among Ohio Appalachian adults, *Prev. Chronic Dis.* **3**(4):1–8; www.cdc.gov/pcd/issues/2006/oct/06 _0032.htm, accessed 14 April 2007.

SECTION IV

ACTING ON HAZARD IMPACTS: EXAMPLES FROM SUB-SAHARAN AFRICA, EASTERN EUROPE, AND CENTRAL ASIA

POVERTY-ENVIRONMENT LINKAGES AND THEIR IMPLICATIONS FOR SECURITY

With Reference to Rwanda

D. H. SMITH*

*Manager—Africa United Nations Development Programme
—United Nations Environment Programme Poverty
and Environment Initiative (UNDP-UNEP PEI)*

Abstract: In many parts of the world the degradation of natural resources is worsening while the social and economic conditions of people are not being improved. In some parts of the world, both poverty and environmental degradation are increasing. One reason for this is the perception held by some that the sustainable management of the environment and economic development are competing priorities. Consequently, national development strategies fail to include the necessary actions and budgets to promote environmentally sustainable resource use. While there are specific cases where measures to end environmentally unsustainable actions will restrict human use of environmental resources in the short term, over the long term environmentally unsustainable natural resource use will reduce the social and economic benefits to humans provided by the environment. Thus, the environment should not be treated as a competitor but as the natural resource base of human social and economic development. In some parts of the world, there are direct and indirect links between environmental degradation, poverty, and security. This paper, which is aimed mainly at policy decision makers, outlines the links between environmentally sustainable use of natural resources and poverty reduction, with their implications for security. It provides examples, especially from Rwanda, to illustrate the nature of these links. It then provides recommendations for environmentally and economically sustainable natural resource use that will also enhance human security.

Keywords: Economic development; environment; poverty; sustainable development, security

---

* Address correspondence to David.Smith@unep.org

P.H. Liotta et al. (eds.), Environmental Change and Human Security, 327–340.
© 2008 *Springer Science + Business Media B.V.*

1. Introduction—Environmental Degradation, Poverty, and Security[1]

Natural resources, such as land and water, are fundamental requirements for human survival and socioeconomic development. While levels of dependency on natural resources for socioeconomic development vary in different parts of the world, in countries like Rwanda, natural resources such as land remain the key economic resource. In Rwanda and most of Africa, agriculture is the dominant economic activity, with, for example about 90 percent of Rwanda's population employed in the agricultural sector. Unfortunately, the degradation of natural resources is a serious problem in significant parts of Africa and also globally. As a result, the economic and social well-being of hundreds of millions of people is being directly or indirectly threatened by environmentally unsustainable natural resource use, particularly unsustainable land and water use, which is resulting in major physical damage and diminished economic benefits.

It is a fact that environmentally unsustainable natural resource use over time reduces the productivity and therefore the social and economic benefits produced by such resources as land and water. Thus, over time, environmentally unsustainable resource use can increase poverty.

Additionally, if environmentally unsustainable use of natural resources in one country has an impact on neighboring countries, increased political tension can result. For example, transboundary waters management issues have been a source of tension in Central Asia and Africa. Within countries such as Kenya, competition over scarce and unsustainably managed pasture and water resources has resulted in armed conflict.

Yet, despite global and national recognition of the problems caused by environmentally unsustainable natural resource use, degradation problems are worsening in many parts of the world. There are a number of reasons for this, including a lack of capacity, inappropriate natural resource policy and management, insufficient financial resources, and a lack of political commitment. A failure to apply integrated environmental-economic management approaches is a key problem.

Underlying the above reasons why insufficient action is being taken to reduce unsustainable natural resource use is that the perception still exists in too many cases that the issue is one of development versus environmental protection. That is, reducing environmentally unsustainable natural resource use comes at the expense of economic development. While in specific cases,

---

[1] Content footnote: Substantive parts of this paper are based on original work by the author carried out as part of his position in UNEP.

there will be instances where human activities will have to be restricted to achieve environmentally sustainable resource use, in general terms over time, unless environmentally sustainable use of key natural renewable natural resources is achieved, people will suffer economic and social costs. Over time, environmentally sustainable natural resource use is an essential factor in achieving sustainable economic and social well-being. Land and water resource use provides clear examples of this.

2. The Costs of Environmentally Unsustainable Land and Water Use

The table below outlines some of the negative economic and social impacts of environmentally unsustainable land and water use (Table 1). It is not exhaustive, but highlights that reduced production, increased input costs,

TABLE 1. Costs of Environmentally Unsustainable Land Use.

Environmentally unsustainable action	Result	Economic impacts
1. Overgrazing	Reduced grass growth	Decreased numbers of livestock
	Soil degradation	Decreased productivity
	Erosion	Reduced incomes
2. Water pollution from pesticide and fertilizer use and/or from farm livestock effluent	Water use restricted, depending on degree of pollution	Reduced farm production due to restricted availability of safe water
		Increased water treatment costs
	Increased incidence of waterborne diseases	Increased cost of obtaining safe water
		Increased incidence of ill-health and medical treatment costs
		Decreased incomes
3. Depletion of water resources through excess withdrawals from surface and groundwater sources	Decreased water availability	Decreased production as less water is available (fewer livestock, less crops)
		Increased costs of obtaining water
		Decreased net incomes
4. Inefficient irrigation	Soil salinization	Larger water bills
	Other soil degradation	Decreased soil productivity, production, and incomes
	Water waste	

(continued)

TABLE 1. (continued)

Environmentally unsustainable action	Result	Economic impacts
		Ultimately, complete cessation of livestock or crop production
5. Deforestation—i.e., failure to manage forests for sustainable yields and complete cutting of areas of forest	Erosion causing increased sedimentation downstream, reduced water retention due to increased runoff, soil degradation, less timber	Decreased productivity downstream from sedimentation
		Decreased rainfall as deforestation impacts on localized climate patterns
		Decreased productivity on site over time if erosion and soil degradation occurs
	Increased vulnerability to flooding	Decrease in timber resources
		Decreased incomes and increased costs

and health costs are real problems—all of which reduce net incomes. Thus, over time, attempts to reduce poverty will be inhibited by environmentally unsustainable land and water use.

The case of the Aral Sea basin is perhaps the best known example of how environmentally unsustainable land and water use can have serious economic consequences. The environmentally unsustainable use of land and freshwater resources for agricultural production in the Aral Sea basin has decreased lake water levels and quality severely. As a result, the fishing industry has collapsed. In addition, the inefficient use of irrigation water in this semiarid region has led to salinization and a subsequent decrease in agricultural production. Serious human health problems have arisen with windblown dust contaminated by agricultural chemical residues (UNDP, 1995; UNEP, 1993).

In Africa, a similar, though not as extreme case is that of the Lake Chad basin. This lake has drastically shrunk in size, with very significant economic consequences for the people whose livelihoods depend on the lake's water and fish resources.

Two more specific examples reinforce the points made above (International Water Management Institute, n.d.):

- Xian County, Fuyang River Basin, North China Plain: Formerly dependent on surface water resources, increasing water demand and upstream developments have led to an increasing dependence on groundwater in this area. However, excess withdrawals have made the water table fall substantially. In order to obtain sufficient water, tubewell depths have increased from 25 m in the 1980s to 200 m plus in the 1990s. The total cost of a 200 m tubewell is about US $3,900, nearly four times the cost of a 25 m tubewell. As water supplies have declined, grain yields have declined by 25 percent.

• Jambul, Kazakhstan: Located in the Aral Sea basin, this area has suffered from increasing salinization due to poor irrigation practices. Salinization is reducing farm productivity and contaminating drinking water supplies. Grain yields are declining as a result of salinization, from 4 t/ha to less than 2, in one case study.

In summary, unsustainable land and water use imposes costs through reduced production and health problems. While there is much focus on the costs of environmental protection, there is not enough focus on the costs of using environmental resources such as land and water unsustainably. This is partly because these costs are often borne by those who have the least say—the poor. Over time, poverty can be increased by unsustainable land and water use. However, this link between poverty and the environment is not sufficiently understood. For example, the link between soil erosion and decreased agricultural productivity is not sufficiently understood. Consequently, environmentally sustainable natural resource use is usually not a priority in national development plans. The long-term impact of this is that achievement of economic development priorities, including food security, will be undermined.

A key to understanding the contribution of environmentally sustainable land and water use to economic development, including poverty reduction, is the concept of ecosystem services.

3. Ecosystem Services: The Basis for the Contribution of the Environment Toward Economic Development and Poverty Reduction

The environment generates social and economic benefits for humans through ecosystem services, which are defined as the conditions and processes through which natural ecosystems contribute to human economic and social well-being. Ecosystem services provide benefits for humans through (UNEP-IISD, 2004):

• Provisioning: natural resources used for economic activities (e.g., food, fuel, plant and animal products, energy, fiber, non-living resources, and water).

- Regulating: life supporting functions for humans (e.g., purification of air and water, mitigation of floods and droughts, decomposition of wastes, generation and renewal of soil).
- Enriching: cultural and religious services (e.g., spiritual components, aesthetic values, social relations, education, and scientific values).

Relating the impacts of land and water degradation listed in the table to the ecosystem services listed above indicates how land and water degradation, over time, reduces the ability of land and water to provide provisioning, regulating, and enriching services. Overgrazing that results in reduced grass cover, erosion, and soil degradation reduces production and incomes, reduces the ability of land to purify itself from pollution, increases the risk of flooding, and inhibits the renewal and regeneration of soil. Land and water degradation also impacts on aesthetic values—for example, tourists do not wish to spend money to view eroded and deforested areas or polluted lakes.

It is also important to note the interdependency between the components of ecosystem services. For example, overharvesting, overuse, misuse, or excessive conversion of ecosystems into human or artificial systems damages regulating services, which in turn reduces the flow of the provisioning service provided by ecosystems.

All people depend on services provided by ecological systems. Yet, the poor are more heavily dependent on these services than the rich, since the rich can buy food or clean water or build appropriate shelters to isolate themselves from economic and social problems caused by land degradation or other environmentally unsustainable behavior.

Unfortunately, as indicated above, the link between environmental sustainability and poverty reduction is not sufficiently understood, nor is this link integrated into economic developing planning. Thus, poverty reduction strategies fail to reflect that ecosystem services are vital contributors to human social and economic well-being. More specifically in the context of land and water use, poverty reduction strategies fail to include policy, management, and capacity building actions to achieve environmentally sustainable land and water use in a manner that reduces poverty.

4. Environment for Development

While the perception that the issue is one of environment versus development, the stress at the World Summit on Sustainable Development (WSSD) on the natural resource base of sustainable development, plus the discussion above, highlight that it is more accurate to refer to how sustainable use of the environment and economic development, over time, reinforce each other—particularly in countries heavily dependent on natural resources such as land and water. That is, it is more realistic to refer to environment for development.

While some may argue that there is no choice but to use the environment in an unsustainable manner—for example, to continue to overgraze because there is no alternative—these arguments are rarely based on an assessment of the costs and benefits of such environmentally unsustainable actions over time. What, for example, are the true costs of overgrazing—in terms of lost productivity in the long term? Or the true costs in terms of the cost of restoring degraded land? (Even if restoration is possible, it is likely to be more expensive than sustainable management would have been in the first place and to take many years to achieve.) Often environmentally unsustainable activities merely postpone the day when alternative economic activities have to be found. In other words, sustainable management now will save money later.

In summary, it is a mistake to assume that environmental sustainability inhibits economic development over time. Rather, the environment is an essential pillar of economic development and thus environmentally sustainable resource use should be a key objective of economic planning.

A key to operationalizing the concept of environment for development is to fully integrate the objective of the achievement of environmental sustainability into national development processes such as poverty reduction strategies and into sectoral development strategies and plans.

In recognition of the importance of integrating environmental sustainability into national development processes such as poverty reduction strategies, the United Nations Environment Programme and the United Nations Development Programme have a joint program—the UNDP-UNEP Poverty and Environment Initiative—which aims at increasing the capacity of developing countries to mainstream environmental sustainability into their development strategies (UNDP-UNEP, 2007). Such mainstreaming will ensure that poverty reduction is not undermined by unsustainable use of environmental resources.

The program assists countries to identify how environmental sustainability and poverty reduction are linked in concrete terms. That is, how specific environmental resources such as land and water contribute to the economy and how environmentally unsustainable use of such resources will make it more difficult to reduce poverty in the long term. Then the country programs demonstrate how such environment-poverty linkages can be integrated into poverty reduction strategies and other national development processes in order to ensure that development and poverty reduction strategies are sustainable. For example, the program demonstrates how countries can ensure that the policies, management tools, legal frameworks, and institutions in their national development processes are consistent with sustainable land

and water use in order to ensure that economic growth and poverty reduction is not undermined by environmentally unsustainable behavior. Rwanda is one of the countries where this program is being implemented.

5. Poverty-Environment-Security Linkages

Links between poverty environment and security have been well described by such authors as Homer-Dixon—including in Rwanda (Percival and Homer-Dixon, 1995). A graphical representation of the links is contained in Figure 1. It shows, for example, how supply- or demand-induced scarcity can lead to conflict. It is also important to recognize that armed conflict can also have environmental impacts—which can in turn have negative socioeconomic impacts. For example, in Rwanda forested areas were destroyed in the aftermath of the genocide by displaced persons.

Armed clashes over scarce water and pasture in Kenya, cross-border armed clashes over grazing between Uganda and Kenya, and the atrocities in Darfur are examples of how poverty and competition over resources like land and water can result in security problems or be a significant contributory factor.

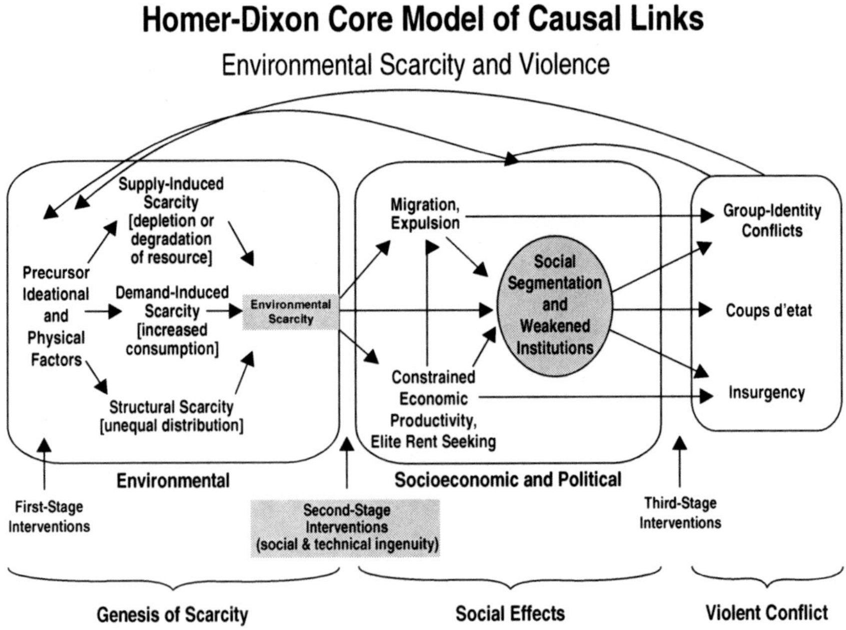

Homer-Dixon Core Model of Causal Links

Environmental Scarcity and Violence

Figure 1. Homer-Dixon Core Model of Causal Links.
Source: Adapted from Homer-Dixon, Environment, Scarcity, and Violence (2001).

Competition over transboundary water resources has potential security implications—for example in Central Asia over rivers that flow into the Aral Sea and in Africa over the Nile—as nations suffer economically from resource use actions by others.

There is also increasing concern that the global environment problem of human-induced climate change will have serious security impacts—as warned of by the United Kingdom's Foreign Secretary at the United Nations Security Council.

In summary, the work by Homer-Dixon and examples such as those referred to above demonstrate that there are links between poverty, environment, and human security. If a key basis for these links is competition over natural resources such as land and water, then it logically follows that efforts to improve the environmental sustainability of resource use will reduce the potential for conflict over such resources. Steps for improving the environmental sustainability of natural resource use will be discussed below after examples of poverty-environment linkages from Rwanda (including any security implications) are outlined.

6. Rwanda Poverty-Environment Linkages and Implications for Security

Many of the figures used in this section are based on ongoing work commissioned by UNDP and UNEP to be published in a report entitled "Economic Analysis of Natural Resource Management in Rwanda" (Musahara et al., forthcoming).

Rwanda is a land-locked hilly country with a population of about eight million and an average population density of about 377 persons per square kilometer. It is a least developed country (LDC) with a low per capita GNP of about US $250. Population growth was estimated to be 2.8 percent in 2002. About 90 percent of the population is dependent on agriculture, which contributes about 41 percent of GDP and 72 percent of all exports. However, the very high and increasing population density has resulted in very high pressure on available land resources, with environmentally unsustainable land use practices including soil erosion, overcultivation, and deforestation increasingly reducing agricultural productivity. The hilly terrain in Rwanda makes it particularly vulnerable to unsustainable land use practices.

Clay and Lewis (1996, quoted in Mushara et al.) estimated that, in one area of Rwanda, farms suffering from serious soil erosion had a 21 percent loss in agricultural productivity compared with land that was not significantly eroded and that soil erosion is moderate to severe on 50 percent of land in Rwanda. The government estimates current rates of soil loss per year causes production losses equivalent to feeding 40,000 people per year. (Government of Rwanda, 2004). Mushara et al. (forthcoming), estimated that soil erosion caused economic losses equivalent to 1.9 percent of GDP per year.

Given the extent of poverty, population density, and the dependence on agriculture, these are very serious losses that are likely to significantly increase pressure on poor rural peoples' economic survival.

In a study of the depletion of the Giswhati Forest, Mushara et al. (forthcoming) documented how a forest of over 280 km^2 in the 1970s has essentially ceased to exist, through a variety of causes including a deliberate conversion program to larger-scale agriculture and the impact of thousands of displaced persons after the genocide. Despite the conversion to agriculture, the loss of the forest has resulted in a number of negative impacts on local poor people, which reinforces the importance of links between poverty and environment and suggests that local rural poor people are worse off as a result of deforestation. Local people used to depend on the forest for wild fruits and vegetables, animals, firewood, and medicinal plants. The shortages of fuel-wood have health impacts as water is not boiled properly and food not cooked adequately. Waterborne diseases and negative nutritional impacts have resulted. Deforestation has resulted in soil erosion, loss of soil fertility, siltation, and increased flooding. The increased siltation in lakes used for hydroelectricity generation stations has reduced electricity production.

A more significant loss in hydroelectric generation has resulted from the conversion of large areas of the Rugezi wetland to agriculture. Population pressure and land degradation elsewhere are prime causes for the conversion of areas of this wetland into agriculture. While agricultural production is a national priority, serious losses of fish, wild animals, and other natural resources have been reported. Increased flooding has also resulted.

The most direct and serious economic impact from the conversion of large areas of wetland is reduced water flows into the main Rwandan hydroelectricity generating reservoirs, a reduction of 50 percent in the average flow compared to before large-scale wetland conversion occurred. Electricity shortages have resulted from the decline in reservoir inflows—with Rwanda having to import generators and fossil fuel to compensate for these production losses. The result has been continued shortages and large increases in electricity prices.

While the prime driving forces of the genocide lay in ethnic rivalries, it is quite possible that aspects of the genocide in some locations, once it had commenced, also reflected the intense competition for scarce land resources (Diamond, 2005). Certainly the mass displacement of people as a result of the genocide has had a serious impact on the environment in some areas—particularly through deforestation. It is estimated that about 30 percent of the population was displaced during the genocide, which also resulted in an estimated 50 percent fall in GDP in one year. Diamond's views about the intensity of competition over land are supported by recent unofficial reports of intra-family disputes over land in Rwanda resulting in husbands killing wives.

Due to the very high population density in Rwanda, the country is particularly vulnerable—physically, socially and economically—to environmental degradation. However, while addressing environmentally unsustainable resource use is critical, the high population density makes it more difficult to do so, because there is less room to maneuver. Traditional soil conservation methodologies include letting some land lie fallow. Sustainable agriculture in hilly terrain involves terracing, tree planting for stability, and not farming some land. These techniques would reduce the intensity of agriculture in some areas and the amount of land actually farmed, and given the dependence of people on agricultural production for survival, this would involve some difficult choices. The short-term pain would have longer-term benefits, but the short-term impacts may not be sustainable without specific measures to compensate for them.

However, every day that action is not taken, the pressure on land and water resources increases, and the room to maneuver decreases even more. Poor people will find it increasingly difficult to survive and resentment will grow. Therein lies the link with security—competition between poor people over a declining natural resource base when there is no real viable alternative for them may lead to conflict. Rich people can buy alternatives; poor people cannot.

If one considers the situation in Rwanda in terms of the Homer-Dixon model of causal links between environment, scarcity, and violence (Figure 1), immediate relevance is seen. For example, a relevant physical factor in Rwanda is the steep terrain that characterizes large areas of the country. Under the "Supply-Induced Scarcity" box, the impact of soil erosion and overcultivation (which, inter alia, depletes soil nutrients) is to reduce the agricultural productivity of Rwanda's land, which effectively reduces the supply of land.

The increasing population increases demand for land and the combination of hilly terrain, plus high and increasing population density, means that environmental scarcity is an increasing issue in Rwanda such that there is worsening competition over an increasingly scarce per capita resource. This combination also results in constrained economic productivity.

In terms of social and institutional factors, the traditional custom of dividing land between sons is a key contributor to the trend of decreasing the average size of plots of land to below minimum economic size. Group-identity conflicts or potential conflicts are key underlying factors in Rwanda, and competition over land can spark or worsen actual physical conflict.

In summary, the situation in Rwanda is a potentially dangerous one—the overwhelming dependence of 90 percent of the population on land, increasing population density, poverty, worsening environmental degradation, a lack of alternatives, and a trigger could combine to lethal effect.

In terms of addressing the problems, the government of Rwanda has demonstrated a clear will to address the country's problems, and issues that have crippled development in other countries, such as corruption, are not an issue in Rwanda. Additionally, the government has made environment a priority issue in its Economic Development and Poverty Reduction Strategy (Government of Rwanda, 2007). The next challenge will be to operationalize that priority in sectoral plans and strategies and their implementation on the ground. It will be very important that the international community does not fail to provide substantive support to Rwanda in its endeavors to address environmental and other problems.

7. Recommendations for Achieving Environmentally Sustainable Natural Resource Use

7.1. IDENTIFICATION OF THE ECONOMIC COSTS OF THE UNSUSTAINABLE USE OF ENVIRONMENTAL NATURAL RESOURCES AND OF THE BENEFITS OF ENVIRONMENTALLY SUSTAINABLE USE

Concrete evidence of the links between environment and poverty reduction is vital to convince government officials and ministers that increased investment in, for example, sustainable land management and cutting water pollution, should be development priorities. Such evidence is likely to be very important in generating political commitment for achieving environmentally sustainable natural resource use. Arguably, the most important factor for achieving environmentally sustainable natural resource use is real commitment by the relevant decision makers at all levels that environmentally sustainable resource use is a core national development objective. With such commitment, it is far more likely that the necessary policies and management plans will be developed and implemented.

In some countries, it would also be relevant to highlight environment-poverty-security linkages. For example, how unsustainable land and water management is increasing tension or actual clashes between different groups.

7.2. FULL INTEGRATION OF ENVIRONMENTAL SUSTAINABILITY INTO POVERTY REDUCTION STRATEGIES, OTHER PRO-POOR NATIONAL DEVELOPMENT PROCESSES, PLUS SECTORAL POLICY AND MANAGEMENT PLANS

As indicated above, the rationale for this is that economic development and poverty reduction are enhanced by environmentally sustainable natural resource use and that using such resources in an environmentally unsustainable manner reduces the economic and social benefits that land and other natural resources generate.

Full integration means that budgetary allocations reflect that environmental sustainability is a core objective. As a practical example in the case of agriculture, it would mean prioritized funding for sustainable land use practices, including measures to combat soil erosion and overcultivation.

7.3. APPLICATION OF INTEGRATED MANAGEMENT APPROACHES BASED ON ENVIRONMENTALLY SUSTAINABILITY AS A CORE GOAL

1. Ecosystem management

   Ecosystem-based management ensures that the relevant ecosystem inter-relationships—plus their economic and social implications—are taken into account when managing the environmental issues under consideration. A key aim of ecosystem management is to ensure that the social and economic benefits generated by ecosystems are not degraded to the extent that human economic development is undermined.

2. Integrated land and water management including integrated environmental-economic-social management

   Integrated water resources management (IWRM) is integrated land and water management, and is the internationally accepted best practice management process for land and water management. Land and water are closely linked in ecological, economic and social terms; thus IWRM incorporates integrated environmental, economic, and social management, and is also consistent with sustainable development, which has three elements—social, economic, and environmental.

3. Multi-objective planning

   A key tool for integrated, inter-sectoral approaches is multi-objective planning that bases management decisions on an integrated assessment of environmental, economic, and social factors. The elements of multi-objective planning are:

   (i) Cost-benefit analysis from the national perspective

   (ii) Cost-benefit analysis from the project or regional perspective

   (iii) Environmental impact analysis

   (iv) Social impact analysis (usually non-monetized)

   Cost-benefit analysis should include economic, social, and environmental impacts. If it does not, then it will not include all the relevant impacts. Some impacts cannot be costed in monetary terms, but they should at least be identified.

8. Conclusions

Environmentally unsustainable use of natural resources is a serious problem in many parts of the world. Environmentally unsustainable resource use is, over time, a threat to economic development—including poverty reduction. Over time, environmentally unsustainable use of natural resources such as land and water reduces incomes, increases costs, and worsens poverty. This is because ecosystems generate social and economic benefits for humans and these benefits are, over time, decreased and even destroyed by environmentally unsustainable activities such as overgrazing, inefficient irrigation, and deforestation. Unsustainable use of natural resources can also have security implications, if it results in increasing competition for such resources that then results in national or subnational political tensions and conflict. Thus, to help achieve economic development, reduce poverty, and reduce the potential for conflict, natural resources should be managed in an environmentally sustainable manner. In doing so, the negative human security implications of the types of environmental change described above will be reduced or eliminated.

References

Clay, D., and Lewis, L. A., 1996, *Land Use, Soil Loss, and Sustainable Agriculture in Rwanda,* Michigan State University/MINAGRI, Chicago/Kigali.
Diamond, J., 2005, *Collapse—How Societies Choose to Fail or Survive,* Penguin, London.
Government of Rwanda, 2004, *National Land Policy,* MINITERE, Kigali.
Government of Rwanda, 2007, *Economic Development and Poverty Reduction Strategy* (draft), Kigali.
International Water Management Institute, n.d., *Comprehensive Assessment of Water Management in Agriculture: Implications of Land and Water Degradation on Food Security,* Colombo.
Mushara, H., Musabe, T., and Kabenga, I., forthcoming, *Economic Analysis of Natural Resource Management in Rwanda,* UNEP/UNDP, Nairobi/New York.
Percival, V., and Homer-Dixon, T., 1995, *Environmental Scarcity and Violent Conflict: The Case of Rwanda,* University of Toronto, Toronto.
United Nations Development Programme (UNDP), 1995, *The Aral in Crisis,* Report presented at Tashkent, Uzbekistan, New York.
United Nations Development Programme/United Nations Environment Programme (UNEP-UNEP), 2007, *Mainstreaming Environment for Poverty Reduction and Pro-Poor Growth,* New York/Nairobi.
United Nations Environment Programme (UNEP), 1993, *Diagnostic Study for the Development of an Action Plan for the Conservation of the Aral Sea,* Technical Report, Water Branch, UNEP, Nairobi.
United Nations Environment Programme/International Institute for Sustainable Development (UNEP-IISD), 2004, *Exploring the Links Human Well-being, Poverty and Ecosystem Services,* Nairobi, Kenya.

HUMAN AND ENVIRONMENTAL SECURITY IN THE SAHEL

A Modest Strategy for Success

COLONEL CINDY R. JEBB* [1], COLONEL LAUREL J. HUMMEL[2], LIEUTENANT COLONEL LUIS RIOS[3] AND LIEUTENANT COLONEL MADELFIA A. ABB[4]

[1-3] *United States Military Academy, West Point, NY 10996*
[4] *Seton Hall University, South Orange, NJ 07079*

Abstract: The human security paradigm is a holistic and empathetic approach to understanding the security environment. The lens of the human security paradigm magnifies the roots of conflict and instability in the Sahel. The analysis of the insecurities that exist in Niger and Chad illuminates the condition of the Sahel's regional security environment. Key to this analysis and the implications of conflict and potential for conflict is understanding the strong connection between the environment and human survival. This understanding is an important first step in crafting effective policy. It is important to recognize that policy making and implementation will take time, intelligence, critical analysis, patience, and empathy. Subsequently, the modest strategy for success will depend on the incremental, tireless, and focused work of a cooperative international community that will best succeed in creating a more peaceful and prosperous world.

* Address correspondence to Colonel Cindy Jebb, Department of Social Sciences, United States Military Academy, West Point, NY 10996; e-mail: cindy.jebb@usma.edu.

The views expressed in this paper are those of the authors. They do not necessarily reflect the official policy or position of the U.S. Army Center for Army Lessons Learned, Department of the Army, the Department of Defense, the U.S. government, or Seton Hall University. We would like to thank the selfless people we met in Niger, Chad, and Djibouti: they include personnel from the U.S. military, Defense Attaché in Chad, Defense Attaché in Niger, EUCOM PAO, U.S. Department of State, United Nation agencies, local and international nongovernmental organizations, African Union, and the people of Niger, Chad, and Djibouti. We also thank the generous support of the Institute of National Security Studies, Foreign Military Studies Office, West Point's Combating Terrorism Center, and Pell Center.

The human security portion of this chapter can also be found in Kennedy and Jebb, forthcoming. Additionally, some portions of this chapter can also be found in Forest, 2005.

P.H. Liotta et al. (eds.), Environmental Change and Human Security, 341–392.
© 2008 *Springer Science + Business Media B.V.*

Keywords: Human security; human insecurities; security environment; terrorism; living systems theory; Niger; Chad; Sahel

1. Introduction

> *No man is an island, entire of itself; every man is a piece of the*
> *continent, a part of the main. If a clod be washed away by the*
> *sea, Europe is the less, as well as if a promontory were, as well*
> *as if a manor of thy friend's or of thine own were: any man's*
> *death diminishes me, because I am involved in mankind, and*
> *therefore never send to know for whom the bell tolls; it tolls*
> *for thee.*
>
> —*John Donne*, "Meditation
> XVII" of *Devotions Upon*
> *Emergent Occasions* (1623)

John Donne's 1623 passage profoundly recognizes the interconnectedness of humanity and the living world. If a bell were to ring for every human being who dies due to human insecurities, the bell would never cease to ring. In the case of many African states, the bell rings constantly. Paraphrasing Donne, every African life that ends diminishes all of us in many and various ways. What happens in Africa affects the rest of the globe, to include the most prosperous and technologically advanced states. The global effects of human insecurities in Africa are real because everything is connected and related. The current War on Terrorism elicits a more narrow view of these connections by focusing on just an outward manifestation of deteriorating security environments. To truly understand the challenges of terrorism perpetrated by radical militants and extremists, we must understand the totality of the security environment, while understanding its unique nature from place to place.

We use a human security approach to accomplish these tasks, that is, understanding the totality and uniqueness of a particular security environment. This chapter's focus on the roots of conflict and instability in the Sahel is timely and critical as the United States establishes a new African Command and the work of the Trans-Sahel Counterterrorism Partnership (TSCTP) continues.[1] We use primarily the case of Niger, and a secondary

---

[1] The TSCTP is an attempt to assist states of the Sahel combat terrorism by providing both military and nonmilitary assistance.

case, Chad, to highlight the importance of understanding security issues in a comprehensive and holistic manner, with added emphasis on understanding the environmental impact on human survival and quality of life. However, we acknowledge that there is no one-size-fits-all policy; instead successful policy requires tireless, smart, and patient work. Specifically, the authors—a political scientist, a biologist, a human geographer, and a climatologist—will highlight the strong connection between human survival and the environment. We acknowledge that, unfortunately, there have been numerous interpretations and definitions of human security, so it has come to mean everything to everyone. For example, when security professionals of powerful states use this term, they arouse suspicions of cloaking interventions and neocolonialism in the name of protection. We contend, however, that the human security paradigm fosters a holistic and empathetic approach toward understanding the security environment. This paper will use on-the-ground experience and interviews, political analysis, and geographical analysis to understand how the environment cuts across all areas of human security, as explained by the United Nations Development Programme (UNDP) and measured by the 2000 Millennium Development Goals discussed below.

This holistic approach is not only critical for understanding the security environment; it is essential for strategists to better anticipate second and third order effects of their decisions and subsequent actions. Moreover, it puts the fight against terrorism in its proper context so that tactical counterterrorist actions do not undermine the strategic goals of mitigating root causes that, in the long term, will help "drain the swamp" of potential recruits. As a first step, we will briefly explain why human security matters by viewing the security environment through the lens of a biologist, using the living systems theory. Second, we will explain human security as an approach and operational concept. Third, we will emphasize the importance of the environment as it relates to all areas of human security. Our case studies on Niger and Chad from the Sahel region will illuminate these points before we conclude with environmental and policy implications.

2. Why Human Security Must Be Explored

In short, human security describes threats that states cannot or will not deter that directly affect individuals. To fully appreciate this paradigm, however, requires a "shift of mind."[2] The living systems theory (LST) drives this shift

---

[2] Senge coined the term Shift of Mind. Shifting our Newtonian understanding of cause and effect to a systemic view allows people to understand that the whole is greater than the sum of its parts and to understand the interrelationships of entities (Senge, 1990).

with its biological view of the world. This perspective sheds light on the nonlinear nature of actions and behaviors. In other words, actions are not constricted to the parameters of cause and effect relationships. Instead, due to myriad connections and relationships, events and activities have second and third order effects that are not always recognizable. Moreover, there are various unintended consequences of even the best planned events. The best we can do is study behavior from multiple perspectives to gain a holistic and empathetic understanding. According to Fritjof Capra, "The behavior of a living [system] organism as an integrated whole cannot be understood from the study of its parts alone" (Capra, 1996: 25). For the whole is more than the sum of its parts.[3] Robert Jervis explains that the whole is indeed different from the sum of its parts.[4] Through the lens of LST, the world is a living system consisting of multiple systems within systems with an almost infinite number of relationships and connections among all systems, living and nonliving.[5]

The foundation of the Living Systems Theory is from the theoretical works of James G. Miller and Fritjof Capra. In his book, *Living Systems*, Miller proposes that complex structures that carry out living processes can be identified at seven hierarchical levels and have nineteen critical subsystems whose processes are essential for life.[6] Types of processes that are essential for life are the processing of matter, energy, and information. According to Miller, there are seven hierarchical levels of living systems: the cell, organ, organism, group, organization, society, and supranational system. The hierarchy of living systems begins at the simplest living system (the biological cell) to the most complex living system (the supranational system that consists of two or more societal systems). Miller painstakingly details the increasing complexities of his nineteen processes for each level of living systems.

Fritjof Capra offers another approach, which we adopt. In his book, *The Web of Life: A New Scientific Understanding of Living Systems*, Capra theorizes that a living system possesses form, matter, process, and

---

[3] According to Capra, Christian von Ehrenfels, a philosopher, coined the phrase "the whole is more than the sum of its parts." This became the basis of all systems thinking approaches. Capra, 1996: 31.

[4] Jervis, 1997: 12–13. Further, Jervis believes "If we are dealing with a system, the whole is different from, not greater than, the sum of the parts." This counters Senge and Capra's view that "The whole is greater than the sum of its parts." The differences in opinion still recognize that the parts of a system only present half of the integrated whole.

[5] Capra, 1996: 158, 160, and 209. In Capra, 2002, he refines his universal living systems theory as he integrates the social and human domains of a living system inherit to a human system. He adds the criterion of meaning.

[6] Miller, 1978: 1. Also refer to Table 1-1: The 19 Critical Subsystems of a Living System on p. 3.

meaning (Capra, 2002: 81). These four criteria of living coexist and enable living systems to adapt, survive, create, and propagate through time. These characteristics reveal that living systems make choices and have unique worldviews. Examples of living systems include societies, governments, terrorist organizations, and other human groupings, as well as environmental elements such as animals, crops, and trees (see Jebb and Abb, 2005). These living systems also interact with nonliving systems such as the soil, air, and water. These insights are important for three reasons: first, they point to the importance of understanding the impact of the security environment, in all its manifestations, on the individual and how that individual decides to interact with other living and nonliving systems around him or her. It also reminds us that groupings of individuals make choices as well: that is governments, societies, ethnic groups, and terrorist organizations. Second, there are myriad connections and relationships among and between these systems. We cannot possibly explore all possible relationships among living and nonliving systems, but we can focus on those relationships between political organizations, societies, and terrorist organizations that are vying for legitimacy. This point leads to our final concern, reflected in the criterion of "meaning."

This fourth criterion, meaning, speaks to the human and social domains of living systems (Capra, 2002: 73). It reflects the rules of behavior, values, intentions, goals, strategies, designs, and power relations, essential to human social life (Capra, 2002: 83–87). In essence, meaning refers to the unique perspective or view of the world a system has, based on its unique qualities. For the purposes of this discussion, this criterion refers to a system's mental map. For example, while the United States may regard Al-Qaeda members as terrorists, Al-Qaeda members view themselves as martyrs. Different societies—based on their unique historical experiences, culture, and norms—have different perspectives on the world and view world events differently. This worldview shapes, for example, the quality of the relationship individuals or groups of individuals have toward a particular state, and legitimacy measures the quality of those state-societal connections. This is a key point as states work to offer their people security in a bid to gain legitimacy vis-à-vis extremists and terrorists. Francis Fukuyama notes the importance of state legitimacy: "the state's institutions not only have to work together properly as a whole in an administrative sense, they also have to be perceived as being legitimate by the underlying society" (Fukuyama, 2004: 26). Security, however, must be interpreted from the individual's perspective because it is the individual who ultimately decides on his/her behavior and whether or not the state has earned his/her loyalty. Subsequently, we now explore human security as a means to gain this holistic and empathetic perspective.

2.1. HUMAN SECURITY

When the UN presents aggregate data such as one billion people who lack access to clean water, two billion people who lack access to clean sanitation, three million people who die from water-related diseases annually, fourteen million (including six million children) who die from hunger annually, and thirty million people in Africa alone who die of HIV/AIDS,[7] it primarily describes the myriad human insecurities of underdeveloped regions and the increasing global gap between the haves and the have-nots (UN General Assembly, 2004: 17). Many of the human insecurities exist in the non-Western world, primarily in underdeveloped regions. These regions consist of weak states with porous borders, overlapping ethnicities, and colonial histories. Many of these weak states entered the international system late, and interestingly, attained legitimacy from the international system but not from their own citizenry. Moreover, the threats facing these states have historically been internal, many times reflecting ineffective institutions, a weak state identity, a lack of state capacity, and a bankrupt economy, as well as dangerous neighbors. Clearly, the formation of many of these states did not follow the western Westphalian path. Some scholars suggest that their development has led to state-nations as opposed to nation-states (See Ayoob, 1995; Buzan, 1983; Tilly, 1985; Rejai and Enloe, 1969).

While there are some commonalities among underdeveloped regions, they are far from all alike. One commonality, however, is that many of these transnational forces are directly harmful to people, and often the state either cannot or will not deter or address them. As the name implies, these transnational threats do not recognize borders and are particularly harmful to susceptible regions that are already weak in myriad ways. Moreover, many of these forces are interrelated and reinforcing, and therefore exacerbate conditions that create widespread suffering.

Human security is a concept that both describes these conditions and provides an approach to better understand their effects. Due to variations among regions and localities, a first step in designing strategy is to understand the dynamics and circumstances at the ground level. The UN recognized this approach in 1994:

The concept of security has far too long been interpreted narrowly: as security of territory from external aggression, or as protection of national interests in foreign policy or as global security from the threat of nuclear holocaust. It has been related to nation-states more than people. ... Forgotten were the legitimate concerns of ordinary people who sought security in their daily lives. For many of them, security symbolized protection from the threat of disease, hunger, unemployment, crime [or terrorism], social conflict, political repression and

---

[7] Note that this section on human security is modified from Kennedy and Jebb, forthcoming.

environmental hazards. With the dark shadows of the Cold War receding, one can see that many conflicts are within nations rather than between nations. (UNDP, 1994: 3, 22–23, as cited in Liotta, 2004: 4–5)

In 2004, The European Union's *A Human Security Doctrine for Europe* used human security as its strategic anchor defining it as:

Individual freedom from basic insecurities ... genocide, widespread or systematic torture and degrading treatment, disappearances, slavery, and crimes against humanity and grave violations of the laws of war as defined in the Statute of the International Criminal Court. (Barcelona Report, 2004: 9)

Even the UN has changed the definition of sovereignty to adjust to today's security environment:

In signing the Charter of the United Nations, States not only benefit from the privileges of sovereignty but also accept its responsibilities. Whatever perceptions may have prevailed when the Westphalian system first gave rise to the notion of State sovereignty, today it clearly carries with it the obligation of a State to protect the welfare of its own peoples and meet the obligations to the wider international community. (UN General Assembly, 2004: 21–22)

Consequently, when a state cannot or will not protect its citizens, then the international community must engage (UN General Assembly, 2004: 22; see also Jebb et al., 2006: 134–136). Scholars have described this forward-leaning definition of international security as the "duty to protect" (see, for example, Halperin, 1993: 105–122; Feinstein and Slaughter, 2004: 136). While human insecurities primarily occur in weak and failing states, due to the nature of weak regions, such insecurities quickly spread. And as we have witnessed in our increasing global world, these insecurities frequently have diffuse global effects, such as migrations, reverberations in Diaspora communities, environmental impacts, and even the exportation of terrorism (see also Gurr, 1997: 123–138; Kaplan, 1994: 44–76). The human security paradigm should remind strategists that they must approach issues holistically and empathetically. This requires painstaking analysis, patience, and tenacity.

2.2. OPERATIONALIZING HUMAN SECURITY: UNDERSTANDING ITS TRANSNATIONAL COMPONENTS

The UNDP report identified components of human security. They are:

- Economic Security—the threat is poverty
- Food Security—the threat is hunger and famine
- Health Security—the threat is injury and disease
- Environmental Security—the threat is pollution, environmental degradation, and resource depletion
- Personal security—the threat involves various forms of violence

- Community security—the threat is to the integrity of cultures
- Political security—the threat is political repression[8]
- The 2000 Millennium Development Goals (MDGs) provide a measure toward reducing human insecurities. They are:
- Goal 1 Eradicate extreme hunger and poverty.
- Goal 2 Achieve universal primary education.
- Goal 3 Promote gender equality and empower women.
- Goal 4. Reduce child mortality.
- Goal 5 Improve maternal mortality.
- Goal 6 Combat HIV/AIDS, malaria, and other diseases.
- Goal 7 Ensure environmental stability.
- Goal 8 Develop a global partnership for development.[9]

Taken together, most areas suffering from human insecurities are facing issues of economic underdevelopment, environmental degradation and resource scarcity, food insecurity, health insecurities, political and/or civil inequalities, and violence to include human trafficking, terrorism, crime, and armed conflict. Many insecure areas also are vulnerable to radical ideologies that stifle development and exacerbate social tensions. Clearly, not all areas face the same insecurities or combination of insecurities, and, therefore, there is no common strategy for all situations. What strategists can count on is some combination, interconnection, and regional effect of these issues. With this in mind, the following is a brief description of these issues.

2.3. ECONOMIC INSECURITIES

The first MDG is to eradicate hunger and poverty. How does one measure poverty? How meaningful is the "one-dollar-a-day" measurement of basic subsistence requirements given that this figure does not capture the full extent of misery that people experience? The dollar-a-day measurement presumes that X amount of capital brings equal access to commensurate commodities, which is a spurious comparison. If that amount does not allow people to have the basic necessities of life: food, clean water, health, then perhaps this measure lacks meaning (Pronk, 2005: 84). The UNDP claims that the best measure of global inequality is the global income distribution model. This model uses national household income data to build an integrated

---

[8] As cited in Liotta and Owen, 2006: 90.
[9] These goals are taken from UNDP, 2005: 39.

global income distribution model. This model suggests a huge gap between the very rich and very poor on a global scale that is greater than the inequalities within any one country. Two-thirds of this gap is caused by income inequalities among countries, while inequalities within countries are responsible for the other third.[10] The conditions of extreme poverty make clear their immediate connection to "disease, drought, and distance from world markets" (Sachs, 2005: 192). Jeffrey Sachs, the world's leading economist, commented that "I began to suspect that the omnipresence of disease and death had played a deep role in Africa's prolonged inability to develop economically" (Sachs, 2005: 194). Further, he states that "good governance and market reforms are not sufficient to guarantee growth if the country is in a poverty trap" (Sachs, 2005: 195).

Unfortunately, globalization has indeed created a poverty trap and hardened the distinctions between the haves and have-nots. Though fewer people actually live below the poverty line of $1 per day, inequalities have, in fact, grown. According to the UNDP, stagnation has been a prevalent aspect of globalization, notably within twenty-five sub-Saharan African countries and ten Latin American countries. In Russia, after the Asian financial crisis, thirty million people slipped below the poverty line, and in Argentina between 2000 and 2003, the number of people below the poverty line tripled. While incomes have been rising in China and India and contributing, therefore, to a more positive aggregate picture of poverty reduction, this has masked the deep inequalities that exist across regions (UNDP, 2005: 35–37). Such chronic poverty corresponds to human misery, as Jeffrey Sachs notes above. When faced with such a situation, people look for relief, whether through migration, rampant urbanization, or other non-state actors. These patterns and social dislocations stemming from economic insecurities may affect social, political, and economic development not only within a particular state, but also within the larger region.

2.4.   FOOD INSECURITIES

While recent estimates claim that about one billion people are undernourished, including over 800 million in the developing world, most projections indicate that by 2050 world population will approach nine to ten billion (Cavalcanti, 2005: 156; Brown, 2004: 178). It is not a coincidence that the first MDG lists the eradication of both hunger and poverty, as they are intricately related. This relationship is increasingly apparent in developing countries, rural areas, and

---

[10] UNDP, 2005: 37–38. The only country that measures a greater gap than the global scale between the very rich and the very poor is Namibia.

growing urban centers (Cavalcanti, 2005: 163). Food security is at the heart of every aspect of human life. Many cultures, notably in the developing world, are centered on food-getting strategies. Though Thomas Malthus in the late eighteenth century warned of the consequences of an unchecked exponentially growing population and a linear increase in food supplies, there may be other factors that require study that were not evident, or at least not considered, in the eighteenth and nineteenth centuries (Bodley, 1996: 83–84). For example, food shortages may also be caused by prevailing societal, political, and economic structural sources that artificially and adversely affect food supplies (Bodley, 1996: 83–112). Henrique Cavalcanti describes three pillars of food security: availability, access, and stability. He defines food security as the

practices and measures related to the assurance of a regular supply and adequate stocks of foodstuffs of guaranteed quality and nutritional value. ... Broadly defined, food security can only be achieved if every person, family, community and nation is considered in the process and at the same time play a responsible role, as sustainable consumers, in ensuring its efficiency and effectiveness. (Cavalcanti, 2005: 153–154)

Policymakers must also consider the regional effects of food security. For example, during the 2005 famine (or what others referred to as the food crisis in Niger), it was clear that food aid was entering the country only to be drifting south to Nigeria. Unfortunately when Nigeria resold the food to Niger, the Nigerians suffering from hunger could not afford to purchase the food that was originally intended for them. Food insecurities know no boundaries, especially among countries that share growing environmental depletion of food sources.

One of the factors that Malthus could not forecast is technology. Between 1950 and 1984, grain production outpaced population increase; however, after 1984, grain harvests declined, especially in Africa. Some of the major causes included soil erosion, desertification, transfer of cropland to non-farm purposes, falling water tables, and rising temperatures. Moreover, shocks such as natural disasters, droughts, and disease can contribute to food insecurities. These conditions may reveal themselves globally in the form of higher food prices, which further emphasizes the disparity between the haves and the have-nots, but at the state level, these conditions can develop into grave instabilities and conflict (Brown, 2004: 4–9; see Cavalcanti reference the effects of shocks, 2005: 154).

2.5. HEALTH INSECURITIES

It should be no surprise that the prevalence of HIV/AIDS in Africa directly exacerbates food insecurities. Clearly, the loss of workers due to AIDS adversely affects the ability to harvest crops, for example (Brown, 2004: 19). In some African countries, HIV/AIDS has infected 20–30 percent of the adult population (Elbe, 2004: 371). But Africa is not alone. According to Nicholas Eberstadt, "major epidemics are already underway in China, India, and Russia, and the local social mores and behavioral practices are set to further spread the disease" (Eberstadt, 2002: 42). Of course, HIV/AIDS is but one aspect of health insecurity. Recently working in Africa, the authors were appalled by the prevalence of malaria. Water-related diseases account for 3.4 million deaths per year and one billion people lack access to clean water. Even in the United States, 218 million people live within 10 miles of a polluted body of water—and that is more than thirty years since the legislation of the Clean Water Act (Durbak and Strauss, 2005: 128–129). The recent scare of bird flu and SARS reveals how quickly disease can spread, and the highlights the need for a responsive, preventative, and proactive public health policy.

The connection between the rise of infectious diseases and the degradation of the environment is clear. According to the World Health Organization (WHO):

Poor environmental health quality is directly responsible for some 25 percent of all preventable ill health, with diarrheal diseases and acute respiratory infections heading the list. Two-thirds of all preventable ill health due to environmental conditions occurs among children, particularly the increase in asthma. Air pollution is a major contributor to a number of diseases and to a lowering of the quality of life in general. (Durbak and Strauss, 2005: 129)

Moreover, the 2005 Millennium Ecosystem Assessment report claims that continued degradation of ecosystems may increase the spread of common diseases, such as malaria, as well as facilitate the evolution of new diseases (Durbak and Strauss, 2005: 129–130). States with inadequate (or no) public health support systems will be most vulnerable and least able to respond. Even developed states that do have such systems will be stressed, and if faced with a true health emergency may be ineffective, possibly losing society's trust. Of course, food and environmental insecurities and poverty exacerbate health insecurities and vice versa. While developed countries are facing declining birth rates and the health challenges associated with aging populations, developing countries' health insecurities, notably high infant and child mortality rates, help to drive high birth rates as parents want to ensure that at least some of their children reach adulthood. Ironically, it is these areas that suffer the most acute array of human insecurities that have the highest population growth rates, a combination of factors that exponentially increase misery and suffering.

2.6. RADICALIZATION

Bruce Hoffman explained that terrorism is dangerous, but the real danger is the potential it has to spark a political movement. It is these radical ideas that pose a great threat to peace and stability (Hoffman, 2004). Why is radical Islam such a threat? First, with the end of the Cold War, democracy's ideological competitor, communism, was defeated. This radical form of political Islam has filled an ideological vacuum. It is dangerous because it appeals to people who are suffering from societal, political, and economic despair and seemingly have nowhere else to turn. Unfortunately, this radicalization prohibits long-term development as measured by the components of human security. For example, in the southern areas of Niger that are influenced by Nigeria's northern Islamic radical elements, women are told not to vaccinate their children because the vaccinations are a conspiracy of the West to sterilize them. Consequently, children are contracting preventable diseases such as polio. The 2002 UNDP Arab Development Report noted the three deficits facing the Arab world: "the freedom deficit; women's empowerment deficit; and, the human capabilities/knowledge deficit relative to income" (UNDP, 2002: 27, as cited in Beitler and Jebb, 2003: 47). A radical agenda does not address these deficits for the Arab region or any other region, and leaves the most marginalized segment of the population in status quo.

Even the radicals themselves put aside ideology for pragmatism. They are able to identify vulnerable groups, and develop loyalties in place of states that cannot deliver basic needs to their population. Herein lies the danger of weak and failing states. Not only do they provide safe havens for terrorists, either wittingly or unwittingly, but they also present extremely vulnerable populations that are susceptible to radicalization in the absence of any alternatives.

2.7. POLITICAL AND CIVIL INEQUALITIES AND VIOLENCE

Conditions that magnify a population's vulnerabilities stem from illegitimate governments that stifle political and civil equalities while failing to protect and enforce rule of law. An inability to enforce rule of law invites a potent mix of criminals, terrorists, smugglers, and human traffickers that have a tendency to reinforce each other's activities. These various criminal and terrorist groups have developed alliances due to both globalization and the availability of weak and failing states. "The absence of rule of law ... provides ideal conditions for the blending of criminal and terrorist activities. Economic hardship in many of these nations leads to corruption and trafficking of illicit goods" (Sanderson, 2004: 2). One type of good that has greatly affected the rise of violence is the accessibility of small and light arms that either finance or are used in terrorist or criminal activities (Sanderson, 2004: 2). And unfortunately there still is trafficking in human beings, though, for example, Niger in 2004 made slavery a criminal offense. However, due to the government's inability to enforce the law, there are a reported 43,000 slaves in Niger (Dixon, 2005: A2–A3). According to Reven Paz and Moshe Terdman, Africa offers opportunities for terrorists:

The political and military conditions in most of the African continent, the broad weakness of its governments, and the internal fighting and corruption of these regimes, ease the ability of the Mujahidn to move, plan, and organize themselves, far from being seen. They enjoy in

Africa easier operational abilities than in other countries, which have effective security, intelligence, and military capacities. (Paz and Terdman, 2006: 2)

Illegitimate and poorly functioning regimes provide a combination of conditions that invite a potent mix of violent groups. This only serves to exacerbate human insecurities.

2.8. ENVIRONMENTAL INSECURITIES

We argue that environmental insecurities have profound effects on all the other human security components. For example, according to Lester Brown, economic development will be meaningless if we do not have a planet that can sustain a global economy (Brown, 2006: 3). Thomas Homer-Dixon warns that

Within the next fifty years, the planet's human population will probably pass 9 billion, and global economic output may quintuple. Largely as a result, scarcities of renewable sources will increase sharply. The total area of high quality agricultural land will drop, as will the extent of forests and the number of species they sustain. Coming generations will also see the widespread depletion and degradation of aquifers, rivers, and other water resources; the decline of many fisheries; and perhaps significant climate change. (Homer-Dixon, 2004: 265)

Many areas in Africa are already witnessing the impact of environmental instabilities such as arable land degradation, drought, and deforestation. Widespread migration, refugee flows, and conflict have set in, putting great pressure on already weak states. According to Homer-Dixon, these types of challenges, in addition to reduced fisheries, will lead to sharp increases in conflict—certainly more immediate increases—more than other grave environmental changes, including ozone depletion and the long-term construct of climate change (Homer-Dixon, 2004: 267).

While the developed world may not directly feel the impact of these environmental scarcities, there may be a more immediate and short-term impact felt in the near future based on energy scarcities. As China, India, and other developing countries continue their economic growth based on the western economic model, the competition for energy resources will grow. For the United States, the economy is a pillar of national security that depends upon vital energy resources. President Clinton in 1999 stated that "Prosperity at home depends upon stability in key regions with which we trade or from which we import critical commodities, such as oil and natural gas."[11] With declining reservoirs of nonrenewable resources, such as oil and gas, and the past empirical evidence demonstrating that such resources are the cause of

---

[11] Klare, 2002: 8–9. See Klare for further discussion on resource driven conflicts.

wars, the possibility of future conflict centered on these resources seems high (Homer-Dixon, 2004: 278). As our cases will demonstrate, environmental insecurities crosscut all components of human security.

3. Case Study 1: Niger

Upon planning the trip to Niger, we were struck by the Defense Attaché's reassurance that Niger is a very peaceful country, unlike many of its neighboring countries. Why was Niger stable, or at least seemingly so, while the Sahel region seemed to be on the precipice of chaos? Moreover, how can a country that had just experienced a food crisis (or famine) be stable? To grasp some understanding both of Niger's commonalities with the Sahel region and its unique qualities, we will begin with a brief background, a description of the food crisis/famine, an environmental analysis of the country, the terrorist threat, and a brief discussion of what is being done.

3.1. BACKGROUND

Niger is three times the size of California, with a population of 11.3 million growing at a rate of 3.3 percent. A continued growth rate at this level means that the population will double in twenty-one years. Niger's ethnic groups include 56 percent Hausa; 22 percent Djerma; 8.5 percent Fulani; 8 percent Tuareg; 4.3 percent Beri-Beri; and, 1.2 percent Arab, Toubou, and Gourmantche. The Hausa, Djerma, and Gourmantche are primarily sedentary farmers, while most of the other groups are nomadic. Ninety-five percent of the population is Islamic. Niger is listed last in the Human Development Index (HDI) with $232 per capita and a literacy rate of only 15 percent.[12]

Niger became a French colony in 1922, and was ruled by a governor general who resided in Dakar, Senegal, and a governor in Niger. Niger had some limited self-governing ability until 1958 when it became an autonomous state and by 1960, an independent state (Niger, Background note, 2006). After independence, Niger was ruled by one party under President Diori. Corruption and drought led to a military coup with Colonel Seyni Kountche ruling until his death in 1987. Colonel Ali Saibou, his chief of staff, took over and tried to liberalize Niger. He released political prisoners and established a new constitution, but the people were not satisfied. A national conference in 1991 paved the way for a transitional government as institutions were installed for

---

[12] Niger, Bureau of African Affairs, 2006; see also Wentling, 2006. Some scholars would argue that the doubling time is closer to twenty-two years.

the Third Republic in 1993. Although this republic attained some successes, including a free press and free and fair elections, the economy declined. By 1996, Colonel Ibrahim Barre overthrew the Third Republic and established the Fourth Republic. However, Barre assumed the presidency under flawed electoral conditions and reversed advances in civil and political liberties. By 1999, Major Daouda Mallam Wanke overthrew Barre and established the Fifth Republic, which currently resembles the French semi-presidential system. Mamadou Tandja won the presidency leading a National Movement for a Developing Society (MNSD) and the Democratic and Social Convention (CDS) coalition and again in 2004 for his second five-year term (Niger, Background note, 2006: 3–4).

A major challenge to peace and stability has been the conflict between the government and the Tuaregs. In the early 1990s, Tuareg rebels championing their nomadic rights challenged the government, risking civil war. In 1995, a peace agreement ended fighting and eased the conflict. Several more agreements were signed with the help of other governments and groups, which essentially has led to greater decentralization of resource control, security management, development, reintegration of rebels, and return of refugees. The goal of the series of arrangements has been to create a climate of trust and mutual understanding among the different groups (Nigerian official, interview, 2006). While these series of agreements have been largely successful, there are still sporadic signs of unrest in the Aïr Mountains located in the north. The Azaouak Liberation Front (FLAA), composed of some former rebels, still confronts government forces intermittently, and unfortunately there are still land mines remaining in border areas (Global Risk Outlook 2006, 2005: 60).

Most of the population lives in the south, where many relief efforts are focused. There are several potential and realized threats emanating from the south. According to one source, local fighting between herdsmen and other local communities is the primary threat (Global Risk Outlook 2006, 2005: 61). The flash point is the declining availability of land and other resources, not ethnicity (Danish Corporation, interview, 2006). The dwindling availability of resources and the problems of desertification have even impacted the capital city of Niamey, and squatters, among them the most recent migrants from the impoverished rural areas, have been evicted from around the city in an attempt to protect the forests from Niger's harsh climate. One resident explained that the areas surrounding the capital have been under siege since the 1980s, and livestock breeders have had to relocate (Issa, 2006: 1). As a USAID official looked back at his first assignment in the 1970s, Niamey had 100,000 people; now it is close to one million (Wentley, interview, 2006). The combination of population growth and dwindling resources makes Niamey and Niger desperate for relief.

The government has not been able to address many of these issues, though there was a brief hope that uranium would be a boost to the economy. In 1980, uranium was responsible for 75 percent of the state's revenue, but uranium's value soon plunged, causing huge deficits. These economic woes contributed to political unrest, which further reduced external aid (Gazibo, 2005: 8–9).

Interestingly, a *Boston Globe* article noted the typical images the West receives as a result of an African devastated economy: bloated bellies, dying children, and rampant diseases. "But to define Niger in such terms does grievous injury to the larger reality of a robust, life-affirming, and religiously tolerant people" (Miles, 2005: A15). This is a key point, perhaps related to the question we first asked: why is Niger seemingly more peaceful than the rest of the Sahel region? However, one of the potential threats to Niger is the influence of northern Nigeria's political and extremist Islamic practices on the southern border region of Niger. For example, as mentioned earlier, some Nigerian women who live in the south have been told by radicals that any vaccination attempts are really attempts by the West to sterilize their children (World Health Official, interview, 2006). Clearly such fallacies are detrimental to society and its subsequent political and socioeconomic development. The subjugation of women as a result of radical Islam further hinders development.[13] Thankfully, Niger, due to its history and culture, is not necessarily widely susceptible to such radical influences.

While people we spoke with noted that there are more women wearing head covers, Islam was still described as "soft." William F. S. Miles describes a key difference between Islam in Nigeria and Niger: "Unlike Nigeria, Islam in Niger does not serve as an identity marker for parts of the population who wish to gain status, prestige, power, or wealth over other regions" (Miles, 2003: 5). The form of Islam emanating from Nigeria is known as Izala. It is a more decentralized, radical form of Islam. Most Nigerians follow a Sufi form of Islam known as Tijanyya, which is more hierarchical and adaptive to the local cultural practices (U.S. Embassy official, interview, 2006). Why this difference in Islamic practices between Niger and the bordering state of Nigeria?

In Niger, the Sufi form of Islam coexists with several different forms of Islam, though in 1990 it was the most prevalent form under the Association Islamique du Niger, a government-sanctioned organization. Since the 1990s, there has been a proliferation of associations. The most notable association

---

[13] Miles further explains how women are absent in Islamic decision-making processes, thus putting into question the volunteerism of supporting radical Islam. All people we interviewed discussed the importance of education for girls as the key to development, as will be explained further in the paper.

is the Jama'at izalat al-bid'a wa Iqamat as-Sunna. This association comes from the Izala movement, "which proclaims the suppression of innovation and the restoration of the sunna, for a return to the original religion" (Office of Dutch-Nigerian Cooperation, 2006). The French legacy in Niger promoted a tangible separation of the state and religion, whereas the British in Nigeria courted the religious leaders as political allies. Upon independence, Niger's leaders continued the French dissociative approach toward religion and the state and continued the state subordination of Islamic leaders and organizations. In Nigeria, Islam emerged as a political force in the north, and the Izala form particularly rejected any Islamic practices influenced by local cultures. However, in Niger Islam continued its adaptive and "soft" nature that encouraged inclusiveness and local cultural practices (Miles, 2003: 5). Clearly, some influence has occurred along the Niger-Nigerian border. But just as Western globalization is maligned and elicits a backlash, so may Islamic globalization. Mills asks: "Will there be a Chinua Achebe, lamenting the passing of the old ways in favor of the 'enlightened' new?"[14] In short, radicalization is not definite for the future of Niger, but there are some danger signs. Probably the most significant aspect is the incessant force of globalization and how the state chooses to respond.

3.2. IS TERRORISM A THREAT?

It is difficult to determine the extent of international terrorism in Niger. There have been reports of the Salafist Group for Preaching and Combat (GSPC) planning attacks against Niger (Laremont and Gregorian, 2006: 27–36). Incidents have been rather isolated, and it is difficult to distinguish between banditry, smuggling, and terrorist activities. What is worrisome is that many foreign fighters in Iraq have been linked to North Africa. Just as the fight against the Soviets in Afghanistan internationalized the jihad in the 1980s, there could be a similar effect once these fighters return to their homes.[15]

What exactly is the GSPC? It is a faction of the Algerian Group Islamique Armé (GIA) and formed in 1998 after the GIA's very violent period between 1996 and 1997. The GSPC, led by Hassan Hattab, continues to fight against the Algerian state, and it is the only group that has operated across all of the Sahelian countries. Although the leaders of the GSPC have pledged allegiance to Al-Qaeda, it is not clear that there is much significance

---

[14] Miles, 2003: 5. For a comprehensive analysis on the different forms of Islam and Islamic associations by region, see Office of Dutch-Nigerian Cooperation, 2006.

[15] The existence of North Africans fighting in Iraq on behalf of Al-Qaeda is taken from Whitlock, 2006: 19.

in this pledge. One of the ties connecting the GSPC to Al-Qaeda is Emad Abdelwahid Ahmed Alwan, who was accused of the attack on the USS *Cole* and suspected of planning a U.S. embassy attack in Bamako. He has since been killed (in September 2002), but there was an affiliation alleged between him and the GSPC. There have been reported clashes between the Nigerian Army and the GSPC. In March 2004, the army fought the GSPC on the border with Chad; in April 2004, the army killed traffickers associated with the GSPC near the Malian border; and in August of 2004, the army clashed with Tuareg bandits (International Crisis Group, 2005: 7–8, 22).

Stephen Ellis believes that the GSPC is a bandit-like criminal group that takes advantage of the ungoverned space of the Sahara, just as outlaws have done for years. Interestingly, they, too, take advantage of any infrastructure improvements for movement of contraband. Unfortunately, they also develop ties with officials such as border guards. Ellis further warns that such a potent mix of relationships could draw in U.S. soldiers such that they unwittingly become pawns as illicit groups vie for power and profit. Furthermore, closing down smuggling activities may not only affect terrorists but may also harm ordinary people by closing down their livelihoods (Ellis, 2004: 463).

Much of the discussion about terrorism in Africa focuses on the ripe conditions for terrorists. Susan Rice, former Assistant Secretary for Africa Affairs said:

Much of Africa has become a veritable incubator for the foot soldiers of terrorism. Its poor, young, disaffected, unhealthy, undereducated populations often have no stake in the government, no faith in the future, and harbor an easily exploitable discontent with the status quo. … These are the swamps we must drain. And we must do so for the cold, hard reason that to do otherwise, we are going to place our national security at further and more permanent risk. (Kraxberger, 2005)

But to drain the swamps, we must understand the security environment from the individual's point of view. Nothing seemed to put the desperate situation in Niger in as sharp a focus as the food crisis of 2005.

3.3. A FOOD CRISIS OR FAMINE?

It was interesting to note that many of the officials from the government and NGOs were adamant to refer to the food crisis as a crisis and not a famine. A famine indicated a government that could not feed its people. According to one official, a famine reflects people and livestock dying of hunger, while a food crisis describes a condition where food reserves are empty and widespread malnutrition exists (WFP personnel, interview, 2006). Whether the situation was a crisis or a famine, it is clear that Niger suffers from a chronic structural condition that leaves it vulnerable to periodic crises/famines that put its population at risk for widespread health risks and death.

There appears to be myriad factors at work that put Niger in such a precarious situation. One observer notes that the problem is chronic poverty, not natural disasters, conflict, or corruption (Loewenberg, 2006: 2). Government and UN officials maintain that the combination of drought and locust infestation caused Niger's "nutritional emergency." They maintain, though, that the emergency was localized and resulted in an 11 percent loss of cereal (Tectonidis, 2006: 1).

Warning signs of a coming emergency seemed visible. Medecins San Frontieres (MSF) noted the wave of malnourished children at their feeding center in Maradi. With low food reserves and a resulting increase in cereal prices, people could not afford food or medical assistance for their families. Subsequently, people sold their livestock, which caused livestock prices to fall (Tectonidis, 2006: 1). Another causal factor was the pace of donor response. The World Food Program (WFP) saw signs of an imminent emergency in 2004, but donor response was not quick enough. The WFP noted that government subsidized food was not enough to combat this problem, so the WFP established free food distribution and food for work programs (WFP personnel, interview, 2006). Unfortunately, much aid did not reach the intended population. One person noted that, as we mentioned earlier, some aid drifted to Nigeria, and then was resold at unaffordable prices in Niger.

The sicknesses that develop in such an emergency have second and third order effects that are not clearly or immediately visible. For example, when mothers bring their sick children to treatment centers, siblings may be left home alone. When large numbers of sick children converge on one location, there is risk of spreading infection (Tectonidis, 2006: 2). According to the UNDP, 3.6 million people were affected by the food shortage, with some places experiencing a daily toll of 4.1 deaths per 10,000 people. According to Thierry Allafort-Duverger of MSF, 30,000 children had been treated for malnutrition and undernutrition midway through the crisis, and he predicted 50,000 by the end of the year (Heath, 2005: 1). Allafort-Duverger noted that "the slightest breakdown, be it a drop of harvests or a rise in prices, is enough to trigger a dramatic rise in the number of children falling victim to severe malnutrition" (Heath, 2005: 1–2).

Moreover, the problem extends beyond Niger's borders. According to UNDP Administrator, Kemal Dervis, "There needs to be an increase of resources allocated to the Sahelian region. A long term commitment is needed for the region, not just in response to crisis." Furthermore, he states that "[O]ur work now has to focus on the complex mechanisms that need to be put in place to really ensure food security, reduce infant mortality, and increase maternal health" (Heath, 2005: 2). Dervis points to the importance of women's empowerment and education to positively affect these conditions (Heath, 2005: 2). Another indicator of the importance of educating girls is the linkage between education and fertility rates. According to the United Nations, women with no education have twice the fertility rate as women with ten or more years of education. In another study, seven years of education seemed to correspond with fertility decline. Not only is education linked to falling fertility rates, but it provides an avenue for increased participation of women in the economy, society, and governance—all factors that lead to human development as opposed to factors that gravely contribute to human misery (Hummel, 2006: 13). Interestingly, a nutritionist from the United Nations Children's Fund (UNICEF) mentioned that even teaching women the simple task of breast feeding instead of giving babies water (usually contaminated), would go a long way in fighting health problems (UNICEF official, interview, 2006). Kofi Annan, after visiting Niger and listening to starving villagers and seeing the effects of hunger on babies, concurred that Niger requires more than free food (Scharnberg, 2005: 1). As a first step in understanding the complexity of Niger's condition, the next section provides an environmental analysis. With the declining environmental conditions and resources combined with an explosive population growth, human survival is precarious. If the state cannot find a way to secure people's basic needs, they will turn to whomever offers relief.

3.4. NIGER: A DELICATE CONTRAST IN CLIMATE ZONES

Like many of its sub-Saharan neighbors, the country of Niger sits land-locked and precariously positioned in a region affected by the ebb and flow of shifting climate patterns. This combination makes its southern region vastly different from the north. Only the extreme southern border area of Niger with Burkina Faso, Benin and Nigeria receives any significant precipitation, which comes in a wet-dry pattern typical of tropical savannas. While the southern 20 percent of the country falls within the definition of a tropical steppe climate, the remaining 80 percent of Niger is considered a true desert climate. Four climographs, Figures 1–4, highlight the winter dry–summer "wet" pattern of precipitation typical to Niger along its southernmost border (Figure 1) and its steady decrease in precipitation as one moves northward (Figures 2–4). For this purpose, the capital of Niamey and the cities of Tahoua, Agadez, and Bilma—the only cities with reliable climate records—are used.

The area closest to Niamey (Figure 1) shows the effects of the northward portion of the intertropical convergence zone's (ITCZ's) seasonal rains. At higher latitude (Figures 3 and 4) it is clear that the majority of Niger is a true desert under the influence of subtropical high pressure (STH).

The late 1960s, 1970s, and 1980s saw a catastrophic, decades-plus drought that afflicted the larger Sahel region of Africa (Niger included). While the

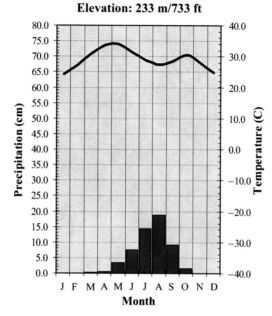

Niamey, Niger
13.4° N / 2.3° E
Elevation: 233 m/733 ft

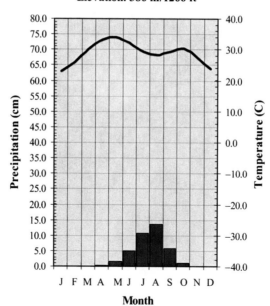

Tahoua, Niger
14.9° N / 5.3° E
Elevation: 386 m/1266 ft

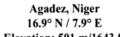

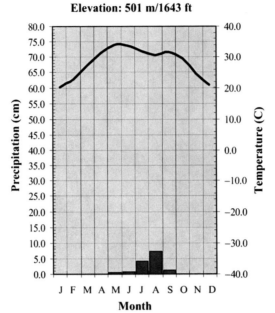

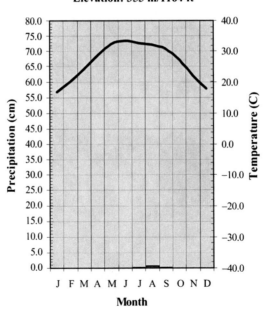

Figures 1–4. Niger Climographs.
Source: Climate data for Figures 1 through 4 collected at http://www.worldclimate.com

catalyst for this drought is still debated among climatologists, one fact is certain: desertification continues to affect the lives of millions of Sahelians. Whether this desertification is anthropogenically induced, a global warming consequence, or a combination of factors, the implications to regional stability and security for this developing part of the world are clear given the area's reliance on changing and increasingly unreliable climate patterns whose end state is unknown. As we have discussed, climate change is not itself the cause of Africa's instability. But it is a very significant component, as "uncontrollable climatic factors complicate Africa's more controllable problems, such as population and poverty ..." and inevitably manifest themselves in the form of ever-increasing tensions between neighboring states (Nyong, 2005).

3.4.1. Climate Change and Desertification Within the Sahel

The notion that climate change and desertification are both under way within the Sahel region is no longer challenged even if the genesis and responsible mechanisms are still the subject of debate. The Sahelian drought that lasted from 1968 though the mid 1980s is generally believed to be an example of climate variability at its most severe, responsible for many deaths and the collapse of agriculture systems throughout the region (Batterbury and Warren, 2001: 1–8). Central to the problem is the very definition of "desertification." It was not until 1992 that the United Nations Environment Programme (UNEP) defined desertification for the benefit of the Intergovernmental Negotiating Committee on Desertification (INC-D) as "land degradation in arid, semi-arid and dry sub-humid areas resulting from various factors including climatic variations and human activities." Accordingly, the very definition of desertification has been refined by some to mean "land degradation in dryland regions, or the permanent decline in the potential of the land to support biological activity and, hence, human welfare" (Hulme and Kelly, 1993: 1).

Niger's relative location within the larger Sahel region means that problems associated with accelerated climate change and desertification directly influence its ability to support its rapidly increasing and very poor population (World Resources Institute, 2006). As of 2005, 38 percent of Niger's fledgling economy was dominated by rural/rain-fed subsistence agriculture and livestock production centered along its only arable stretch of land, the southern border (World Bank, 2006). The wide range of economic stressors affecting Niger combined with its dependence on a marginal climate make small variations from the expected climatic norm able to significantly and deleteriously affect the lives of most of its people. The future of the Sahel is rather uncertain, especially when viewed in light of climate change considerations and the almost unimaginable set of interactions involving possible global warming, increased land degradation, and other inhibitors of human security.

The climate record over most of the African continent shows a warming of 0.7°C during the twentieth century along with an episodic decrease in precipitation over significant portions of the Sahel and Central Africa (McCarthy et al., 2001: 489). Global Circulation Models (GCMs) run by climatologists working for the Intergovernmental Panel on Climate Change (IPCC) have attempted to draw a link between an increasingly warmer planet (and thus, a warmer ocean) and precipitation patterns (Hulme and Kelly, 1993: 7–8). The findings show that although an increase in overall rainfall, when globally averaged, is likely, this is not the case throughout, with drier than normal conditions expected to prevail over the southwestern Sahel, specifically Mali and Niger (Hulme and Kelly, 1993: 7). While computer model reliability in forecasting long-term rainfall patterns remains uncertain, most consistently show a decrease in rainfall over the region that accompanies a continent-wide rise in temperature ranging from 0.2°C to 0.5°C per decade through the end of the century (McCarthy et al., 2001: 489–490). This translates to an increase in temperature ranging between 1.8°C and 4.5°C over the next ninety years, a conclusion whose magnitude and implications are more than merely academic. Numerical model output suggests that for every 1°C of mean continent-wide warming, a corresponding 6 percent reduction in rainfall will follow, bringing the potential reduction in rainfall over the Sahel region of the African continent to anywhere from 10 to 30 percent by the year 2100 (Hulme and Kelly, 1993: 7). Understanding the potential impacts of such dire climate simulations is critical to assessing what might happen to regional stability. If continent-wide drying occurs as a slow and steady process, the ecosystem might have time to adapt and cope with the consequences. If the rate of drying, however, is more sudden and concentrated, like the 20–40 percent declines observed during the 1960s through 1980s, then the whole system might be severely stressed (Hulme and Kelly, 1993: 7). If the latter is the case, there will likely be a large-scale failure in the region to meet the demand for subsistence food and water.

3.5. SECURITY IMPLICATIONS OF CLIMATE CHANGE AND DESERTIFICATION: A NIGER PERSPECTIVE

The long-term implications of climate change and desertification within sub-Saharan Africa will continue to manifest themselves in the form of widespread intra- and interstate conflict. As of 2002, it was estimated that over 50 percent of the population was fifteen years old or younger (World Resources Institute, 2006). As limited and shifting resources come under an ever-increasing amount of stress, the volatile combination of a young and predominately poor population will make Niger vulnerable to continued deterioration of human security. Specifically, we focus on concerns about the integrity of water resources, food security, natural resources, and continued or accelerated desertification as factors affecting the overall human security of Niger.

3.5.1. *Water Resources, Food Resources, and Their Impact on Human Security*

Access to sanitary and reliable water resources is unavailable to approximately two billion people around the world, but is especially problematic in the Middle East and Africa. These regions considered water-scarce or water-stressed are likely to double in area by the year 2025. Changes in climate are theorized to magnify both episodic and longer-term shortfalls in water availability within already arid and semiarid locations. Since many parts of Africa, for example, are dependent on simple, nonredundant, and single-point water sources for survival, any significant dry period can be catastrophic. Within the Sahel, the ability to respond to the many potential second and third order effects brought about by climate change will depend on the region's ability to manage water resources from an intrastate and interstate perspective. This means placing emphasis on integrated, cross-state concerns and properly managing river basins to the greatest extent possible (Watson et al., 1997: 3).

Generally speaking, close to 800 million people around the world are considered undernourished or malnourished, with many of them in Africa. Climate change is likely to affect the overall agricultural production cycle by inducing changes in the growing season as a result of shifting temperature and precipitation patterns. While some changes might be beneficial, such as longer growing seasons, many other interactions may offset any gains. In regions where agricultural practices are robust and well-established, the impacts of climate change might be easier to overcome. However, in places where agriculture is already at or near the climatic tolerance of staple crops, or where subsistence practices dominate, any change to the climate norm is likely to lead to complete failure (Watson et al., 1997: 3–5).

Niger is a state with only one active river basin, the Niger River. All other stream channels within its boundaries are ephemeral, making the Niger River a vital enabler to the nation's otherwise fragile economy, as well as to the entire Western Sahel region. The river's headwaters originate in the highlands of Guinea near the Sierra Leone border. It runs through Mali, the border between Benin and Niger, and finally through Nigeria, where it empties into the Gulf of Guinea (Figure 5). With few interruptions, approximately 5,000 km of its length (including tributaries) are navigable, making the Niger the lifeblood of a sustainable fishing, trading, and agricultural economy throughout the Western Sahel. However, the Niger River is a stream that, like the Sahel itself, traverses a changeable environment from humid to arid and back to humid.

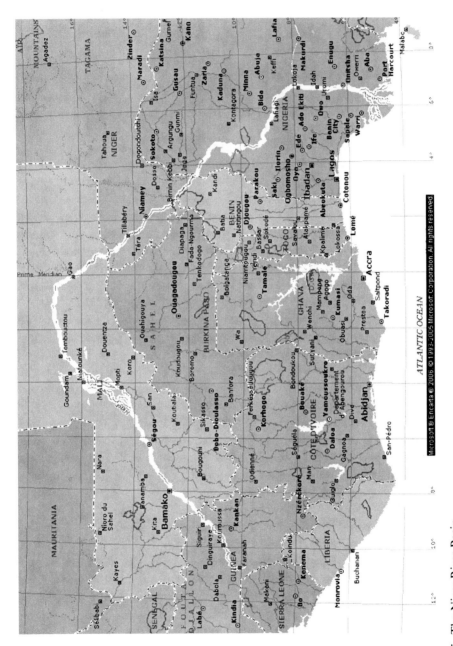

Figure 5. The Niger River Basin.

Thus, it is very vulnerable to slight changes in environmental conditions (Microsoft Encarta, 2006). Far from the river's headwaters and downstream from the point of minimal flow (Central Mali), the Niger is especially susceptible to the whims of variable precipitation patterns and anthropogenic forces. Irrigation schemes dating back to the early 1930s have successively dammed and diverted water away from its main channel, while additional flow from tributaries has been impeded due to smaller water works projects. What water does flow into the state of Niger remains completely unregulated and largely untapped except for extremely small-scale, local-use purposes.

Hydrologic records dating back to the early 1960s show that many of Africa's major river basins have undergone a steady decrease in water discharge. The historical record for Niamey, Niger, for example, shows a decrease in Niger River flow volume by more than 10 percent. This is likely due to a combination of less precipitation, more evaporation, and higher stress on upstream water flow (McCarthy et al., 2001: 498). In places like Mali or Nigeria, where hydroelectric plants benefit from a more reliable river discharge and reservoir level, declines to the Niger River basin flow are less problematic than in Niger, where basic irrigation needs can be obliterated over multiple seasons, affecting the livelihood of farmers and pastoralists engaged in basic subsistence practices.

As stated, Niger's relative location within the Sahel exerts an enormous amount of pressure on already weak pastoral and agricultural activities. Pastoralists from Niger, for example, tend to move their herds north or south with the seasonal rains. Often, this takes them into Nigeria across a largely unregulated and ungoverned border where their animals can feed and survive on, literally, greener pastures. These cross-border activities aggravate existing tensions between the states as competition for scarce resources intensifies. Another complicating factor for pastoralists and subsistence farmers is the greater region's desire to engage in the cultivation of cash crops for the global market. This endeavor drives subsistence farmers and pastoralists into progressively marginal lands, increasing the area's susceptibility to conflict. These human and environmental competitions set up a downward spiral that further exacerbates existing human security concerns. This insecurity may prevent the institution of appropriate management practices and techniques that might otherwise prevent the loss and/or waste of valuable natural resources like water and food. The end result would be a cycle of increased competition and continued conflict (Nyong, 2005: 1).

3.6. WHAT IS BEING DONE?

One of the authors was able to meet with several members of the UN, NGO, host nation, and U.S. government officials in August 2006. Observations during this trip support many of the points addressed above. We had tremendous support from Lieutenant Colonel Stephen Hughes, the Defense Attaché, who coordinated the meetings as indicated throughout our discussion on Niger. At this point, we will highlight some persistent themes.

First, we would be remiss if we did not mention the tremendously selfless, compassionate people we met. It is inspiring to meet people who want to make a difference, have chosen to work hard under harsh conditions and under what appear to be insurmountable odds. Herein lies the modest strategy. For the most part, there is good work happening in small doses, but over time, these efforts will make a difference. As an example, one gentleman who served as a translator regularly assisted his home village, primarily in the field of education, even though he no longer resides there. Having said this, a great deal more can be accomplished to assist this type of individual work. A far more coordinated effort among NGOs, the UN, host nation, the U.S. government, and neighboring countries is required. This is more easily said than done, but unfortunately, each organization has a specific charter that narrows the focus of effort while creating critical blind spots that prevent anticipation of unintended consequences and negative second and third order effects.

Some critical themes that crosscut many of the current efforts we find are significant. First, as the gentleman in charge of a human rights NGO explained, there must be rule of law and justice. Second, education is critical to producing a literate citizenry with a more adaptable and resilient mindset, which in turn can produce leaders and institutions necessary for a functioning, legitimate government. Third, education will provide tools for cultural change, specifically concerning ways of life that depend on scarce land and in terms of elevating girls' and women's participation in social, economic, and political life. The current conflicts between pastoralists and agriculturalists seen as land becomes scarcer may be abated by the diffusion of more sustainable practices that could both increase yield and decrease land degradation (NGO head, interview, 2006).

There are local practices that can be leveraged to better address Niger's complex threats. For example, a cultural practice, Habbanae, was recognized by the Harmony List in 2006 as one that provides sustainable development and enriches quality of life. In short, the cultural practice occurs when a herdsman falls on hard times. The community responds by providing a pregnant cow for three years. This practice not only helps those in need; it also strengthens the bonds of the community (U.S. Federal News Service, 2006: 1). There are cultural practices that also prevent adaptability. For example, people have come to accept famine as a part of life. People still grow millet, which does not provide adequate nutrition. Crop diversification, which would introduce a more nutritious, protein enriched diet, is key to survival. Such change is difficult in a society that is uneducated and continues to experience a death rate of 262 per 1,000 children under the age of five. It is difficult to talk about family planning under these conditions, so women on average continue to have seven babies each (Scharnberg, 2005: 2–3). But women need to be a target for programs that empower and educate. For example, a UNICEF program helps women with malnourished children by providing them with goats, especially valuable during periods of drought. Focusing on projects for women provides a "direct advantage to the children" (Scharnberg, 2005: 2).

In fact, this focus on women is one of three strategic points as outlined by Mark Wentling, a career USAID officer with multiple assignments in Niger. Overall, his strategic plan calls for:

- Conserving and Making Land More Productive
  - Assist farmers to use and manage natural regeneration methods in their fields
  - Transfer, as possible, to farmer fields Sahelian Eco-Farm Practices
  - Work with communities to draft land use maps of their geographical areas
  - Help communities obtain improved seeds, seedlings, and cuttings
  - Establish where possible, "off-season" small irrigation schemes
- Improving the Well-Being of Women and Children (Women's Groups)
  - Provide nutritional, reproductive health, and family planning education
  - Arrange for the identification and treatment of all children suffering from malnutrition
  - Support life skills (asset management) and literacy training (critical thinking)
  - Organize micro-credit schemes for women
  - Train women to be good managers of cereal banks
  - Help women improve their livestock holdings
  - Increase the number of children in school, especially girls
  - Introduce tree education programs and increase the number and variety of trees women and children possess
- Improving Community Well-Being
  - Help provide and maintain adequate water, sanitation, and health services
  - Mount and maintain a malaria prevention and control program
  - Assist with improving animal health and growth
  - Develop marketing plans to obtain highest prices for community/farm products

- Provide training to men in the good management of their assets and incomes
- Arrange for Food-for-Work activities during the hungry seasons[16]

Wentling provides guidance for this strategy. First, it must be viewed holistically, which requires understanding how one aspect of the strategy affects other portions of it. Second, the community must be empowered and be a stakeholder in any project (Wentling, 2006: 3). He acknowledges that "Some of these choices [about changing ways of life and thinking] will be very hard to make but, in the end, it is the communities and their leaders that must make them" (Wentling, 2006: 32).

Homer-Dixon reminds us "that resource scarcity is, in part, subjective; it is determined not just by physical limits, but also by preferences, beliefs, and norms" (Homer-Dixon, 2004: 269–271). As we explore the environment as a critical component of human security, it is important to note that there are possible cultural and technological adaptations that could provide relief in short-term and perhaps even long-term sustainability. The first step is acknowledging the short- and long-term impact of environmental change. Human insecurities know no boundaries, so we move onto our next case and neighbor of Niger: Chad.

4. Case Study 2: Chad

Chad, our secondary case, provides another excellent opportunity to understand the political and socioeconomic situation through a human security lens.[17] Clearly, it suffers from many of the same environmental conditions outlined earlier. As a result, it, too, presents vulnerable populations to dangerous radicals, extremists, and terrorists. Our trip to Chad, however, allowed us to pay attention not only to the extremely vulnerable refugees as a result of the crisis in neighboring Sudan, but also gave us a unique opportunity to witness the tremendous work by many different organizations who are trying to alleviate the situation. Conflict prevention will be key to managing the potential crisis in eastern Chad, which is mounting as a result of the refugee crisis that looks more like a permanent situation than a temporary one. This case will focus on the refugee situation, which has only exacerbated the fragile environmental conditions with added competition for scarce resources.

---

[16] Taken directly from Wentling, 2006: 2–3.

[17] As mentioned in the introduction, this portion on Chad comes from our previous paper with some modifications: Jebb and Abb, 2005.

4.1. BACKGROUND

At first glance, Chad looks similar to Niger. It, too, is landlocked. And though we are long past the days of ascribing to environmental determinism, geography still matters. Jeffrey Sachs notes Harvard's Center for International Development's studies that demonstrate the link between proximity to the sea and wealth. The poorest countries in the world are tropical and landlocked, and those countries are Chad, Mali, Niger, Central African Republic, Rwanda, Burundi, and Bolivia (Sachs, 2000: 102–103). He attributes two challenges to international development: climate and geographical obstacle to transport. In the age of globalization, these challenges have real effect on success or failure (Sachs, 2000: 103). Vast gaps in wealth between countries emerged when the world economy became industrial and science-based starting two hundred years ago (Sachs, 2000: 93–105, 101–102). These inequalities have only grown. It is in these disadvantaged countries that we see the gravest humanitarian and social tragedies (Sachs, 2000: 104).

Chad is twice the size of Texas, with a population of 9.5 million. The growth rate is 3.0 percent—which equates to a doubling time of twenty-three years—and there exist over 200 ethnic groups. They include, in the north and center, the Gorane, Zaghawa, Kanembou, Oaddai, Arabs, Baguimi, Hadjerai, Fulbe, Kotoko, Hausa, Boulala, and Maba, and in the south, the Sara, Moudang, Moussei, and Massa. There are over 120 indigenous languages, but the official languages include French and Arabic (Chadian Arabic). Fifty-one percent of the population is Muslim, 35 percent are Christian, 7 percent are animist, and 7 percent are various other religions (Chad, Background note, Bureau of African Affairs, 2006).

To even begin to understand the mosaic of cultures, one must first understand the climate, political strife, and Sudanese crisis. As discussed earlier, the climate in Chad and the rest of the Sahel is one of extreme rainfall variability. Over the last forty years, overall rainfall amounts have decreased, which has had an adverse effect on the area. The region experienced a wet period in the 1950s and 1960s, followed by twenty years of low rainfall in the 1970s and 1980s, and then some increase of rainfall in the 1990s. There is not complete consensus on the reasons for these changes, which highlights the importance of further environmental research. However, it is clear that a drought in a populated area produces food crises or famines. Extreme climate fluctuations have an effect on food collecting strategies and other economic decisions. Populations travel and may experiment with other methods for survival (de Bruijn and van Dijk, 2002: 5). The profound cultural changes that accompany such changes in food systems have great impacts. According to John H. Bodley, "Food systems are cultural mechanisms for meeting basic nutritional needs" (Bodley, 1996: 83). When local coping mechanisms are overrun by maladaptive and counterproductive structures of the state, then there is potential for more widespread famine and other cultural dislocation.[18] Added to these environmental changes is Chad's civil war, which began in 1965 and continued on and off into the 1990s. While the conflicts seem to have common themes of north versus south, Islam versus Christianity and animism, the food gathering cultures of pastoralism and agriculture have played a key role as well. These civil outbreaks have come at great cost. Conflicts cause people "to migrate to other areas; it destroys cultures, social networks, and leads to an insecure and uncertain economic and social situation" (Bruijn and van Dijk, 2002: p. 6). It is no wonder that the per capita income is only $237 (Chad, Background note, Bureau of African Affairs, 2006: 2).

The political system since Chad's independence from France has been marked by strife, war, ineffectiveness, and illegitimacy. Civil war broke out in 1965 that mainly pitted the Muslim north and east against the Christian south; a coup occurred in 1975, and widespread fighting broke out again in 1979. Hissein Habre emerged as the leader until 1989 when the current president, Idriss Deby, overthrew him. In 1991, Deby became president and subsequently won the 1996 and 2001 elections. In 2004 the National Assembly approved his desire to remain in office beyond the constitutional limit of two terms (Chad, Background note, Bureau of African Affairs, 2006: 3–5).

Two of the authors had just arrived in country in June 2005, when Chad held a referendum on Deby's third term. Interestingly, hardly anyone went to the polls, which perhaps indicated complete apathy or defiance of the president. Key to understanding the complexity of Chad is the president's close ties to his ethnic group, the Zaghawa, of which many members inhabit refugee camps in eastern Chad. Not surprisingly, many of his top advisors, including the military, are also from the Zaghawa clan (Chad, Background note, Bureau of African Affairs, 2006: 6). Interestingly, we had a chance to speak with a leader of an opposition group, the Justice and Equality Movement or JEM from Sudan, who had protection in N'djamena, the capital city. According to this rebel leader, one of the platforms the JEM stands for is security for the greater Sudan. Yet, he identifies himself first as a Zaghawan and secondarily as Sudanese. It seems difficult then to advocate for and promote a Sudanese nationalistic agenda. This situation is reflective of weak states that have yet to elicit national loyalty and legitimacy from their citizens, whose primary allegiance is to their own ethnic group rather than to the government of a state whose motivations are suspect.

---

[18] For more discussion on the structural maladaptation introduced by global and political systems, see Bodley, 1996: 83–89.

4.2. REGIONAL SPILLOVER EFFECTS: SUDAN

There are many estimates coming out of the Darfur region regarding the number of civilians killed and displaced. One estimate has the total at more than 2.5 million as of December 2004. Many of the reports describe unbelievable cruelty and brutality. There is local and international consensus that the perpetrators are the Sudanese military in conjunction with the Arab militias, known as the "Janjaweed" or "evil horsemen" (Kasfir, 2005: 195). Reports indicate that the attacks began in February of 2003 by the Sudan Liberation Movement/Army (SLM/A) followed by the JEM's attacks. Then in April 2003, the SLA launched an attack on the airport in the capital of the North Darfur state, destroying seven planes and killing approximately 100 soldiers. Immediately following this attack, the Janjaweed formed, armed by the government, and conducted brutal retribution attacks against the civilian population (Kasfir, 2005: 196). By January 2005, the UN reported that there were approximately "1.65 million internally displaced persons (IDPs) living in eighty-one camps and safe areas, plus another 627,000 'conflict affected persons,' and 203,000 refugees in Chad" (Kasfir, 2005: 196).

Darfur consists of roughly 20 percent of Sudan and holds about 14 percent of Sudan's population. The population is a mixture of Muslim Arab and Muslim non-Arab ethnic groups. One of the largest non-Arab groups is the Fur (Darfur translates to homeland of the Fur); significantly, the nomadic non-Arab group is the Zaghawa (President Deby's group), and the Arab group, known as Meidab, live predominantly in the North Darfur state. In the Western Darfur state are sedentary non-Arabs from the Fur, Massalit, Daju, and other smaller groups. In a familiar theme, the climate and scarce resources have always been a source of periodic conflict mainly between nomadic Arab herders and non-Arab farmers. Land has been a source of conflict, especially with the continued desertification over the years (Documenting Atrocities in Darfur, 2004). With a neighboring country suffering from what some call genocide, it is no wonder that the border region is tense. Before 2006, the Janjaweed's border activities had decreased with the increase of Chadian, French, and African Union's forces' surveillance (UNHCR, 2004).

However, there have always been reports of banditry along the border and in Chad, and in 2006, the fighting in the Sudan actually spilled over into Chad. An ongoing security concern centers on the tensions between the Sudanese refugees in Chad and the east Chadians living in the proximity of the camps. With the increased number of refugees, there is more competition for the already scarce resources in the area. We will come back to this point, but the situation will only worsen as the strain on resources in combination with the already desperate straits of both native Chadians and refugees continues.[19]

4.3. TERRORISM

Is terrorism a real threat in such a situation? Clearly, the Sahel is being pro-claimed by the U.S. military as "the new front on terrorism" (ICG, 2005). There is some question, however, concerning how real a threat terrorism poses. Chad, and in general the Sahel, suffer from unhealthy societal-political relations. Governments, especially in Chad, are corrupt and ineffec-tive, and most importantly, are illegitimate in the eyes of the people. There should be concern over other dangerous actors exploiting this vulnerability. It is important that the United States and other members of the international community analyze terrorism in context of the larger security situation facing the local people.

According to the International Crisis Group, "the Sahel is not a hotbed of terrorist activity. A misconceived and heavy handed approach could tip the scale in the wrong way" (ICG, 2005). For over sixty years, the Sahel has experienced fundamentalist Islam that did not have ties to anti-Western violence. In fact, the much cited Salafi group, Preaching and Combat (GSPC), which has its origins in Algeria, has suffered a grave defeat by Algerian and Sahelian forces. It had crossed into Chad in 2004, chased by the Pan-Sahelian Initiative's trained forces, and battled the Chadian Army. In the battle, the GSPC lost forty-three of its members. Thus, it would seem that the proclivity of the Sahel toward terrorist activity has been overstated.

Nonetheless, it appears that in the absence of other U.S. government agencies, the U.S. military has the resources to engage in the region and focus on the potential for terrorist activity, and that it would be prudent to do so. It is harmful if U.S. military planners understand the region's problems only by viewing the complex issues of the Sahel through a counterterrorism lens.[20] This overly simple view is not helpful, and fails to consider the many factors of human and environmental security at work here.

The ICG report warns that the governments of the Sahel may be manipu-lating U.S. fears of terrorism to leverage benefits, reminiscent of the Cold

---

[19] UNHCR, 2005a: 1. Note that reports in the fall of 2006 indicated that fighting in Sudan spilled into Chad. The town we visited, Abeche, though unharmed, reported rebel elements in the town.

[20] ICG, 2005: i. In our travels, the military planners we met understood the complexity of the issues affecting human survival and did not use only a counterterrorism lens.

War days (ICG, 2005: 2). The United States must be careful not to over-generalize about Islamic militants or extremists. For example, in the Sahel, the clerics, scholars, and political leaders focus on the inclusivity of Sahelian Islam, "pointing to the role that Sufi Brotherhoods have played in forming the regional culture of tolerance" (ICG, 2005: 4). This culture of tolerance should be valued and could assist with improving human security conditions.

4.4. WHAT IS BEING DONE

We were in Chad in June 2005, and our observations support many of the points addressed above. We had tremendous support from the Defense Attaché, Lieutenant Colonel Tim Mitchell, who enabled meetings with several members of NGOs, UNHCR, host nation, State Department, African Union, and the rebel leader from the Justice and Equality Movement. Additionally, we had follow-up conversations in Washington, DC with members of U.S. Agency for International Development and the State Department's Conflict, Reconstruction and Stabilization office. We had the opportunity to travel outside the capital of N'djamena, to include the town of Abeche and refugee camps in Farchana and Gaga (see Appendix 1).

One cannot address Chad without addressing the Sudanese crisis. As already mentioned, our observations confirm our research. The refugee crisis has exacerbated the conditions for the Chadians, and interestingly, the UNHCR, NGO, U.S. Department of State, military, and other efforts have both helped and hurt the Chadians. The main problem is that with a corrupt Chadian government and an illegitimate Sudanese government, the organizations that are trying to alleviate the plight of both refugees and eastern Chadians cannot engage in sustainability or capacity-building activities. Instead, they are forced to provide aid directly to the people, which is beneficial but is not providing a long-term systemic solution. In other words, the crisis seems to be the status quo. Refugees do not want to go home since it is not yet safe to do so, and the Chadian government cannot be trusted to use support that would benefit society.

What is remarkable is that we found tremendously dedicated UNHCR and NGO professionals working under extremely harsh conditions, knowing that their efforts are not solving the crisis; yet they are not deterred from providing direct aid to refugees who find themselves in desperate situations. The UNHCR has a charter to address refugees, while other organizations are designed to address the local population. Unfortunately, human insecurities do not discriminate among refugees, Internally Displaced Persons (IDPs), and other categories of people. The problem that the UNHCR inadvertently created was the resentment felt by the local population suffering from human insecurities but who do not qualify under the UNHCR charter to receive aid.

Fortunately, the UNHCR recognized this problem and was able to earmark some funds for the local population, while seeking help from other organizations (UNHCR, 2005b).

The Chadians who live in desolate regions of the country suffer from some of the same insecurities as the refugees without the assistance of their government. They are left to cope on their own as the extent of the Chadian government's capability lies within the city limits of its capital, N'djamena. These ungoverned spaces are ripe with human insecurities described earlier, and unfortunately, the people suffer.

Sustaining refugee camps is an enormous challenge. Within the refugee camps, there are security, social, education, and myriad other issues that must be dealt with on a daily basis. With everyone with whom we talked, physical security was a priority. The African Union's military forces from various African nations are charged with providing the external physical security of the refugee camps. The Chadian military units provide the internal physical security. One of the biggest challenges these military units face is the distrust or reluctance to trust by the refugees. From the refugees' perspective, how can they trust someone in a military uniform when the very people from whom they fled were wearing military uniforms? Slowly, the African Union is learning how to deal with such perceptions and taking steps to build trust (African Union official, interview, 2005). Though their strengths and capabilities are very limited, this is an important first step. The UNHCR, as the primary coordinator of all the formal and informal efforts, does everything it can to work with the African Union. Cooperation between the two organizations is mutually beneficial and in the best interest of the refugees. Refugees need to feel protected, as do the UNHCR, NGO, and other personnel.

Most of the refugees in the camps are children (64 percent) and women. A problem arises as refugees become reliant on this secure environment and their desire to return home disappears. This becomes even more evident when the refugees are provided the support to confront poverty, famine, malnutrition, diseases, and illiteracy. Many refugees will decide to stay in the camps. In fact, if more permanent structures are erected, then the expectation for a prolonged stay increases. The reliance on receiving rations and shelter increases with each passing day. Parents willingly accept free vaccinations and education for the children. To protect the women from the perils of the surrounding Chadian villages, women are trucked to areas where they collect firewood and other resources for cooking. Other services coordinated by UNHCR include food, health, veterinary services, sanitation, water, shelter, and other critical services. With each service, cultural considerations must be understood and integrated. Some examples include the number of people culturally acceptable to live in a tent, use of centralized latrines, and modern medicine versus traditional healers (UNHCR, 2005b). For many refugees, there is no home to which to return. Their whole social fabric is gone, and the new social network is the camp. Wisely, the camps try to preserve some tribal unity in the way the living quarters are arranged. With each passing day, the refugees reach a sense of routine normalcy where they feel secure from the physical and nonphysical threats to security. The problem is that the mental attitude required for sustainment and self-help is not nurtured under these conditions. While the UNHCR and NGOs are effective in providing crisis assistance, they are aware of the long-term maladaptations that emerge.[21]

Indeed, cooperation between UNHCR, NGOs, the African Union, the U.S. Department of State, the host nation, and the military are necessary and vital to the human security of the refugees. They all recognize the importance of addressing human insecurities and the protection and preservation of life as critical components of regional stability (with the exception of some of the host nation elites, as described earlier). Each organization contributes a specific set of skills and knowledge that are critical for accomplishing the common goal. What helps in the synchronization of effort is UNHCR's role as the primary coordinator. It works with the Chadian government as well to resolve issues at the national and local levels. The director of the UNHCR effort in Chad routinely involves all available NGOs in the decision-making process whenever possible. She fully understands the diversity of skills and resources NGOs can contribute. Together, they form networks of networks able to provide the flexibility and responses needed in a complex and challenging refugee and humanitarian assistance operation.

The short-term success of the efforts in the camps and outlying areas is easy to measure. One can easily determine numbers of people receiving medical aid, schooling, and food. However, what are not easily measurable are the long-term impacts, especially when host nation support is minimal or even working at cross-purposes. A gentleman from the UNDP stressed the need for elite accountability. Without changing the elite mindset, there is little hope of permanent change (Amaning, interview, 2005). Another key indicator for long-term change is education for girls. Education, especially for girls, as with the previous case, has potential for long-term cultural changes that could generate positive social, economic, and political development (Mitchell, interview, 2005). What was evident through our discussions is that there is an emerging consensus that human insecurities must be addressed with a holistic approach and that real focused effort is attainable by all agencies and organizations, both military and nonmilitary, if there is

---

[21] These points were discussed with the director of the UNHCR effort and member of the UNHCR, Child Protection Office, June 2005, Abeche, Chad.

pre-deployment planning so that everyone understands each other's roles
and missions.

Our focus was on eastern Chad due to the effects of the Sudanese cri-
sis; however, Chad experiences human insecurities throughout the country.
When we discussed the northern border with one official, he recounted how
U.S. forces helped the Chadian military conduct border patrols along its
northern border with Libya. Unfortunately, the border patrols also stopped
the smuggling upon which many ordinary Chadians rely for their livelihood.
In the south, Chad also was affected by refugees of the crisis in the Central
African Republic. How do these situations and conditions relate to terrorism? In truth, it is not clear; however, they present tremendous vulnerabilities
within society in the midst of an illegitimate regime. As we discussed earlier,
when people are faced with desperate situations that the state cannot or will
not address, they may turn to alternate sources of assistance. Currently, the
international community is working to provide some relief. Unfortunately,
without a legitimate and effective government, this relief lacks sustainment
at the national level. In a harsh environment with dwindling resources and
bad governance, the added pressures of refugees only makes the competition
for scarce resources more intense and conflict prone. This situation ought to
cause the world concern on many levels.

5. Environmental Policy Implications: The Continued or Accelerated
Desertification and Its Impact on Human Security

For both Chad and Niger, the continued degradation of the environment
will only exacerbate current human insecurities. For Chad, the refugee
situation is at best a semipermanent situation that competes for already
very scarce resources, and for Niger, its food supply is intricately related to
rainfall variability, among other factors. Of course, the earth's climate has
been under constant change. For example, the area of the southwest United
States known as the four corners region used to be a much more humid,
savanna-like climate approximately eleven hundred years ago. The native
Pueblo peoples who once thrived in this region by cultivating a set of staple
crops suddenly vanished, due to a sudden and unrecoverable change in
the expected environment. The region was dramatically transformed and
became the large desert that is today's Colorado plateau (PBS, 2006). The
Sahel faces a potentially similar fate as deserts continue to expand and
climate change inexorably changes temperature and precipitation in ways
that will be less sustainable for life.

Continued or accelerated desertification has both natural and human
components, each of which can diminish or reinforce the other. Basic
temperature and precipitation parameters not only drive vegetative cover,

but, in turn, soil formation, types of agriculture possible and a region's over-all carrying capacity. Climatic factors can induce positive feedback cycles resulting from, for example, the loss of vegetative cover and a subsequent loss of moisture and increase in solar energy reflection (albedo), leading to progressively drier conditions. For example, loss of vegetative cover can lead to loss of soil through erosion and a decrease in precipitation, which can then lead to further loss in soil (McCarthy et al., 2001: 517–519).

Precipitation patterns in the Sahel can be affected by factors geographi-cally far removed. For instance, sea surface temperature and atmospheric pressure anomalies in the Pacific and Atlantic oceans have been shown to alter the amount of rainfall over the African continent (McCarthy et al., 2001: 519–520). These teleconnections clearly show that the problem is indeed a global one, and that changes to environmental conditions far from the Sahel have tremendous impact on whether the climate there is favorable or hostile to humans (McCarthy et al., 2001: 519–520).

Anthropogenic forces driving desertification include overreliance on unsustainable agricultural practices, overgrazing, and deforestation—all of which can radically alter the Sahel's delicate ecosystem through excessive erosion as well as loss of moisture via destruction of protective tree canopy and productive arable land. The United Nations has noted that, on average, two-thirds of the already desertified land in Africa became that way due to overgrazing while the remaining one-third was due to improper agricultural practices and deforestation (Watson et al., 1997: 4).

Whether anthropogenically or naturally induced, desertification reduces soil fertility, livestock and agricultural yields, and water-retention capacity. If the Niger portion of the Sahel is rendered useless, the net result will be an increase in interstate as well as rural to urban migration, placing more pressure on local, city, and state-level governments already stressed by the steady arrival of the dispossessed to places without the economic bases and infrastructure to support them (Nyong, 2005: 1; Watson et al., 1997: 4).

5.1. METHODOLOGIES FOR IMPROVEMENT AND CHANGE: A MULTI-STATE APPROACH

Many factors place the large majority of the African continent at an elevated risk to the impacts of climate change and desertification. Given the Sahel region's weak economic and political standing, it might be tempting to argue that inte-grated solutions are simply beyond its means (Watson et al., 1997: 1–4).

As a result of the known vulnerability of the region to small changes in environmental conditions, one possible solution is the use of existing climate models to enable the African Union in assisting states within Africa. These models, although far from perfect, are potentially valuable to agents within Sahelian governments. Potential uses might include the public broadcast of climate outlooks that could enable pastoralists and agriculturalists to better prepare for upcoming seasons. Some of the teleconnections mentioned earlier can be forecast with some degree of accuracy, thereby pointing to the right call to action needed. These products are often used within the United States and can help prepare populations if implemented in a timely manner. Specifically, climate models run by the National Oceanographic and Atmospheric Administration (NOAA) assist U.S. farmers by predicting where a drier/wetter than normal pattern is likely. In the western Sahel, where monsoonal weather patterns are easier to predict, using such tools might be quite beneficial and adaptable to their needs. Many climate models can, and often do, take into account large-scale teleconnections, to include vegetative cover, sea-surface temperature, soil moisture, and other variables in a comprehensive attempt to obtain the most accurate long-term forecast (McCarthy et al., 2001: 519). Although the infrastructure needed to undertake, deploy, and manage such a lofty technical and computational goal within the Sahel is intensive in terms of staffing and capital, such tools already exist within government agencies and educational institutions of the United States and the European Union. These items could be included as part of a "climate aid package" potentially kick-starting both interstate cooperation and a larger scientific presence by climatologists, soil scientists, and others willing to lend their expertise.

Managing the already fragile water resources is a potentially contentious proposition, but it is critical. Niger, which has no ambitious water projects, must coordinate with Mali, its upstream neighbor, to establish a compact that would allow for a more reliable water flow into Niger. Without greater flow, no amount of aid or technology will allow for even the most modest irrigation projects to be emplaced. Of course, anything that happens in Niger will affect its downstream neighbor Nigeria. At a regional level, states within the Sahel ought to consider a project similar to one proposed in southern Africa, which suggests diverting Zambezi River water from a surplus region to a deficit region (McCarthy et al., 2001: 496–500). Of course, such a compact, whether undertaken in the Sahel or in southern Africa, would require careful interstate cooperation, and thus is fraught with potential problems— but also possibilities.

While an agreement to divert and more formally manage the Niger's flow may seem too ambitious, the larger Sahel region, including Chad, is home to several other river basins such as the Senegal, Black Volta, White Volta, Benue, Chari, and Hadejia. Each of these basins, in some way, traverses changeable climate regimes. The idea of closely monitoring and regulating water flow as a commodity to be sold and bought across the Sahel would seem to be a reasonable and cost-effective way of diminishing future water tensions, especially if climate change forecasts are correct and the area is about to undergo a significant amount of drying. This project, if realized, might have a similar effect to that of the many Colorado River projects since the early 1900s—the growth of year-round agriculture in regions of California and Arizona that otherwise receive very little precipitation.

Once the larger scale issues of water conservation and management are addressed, then local and state governments can work to institute other less invasive techniques that promote conservation and careful management of water resources. Many of the tribal regions within the Sahel have long-established patterns and techniques of dealing with the seasons. It seems appropriate and prudent that an approach combining the state of the art in scientific thought, sound governmental policies, and local ingenuity congruent with indigenous knowledge might be the best integration tool for dealing with the ongoing changes to the environment at all scales of action. The long record of computer model runs and historical data strongly suggests that climate change not only increases uncertainty but also the chance of greater short-term variability, implying that wet regions might experience too much rain while dry regions might experience more severe drought. If one lives in the Sahel, this level of uncertainty can delineate the difference between life and death.

6. Policy Implications

The patterns and insights revealed through a human security paradigm offer considerations for policymakers.[22] A key discovery is the recognition of the interrelationships and connections among these various human insecurities and actors in Chad and Niger. Addressing the issues of basic survival at the individual level highlights the security implications at the state, regional, and global levels. Most notably are the acute implications for weak or developing states and the regional neighborhoods in which they reside. As states are unable or unwilling to provide security at the individual level in the ways just described, people lose faith in their governments and look for other ways to ensure survivability. While sometimes people turn toward ethnic groups, kin, and other substate entities, they may also be vulnerable to terrorist groups who may offer hope and tangible support. Security policy requires a holistic approach that views human security as a critical linchpin to

---

[22] Portions of this section related to the Horn of Africa can be found in Forest, 2005.

building state capacity and legitimacy vis-à-vis these other groups. Our cases further demonstrate how the environment directly affects human survival. Conditions of dwindling resources and growing populations can no longer be ignored or deferred to another day.

Legitimacy, as a form of interdependence and interconnection between political systems and societies, is central to fostering good governance. Good governance provides the foundation for building state capacity to address these human insecurities. Capacity building must be inherent in any policy that will have long-term effects. In Chad, the government is so corrupt that organizations work directly with the people to ensure that they receive support. The good news is that people get aid, but the bad news is that there is no capacity-building effort in effect. Consequently, it appears that Chad, especially concerning the Sudanese refugees in Chad and the local population surrounding the camps, will be in a permanent crisis mode. For Niger, the food shortages cannot only be addressed when the cameras are rolling; there must be a renewed effort toward structural change that will foster real sustainability.

For change to occur, not only must there be significant focus at the grass-roots level, but there must also be focus at the elite level. In other words, at the political level, the international community cannot turn a blind eye toward rampant state corruption and self-declared dictators. Diplomats must be empowered to courageously speak truth to power, and state officials and politicians must be held accountable. Measuring effectiveness toward the fight against terrorism will require a more nuanced standard. The United States must be careful not to fall victim to the Cold War manipulations of the past. Respect for human rights, efforts toward addressing human insecurities, and other efforts toward real political reform must be considered as standards for fighting terrorism. If not, aid, especially military aid, may mistakenly be provided, which may only further exacerbate the problem. In effect, such aid only helps militarize a regime, not democratize it.[23]

The issue, then, is how to break the mindset that condones extreme state corruption, perpetuates human insecurities, and continues to present vulnerabilities that terrorists can exploit. How does one craft policy that empowers people and governments to take responsibility and move forward, especially in extremely harsh environments and climates? The key will be education, and it is important to realize that although

---

[23] The point about speaking truth to power was based on an interview with Mr. Amaning, Resident Director of UNDP.

the effects of education will be long-term, it will be at least a generation before those effects are realized. Key to the education process will be the inclusion of women for many reasons, not least of which is that it doubles the talent pool of a society. It is important for the international community to empower local people and governments and cultivate stakeholders on all projects by working collaboratively during the decision-making and implementation process with local and national politicians and informal leaders.

Our discussion of human insecurities highlights the importance of understanding the environment. Unfortunately, the regions that are gravely suffering from human insecurities exist in harsh and extreme climates. The prognosis for Niger and Chad is that desertification will continue, thus the urgency for finding ways to mitigate its effects is paramount. Many people living in these harsh climates face the possibility of losing their culture and their way of life. In Niger, land scarcity endangers both pastoralists and agriculturalists. This is important on several counts. First and foremost, as natural environments change, they contribute to the dynamic human insecurity challenges. These challenges put extreme pressure on governments to respond. A government's failure to respond may precipitate other adverse effects. For example, increasing scarcity of resources may lead to conflict, while creating further societal, economic, and political pressures. Environmental studies that provide policymakers with forecasts and recommendations will become increasingly important as people adapt. Moreover, infrastructure becomes a key component in any policy that seeks to develop an economy. It is hard enough growing crops, but just providing the goods to market in a timely and safe fashion becomes an insurmountable task. Understanding the environment is a first step toward understanding the totality of challenges the population faces as well as the state and external actors.

Understanding these challenges will better enable us to anticipate second and third order effects of external actors' actions within a state, and their spillover effects on the larger region. There will be unintended consequences, but by understanding these complexities, we may be able to better adapt. Our cases highlighted some of these consequences: in Chad, setting up refugee camps increased tension between eastern Chadians and the refugees as they compete for scarce resources; after the United States trained the Chadian Army on effective border control techniques along the Libyan-Chadian border, unanticipated tension arose as many Chadians had been relying on smuggling activities for their livelihood that were now shut down; and food aid targeted for Niger ends up in Nigeria, which in turn sells the aid back to Nigerians at unaffordable prices. Understanding the integrated and complex nature of all aspects of an environment requires in-depth knowledge at all levels to mitigate these and other unintended consequences.

This enormous task requires culturally knowledgeable and aware security professionals. For the U.S. military, which is currently transforming, it is clear that the most important asset that requires transforming is the intellect. The military cannot transform effectively without ensuring proper education and training of its service members, and true "buy-in" to the components of human security by its leaders. The foreseeable future will require adaptable, innovative, and culturally aware military members. The U.S. military has excellent professionals, but it must continue to invest at a more rapid rate on their intellectual capital. In addition, training and education must have interagency, NGO, and IGO involvement. While it is important to gain cultural perspectives of a region, it is just as important to understand the different organizational cultures among the suite of actors involved in addressing human insecurities. Furthermore, an interagency, international, and military planning cell prior to arriving in country would be invaluable. At the very least, people would better understand the different players' roles and capabilities, and many misunderstandings could be resolved.

What we discovered in Chad is that there was consensus among members of various NGOs, IGOs, and host nation officials that in fact, the military does play an important role in addressing human insecurities. The most important role is providing security where none exists. However, the military can also play a critical preventative conflict role. The Trans-Sahel Counterterrorism Partnership (TSCTP) is designed with this preventative role in mind. Interestingly, the Combined Joint Task Force (CJTF) in the Horn of Africa may be a good model, as it demonstrates how fighting terrorism with schools, bathrooms, and roads can prove effective in creating and sustaining a relationship and connection with the people in the regions of the Horn of Africa (Jebb and Abb, 2005). The key is basing all actions on the premise that the United States is not in the Sahel to impose an American culture. Instead, the key to success is providing empowerment to help the people to self-actualize, self-determine, and self-govern. Second, the best type of military training is professionalizing host nation militaries. In other words, educating militaries on standards of behavior will go a long way in areas where people are socialized to fear the military uniform—and for good reason. Professionalization in the long term will be more important than teaching perishable skills at the tactical level.

Fortunately, we are starting to see the interagency play a more prominent role. In Niger, USAID opened its first office in several years, though while we were in Chad, USAID had no presence. However, the interagency elements must be resourced and empowered to bring to bear U.S. diplomatic, economic, and informational elements of power. It may be more helpful for all agencies to be organized based on regions, not states, as many of the issues we described easily spill over state borders. Perhaps it is time to create regional-level ambassadorial positions. Second, we were overwhelmingly impressed by the dedicated, selfless UNHCR and NGO professionals on the ground. However, they, too require more resources, and unfortunately, they have been put in a crisis situation that appears never ending unless the international community puts pressure on dysfunctional regimes. The UNHCR and the NGOs that it coordinates have been unable to create sustainable capacities in Chad because of the corrupt regime. Third, all agencies including NGOs must be held accountable. William Easterly criticizes the Millennium Development Goals (MDGs) because they require collective responsibility, which, in truth, means that there is no one to hold accountable. Instead, he calls for feedback and accountability on discrete tasks conducted by each organization (Easterly, 2006: 15–17). Mechanisms for accountability are essential.

Policymakers should not use only a counterterrorism lens to view the security environment. Instead, understanding the security challenges from a human security lens provides policymakers with insights that are vital for effective policy. Attacking extremists and radicals by addressing human insecurities can be effective, although it will take time and results will not be immediately visible. Cultivating stable states that demonstrate good governance and legitimacy takes time. Many of these states are still in the early steps of state-making and struggling with enormous pressures as discussed throughout this paper. World leaders need a better way to measure success and incorporate strategic patience in their planning and assessments. The Western concept of time will not do. For the United States and allies, this new orientation requires political and societal support as well as politically courageous and enlightened political leaders.

Policymakers must also realize that they cannot make choices for people or governments. The communities in Niger must decide their path, and for some that may mean a shift in their thinking and way of life, especially as they develop ways to adapt to the environment. Indeed, policymakers can help create conditions that will assist people in making sound choices. These efforts will take time, intelligence, critical analysis, patience, and empathy. Subsequently, the modest strategy for success will depend on the incremental, tireless, and focused work of the international community that will best succeed in creating a more peaceful and prosperous world.

APPENDIX 1.

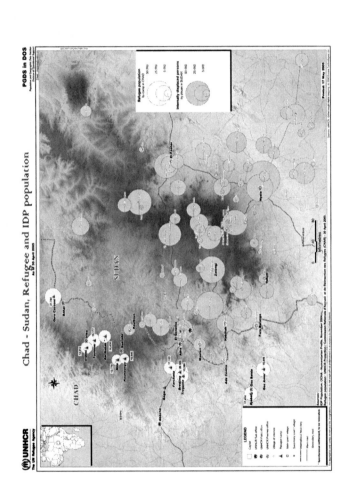

APPENDIX 2.

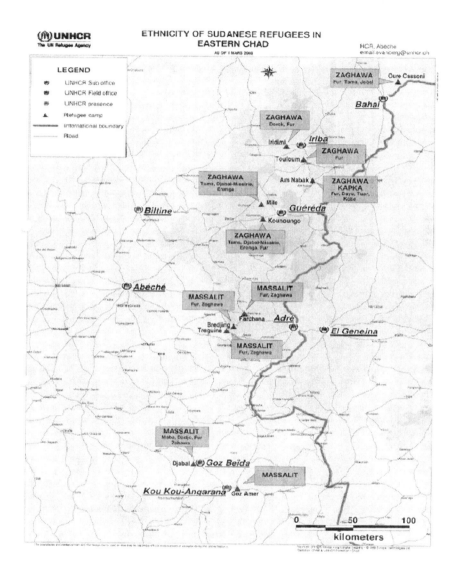

APPENDIX 3.

ACTIVITÉS D'ASSISTANCE AUX REFUGIÉS SOUDANAIS A L'EST DU TCHAD
2004

OPERATIONS EN FAVEUR DES REFUGIES A L'EST DU TCHAD 2004 - TABLEAU DES PARTENAIRES OPERATIONNELS ET D'EXECUTION

SITUATION AU 01 FEVRIER 2005

ZONES	SUD			CENTRE			CENTRE NORD			NORD			HCR BAHAI			
Bureau de terrain HCR	HCR GOZ BEIDA			HCR ADRE			OUREDA			HCR IRIBA						
Gestionnaires de camps	InterSOS	InterSOS		CARE → HCR						CARE			IRC			
Nombre de camps	1	2		3	4	5	6	7		8	9	10	11	Site 1	Site 2	
Camps de réfugiés	GOZ AMER	DJABAL	FARCHANA	BREDJING	TREGUINE	KOUNOUNGO	MILE	AM NABAK	TOULOUM	IRIDIMI	OURE CASSONI					
Population planifiée	20,000	18,000	20,000	24,689	18,000	12,000	18,000	18,000	18,000	16,000	24,000					
Population existante	19,303	17,256	19,389	32,475	14,490	12,991	14,982	18,450	21,243	17,508	28,194					
INFRASTRUCTURES																
DANS LES CAMPS																
Installation et construction	PAM/InterSOS	InterSOS	GTZ	EEMET	FICR-CRT	GTZ	NCA	CARE	NCA	CNAR	CNAR		OXFAM			
Système d'adduction d'eau	InterSOS	InterSOS	Secadev/OXFAM	OXFAM	OXFAM	OXFAM	NCA	OXFAM/HELP	NCA	IRC						
Abris, Radiers des tentes	PAM/InterSOS	InterSOS	FICR	EEMET/CARE	FICR-CRT	GTZ	NCA	CARE	NCA	IRC						
Construction de latrines	PAM/InterSOS	InterSOS	GTZ/MSF Hollande	OXFAM	FICR-CRT	GTZ	NCA	OXFAM	NCA	IRC						
ACTIVITES																
Enregistrement	CNAR	CNAR	CNAR	CNAR	CNAR	CNAR	CNAR	CNAR	CNAR	CNAR	CNAR					
Accueil et information	InterSOS	InterSOS	CRT	CRT	CRT	CRT	CRT	CRT	CRT	CRT	IRC					
Services, Communautaires	InterSOS	InterSOS	CORD		FICR-CRT	CORD	CARE	CARE	CARE	CARE	IRC					
Activités génératrices de revenus	NAS	NAS	MSF Hollande	FICR	Première Urgence	IRC	IRC		NCA	NCA	?					
Protection de l'enfant						CCP	CCP		COO	COO						
Approvisionnement en vivres	PAM	PAM	PAM	PAM	PAM	PAM	PAM	PAM	PAM	PAM	PAM					
Distribution de vivres et de non-vivres	InterSOS	InterSOS	SECADEV	FICR	FICR-CRT	SECADEV	CARE	CARE/HELP	CARE	CARE	ACTED					
Basket Supplementary Feeding	InterSOS	InterSOS	SECADEV	FICR	FICR-CRT	SECADEV	CARE	CARE/HELP	CARE	CARE	ACTED					
Gestion de l'eau	InterSOS	InterSOS	SECADEV	OXFAM	OXFAM	SECADEV	NCA/CARE	OXFAM	NCA/CARE		IRC					
Assainissement/construction de l'hygiene	InterSOS	InterSOS	SECADEV/ICRS OXFAM*	OXFAM	FICR-CRT EEMET	CARE	NCA	OXFAM	NCA	CARE	IRC					
Soins de santé primaires	COOPI	COOPI	MSF Hollande	MSF Hollande	FICR-CRT	IRC	IRC	IRC	MSF Luxembourg	MSF Luxembourg	IRC					
Nutrition	COOPI	COOPI	MSF Holl.	MSF Holl.	MSF Hollande	IRC	IRC	IRC	MSF Luxembourg	MSF Luxembourg	IRC					
Food Basket monitoring	Première Urgence	Première Urgence	Première Urgence	Première Urgence	Première Urgence	ACF	ACF	ACF	ACF	ACF	ACF					
Habitat & références	Hôpital de Goz Beida - COOPI		Hôpital d'Abre - MSF France		Hôpital d'Iriba	Hôpital de Guéréda - IRC			Hôpital d'Iriba - MSF Belgique		Hôpital BAHAI-IRC					
Education	InterSOS	InterSOS	SECADEV/UNHS CORD/POUSSANT ROUSSE BACKDO	CORD	COO	SECADEV	SECADEV	CARE	SECADEV/IHS	CARE	IRC					
Environnement - ONG Responsable	InterSOS	InterSOS	SECADEV/BCI	ACS Japon	SECADEV	SECADEV	CARE	CARE	CARE	CARE	ACTED					
Environnement - Activités d'appui	ACS/BCI	ACS/BCI	SECADEV/BCI	Première Urgence/BCI	Première Urgence	BCI	CAMERON	CAMERON	CARLISIC	CAMERON						
Logistique	GTZ	GTZ	GTZ	GTZ	FICR-CRT	GTZ	GTZ	GTZ	GTZ	GTZ	GTZ					
Renforcement des capacités			CRS et OXFAM*			ICRS et OXFAM				CRS/OXFAM						

* Camps où il y a des activités de renforcement de capacités
→ Transfert en cours

N.B.
Pour toute suggestion, correction, mise à jour, prière contacter HCR, Section Programme 00 85 10, Tchurg 00882 164 112 9435, email: mahamag@unhcr.ch

References

African Union official, 2005, interview, 13 June, Abeche, Chad.

Amaning, 2005, interview, Director of UNDP, 10 June, N'djamena, Chad.

Ayoob, M., 1995, *The Third World Security Predicament,* Lynne Reinner, Boulder, CO.

Batterbury, S., and Warren, A., 2001, The African Sahel 25 years after the great drought: Assessing progress and moving towards new agendas and approaches, *Global Environmental Change* **11**(1):1–8.

Beitler, R. M., and Jebb, C. R., 2003, Egypt as a failing state: Implications for U.S. national security, INSS Occasional Paper 51, July, p. 47.

Bodley, J. H., 1996, *Anthropology and Contemporary Human Problems,* Mayfield, Mountain View, CA, pp. 83–84.

Brown, L. R., 2004, *Outgrowing the Earth: The Food Security Challenge in an Age of Falling Water Tables and Rising Temperatures,* W.W. Norton, New York, pp. 4–9.

Brown, L. R., 2006, *Plan B 2.0: Rescuing a Planet Under Stress and a Civilization in Trouble,* W.W. Norton, New York, p. 3.

Buzan, B., 1983, *People, States, and Fear, 2d edition,* University of North Carolina, Chapel Hill, NC.

Capra, F., 1996, *The Web of Life: A New Scientific Understanding of Living Systems,* Anchor Books Doubleday, New York.

Capra, F., 2002, *The Hidden Connections: Integrating the Biological, Cognitive, and Social Dimensions of Life into a Science of Sustainability,* Doubleday, New York.

Cavalcanti, H. B., 2005, Food security, in: *Human and Environmental Security: An Agenda for Change,* F. Dodds and T. Pippard, eds., Earthscan, London, p. 156.

Chad, Background note, Bureau of African Affairs, 2006, U.S. Department of State; http://www.state.gov, p. 1.

Danish Corporation, 2006, interview with Vanessa, 27 August, Niamey, Niger.

de Bruijn, M., and van Dijk, H., 2002, Climate variability and political insecurity: The Guera in central Chad, August, p. 5.

Dixon, R., 2005, The world; Secret lives of servitude in Niger; The government has banned slavery and denies it exists. Though few speak of it, the practice is a tradition many do not question, *Los Angeles Times,* September 3, pp. A2–A3; http://0-proquest.umi.com, accessed 4 August 2006.

Documenting Atrocities in Darfur, 2004, Department of State Publication 11182, Bureau of Democracy, Human Rights, and Labor and Bureau of Intelligence and Research, September.

Durbak, C. K., and Strauss, C. M., 2005, Securing a healthier world, in: *Human and Environmental Security: An Agenda for Change,* F. Dodds and T. Pippard, eds., Earthscan, London, pp. 128–129.

Easterly, W., 2006, *The White Man's Burden: Why the West's Efforts to Aid the Rest Have Done so Much Ill and So Little Good,* Penguin, New York, pp. 15–17.

Eberstadt, N., 2002, The Future of AIDS, *Foreign Affairs* **81**(6):22–45).

Elbe, S., 2004, HIV/AIDS and the changing landscape of war, in: *New Global Dangers: Changing Dimensions of International Security,* M. E. Brown et al., eds., MIT, Cambridge, p. 371.

Ellis, S., 2004, Briefing: the Pan-Sahel Initiative, *African Affairs* **103**(412):459–464.

Feinstein, L., and Slaughter, A., 2004, A duty to prevent, *Foreign Affairs* **83**(1):136.

Forest, J., ed., 2005, *The Making of a Terrorist,* vol. 3, *Root Causes,* Praeger Security International, Westport, CT.

Fukuyama, F., 2004, *State-Building: Governance and World Order in the 21st Century,* Cornell University Press, Ithaca, NY, p. 26.

Gazibo, M., 2005, Foreign aid and democratization: Benin and Niger, *African Studies Review* **48**(3):8–9; http://0-proquest.umi.com.usmalibrary.usma.edu, accessed 4 August 2006.

Global Risk Outlook 2006, 2005, Executive Analysis, London, p. 60.

Gurr, T., 1997, Why minorities rebel: Explaining ethnopolitical protest and rebellion, in: *Minorities and Risk: A Global View of Ethnopolitical Conflict,s* United States Institute of Peace, Washington, DC, pp. 123–138.

Halperin, M. H., 1993, Guaranteeing democracy, *Foreign Policy* **91**:105–122

Heath, D. E., 2005, Niger: More than a food shortage, *Environment* **47**(10):1; http://0-proquest.umi.com.usmalibrary .usma.edu, accessed 4 August 2006.

Hoffman, B., 2004, Lecture to Terrorism and Counterterrorism class at USMA, West Point, April.

Homer-Dixon, T. F., 2004, Environmental scarcities and violent conflict, in: *New Global Dangers: Changing Dimensions of International Security,* M. E. Brown et al., eds., MIT, Cambridge, p. 265.

Hulme, M., and Kelly, M., 1993, Exploring the links between desertification and climate change, *Environment* **35**(6):5–11, 39–45.

Human security doctrine for Europe, 2004, The Barcelona Report of the Study Group on Europe's Security Capabilities, 15 September, p. 9.

Hummel, L. J., 2006, Lowering fertility rates in developing states: Security and policy implications for sub-Saharan Africa, United States Army War College, Carlisle Barracks, PA, p. 13.

International Crisis Group (ICG), 2005, Islamist terrorism in the Sahel: Fact or fiction, *Africa Report* 92, 31 March, pp. 7–8, 22.

Issa, O., 2006, Niger: Forest squatters demand new homes before eviction, Global Information Network, New York, 3 August, p. 1; http://0-proquest.umi.com.usmalibrary.usma.edu, accessed 4 August 2006.

Jebb, C., and Abb, M., 2005, *Human Security and Good Governance: A Living Systems Approach to Understanding and Combating Terrorism*, INSS, 31 December.

Jebb, C. R., Liotta, P. H., Sherlock, T., and Beitler, R. M., 2006, *The Fight For Legitimacy: Democracy Versus Terrorism,* Praeger Security International, Westport, CT, pp. 134–136.

Jervis, R., 1997, *System Effects: Complexity in Political and Social Life,* Princeton University Press, Princeton, NJ, pp. 12–13.

Kaplan, R., 1994, The coming anarchy, *Atlantic Monthly* **273**(2):44–76.

Kasfir, N., 2005, Sudan's Darfur: is it genocide? *Current History* **104**(682):195–202.

Kennedy, L., and Jebb, C., forthcoming, Non-state actors and transnational issues, in: *American National Security,* A. A. Jordan, W. J. Taylor Jr., M. J. Meese, and S. C. Nielsen, eds., Johns Hopkins University Press, Baltimore, MD.

Klare, M. T., 2002, *Resource Wars: The New Landscape of Global Conflict,* Henry Holt and Company, New York, pp. 8–9.

Kraxberger, B., 2005, The United States and Africa: Shifting geopolitics in an "Age of Terror," *Africa Today* **52**(1); http://0-proquest.umi.com.usmalibrary.usma.edu, accessed 1 December 2006.

Laremont, R., and Gregorian, H., 2006, Political Islam in West Africa and the Sahel, *Military Review* **86**(1):27–36; http://0-proquest.umi.com.usmalibrary.usma.edu, accessed 1 December 2006.

Liotta, P. H., 2004, *The Uncertain Certainty,* Lexington Books, Lanham, MD, pp. 4–5.

Liotta, P. H., and Owen, T., 2006, Sense and symbolism: Europe takes on human security, *Parameters* **36**(3):85–102.

Loewenberg, S., 2006, Millions in Niger facing food shortages once again, *The Lancet* **367**(9521):2; http://0-proquest.umi.com.usmalibrary.usma.edu, accessed 8 August 2006.

McCarthy, J. J. et al., 2001, Climate change 2001: Impacts, adaptations and vulnerabilities, Contribution of Working Group II to the Third Assessment Report of the Intergovernmental Panel on Climate Change, p. 489; http://www .grida.no/climate/ipcc_tar/wg2/001.htm, accessed 30 October 2006.

Microsoft Encarta, 2006, Niger River Article, accessed 7 November 2006.

Miles, W. F. S., 2003, Shari'a as De-Africanization: Evidence from Hausaland, *Africa Today* 50(1):5; http://0-proquest.umi.com.usmalibrary.usma.edu, accessed 1 December 2006.

Miles, W. F. S., 2005, The Niger we should know, *Boston Globe*, August 23, p. A15; http://0-proquest.umi.com, accessed 4 August 2006.

Miller, J. G., 1978, *Living Systems,* McGraw-Hill, New York, p. 1.

Mitchell, T., 2005, interview, Defense Attaché, Chad, 10 June, N'djamena, Chad.

NGO head, 2006, interview, 28 August, Niamey, Niger.

Niger, Background note, Bureau of African Affairs, 2006, U.S. Department of State; http://www.state.gov, accessed 4 August 2006.

Nigerian official, Interview, 27 August 2006, Niamey, Niger.

Nyong, A., 2005, Drought and conflict in the Western Sahel: Developing conflict management strategies, Event summary compiled by Ms. Alison Williams of the Woodrow Wilson International Center for Scholars, 18 October.

Office of Dutch-Nigerian Cooperation, 2006, Study of the practices of Islam in Niger, *Provisional Report*, April.

Paz, R., and Terdman, M., 2006, Africa: The gold mine of Al Qaeda and global jihad, Global Research in International Affairs Center, The project for the Research of Islamist Movements, *Occasional Papers* 4(2):2.

Pronk, J., 2005, Globalization, poverty, and security, in: *Human and Environmental Security: An Agenda for Change,* F. Dodds and T. Pippard, eds., Earthscan, London, p. 84.

Public Broadcasting System (PBS), Nature Documentary, 2006; http://www.pbs.org/wnet/nature/grandcanyon.

Rejai, M., and Enloe, C. H., 1969, Nation-states and state-nations, *International Studies Quarterly* 13(2):140–158.

Sachs, J. D., 2000, The geography of economic development, *Naval War College Review* 53(4):93–105, 101–102.

Sachs, J. D., 2005, *The End of Poverty: Economic Possibilities of Our Time,* Penguin Books, New York, p. 192.

Sanderson, T. M., 2004, Transnational terror and organized crime: Blurring the lines, *SAIS Review* 24(1):2; http://0-proquest.umi.com, accessed 1 December 2006.

Scharnberg, K., 2005, Do-it-yourself famine fight, *Chicago Tribune*, August 25, p. 1; http://0-proquest.umi.com, accessed 4 August 2006.

Senge, P., 1990, *The Fifth Discipline: The Art & Practice of The Learning Organization,* Currency Doubleday Books, New York, p. 73.

Tectonidis, M., 2006, Crisis in Niger—Outpatient care for severe acute malnutrition, *New England Journal of Medicine* 354(3):1; http://0-proquest.umi.com.usmalibrary.usma.edu, accessed 4 August 2006.

Tilly, C., 1985, War making and state making as organized crime, in: *Bringing the State Back In,* P. B. Evans, D. Rueschmeyer, and T. Skocpol, eds., Cambridge University Press, Cambridge.

U.S. Embassy official, 2006, interview, 29 August, Niamey, Niger.

U.S. Federal News Service, 2006, "Habbanae" loans of North Niger for harmony list, Washington, DC, June 20, p. 1; http://0-proquest.umi.com.usmalibrary.usma.edu, accessed 4 August 2006.

United Nations Children's Fund (UNICEF) official, 2006, interview, 28 August, Niamey, Niger.

United Nations Development Programme (UNDP) Report, 1994, pp. 3, 22–23.

United Nations Development Programme (UNDP), 2002, Arab human development report 2002: Creating opportunities for future generations, UNDP, New York, p. 27.

United Nations Development Programme (UNDP), 2005, *Human Development Report, 2005: International Cooperation at a Crossroads: Aid, Trade and Security in an Unequal World*, p. 39.

United Nations High Commissioner for Refugees (UNHCR), 2004, Briefing Note on Chad, December 6.

United Nations High Commissioner for Refugees (UNHCR), 2005a, Sudan Situation Update, February, p. 1. Note that recent reports in the fall of 2006 indicated that fighting in Sudan spilled into Chad. The town we visited, Abeche, though unharmed, reported rebel elements in the town.

United Nations High Commissioner for Refugees (UNHCR), 2005b, Packet developed for Senator Biden's trip, 31 May.

Watson, R. T. et al., 1997, Summary for policymakers—The regional impacts of climate change: An assessment of vulnerability, Intergovernmental Panel on Climate Change Special Report, November, p. 3; http://www.ipcc.ch, accessed 30 October 2006.

Wentley, M., 2006, USAID, interview, 28 August, Niamey, Niger.

Wentling, M., 2006, Draft paper: Elements of an assistance strategy.

WFP personnel, 2006, interview, 28 August, Niamey, Niger.

Whitlock, C., 2006, Terror group expands in N. Africa/Faction backed by al-Qaida runs training camps in the region, *Houston Chronicle*, October 6, p. 19; http://0-proquest.umi.com.usmalibrary.usma.edu, accessed 1 December 2006.

World Bank Web Site; http://www.worldbank.org/niger, accessed 7 November 2006.

World Health Official, 2006, interview, 28 August, Niamey, Niger.

World Resources Institute, 2006; http://earthtrends.wri.org/text/population-health/country-profile-136.html, accessed 15 November 2006.

ENVIRONMENT AND SECURITY IN EASTERN EUROPE

OLEG UDOVYK*

National Institute for Strategic Studies, Kyiv, Ukraine

Abstract: This paper examines the links between environment and security in three East European countries—Belarus, Moldova, and Ukraine. It is based on the report of the Environment and Security Initiative and incorporates results from more recent research. The paper highlights the importance of recognizing the region's geopolitical positioning between the European Union and the Russian Federation, improving energy security without jeopardizing the environment, cleaning up obsolete military infrastructure, chemicals, and metallurgical stocks, addressing frozen conflicts, and strengthening cooperation over shared rivers and ecosystems.

Keywords: Belarus; Moldova; Ukraine; energy dilemma; Chernobyl legacy; frozen conflicts

1. Introduction

It is increasingly recognized today that security is not just a military issue, and that the destruction and over-exploitation of natural resources and ecosystems can also threaten the security of communities and nations. Here we describe such challenges in Eastern Europe. This region extends from the northern shore of the Black Sea in Ukraine up to the Baltic Sea basin in Belarus. It covers 845,000 square kilometers and is home to almost sixty million people. These nations share common borders, watersheds, and infrastructure and have many similarities in their geography, history, culture, and economy.

Belarus, Moldova, and Ukraine are nations with recent sovereign statehood. They are positioned between an enlarging EU and a historically influential Russia. The area's unique position and history have played a large part in the overlapping of environmental and security issues, which have evolved over three distinct periods: the Soviet years of intensive industrialization, a

---

* Address correspondence to Oleg Udovyk, National Institute for Strategic Studies, Kyiv, Ukraine

P.H. Liotta et al. (eds.), Environmental Change and Human Security, 393–405.

difficult period of political and economic transition, and the recent economic recovery with its new challenges.

2. The Regional Context

Following the sudden disintegration of the USSR, Belarus, Moldova, and Ukraine immediately faced a historic challenge for which they were ill equipped. Outsiders often fail to appreciate their problems but are quick to notice poverty, corruption, and other negative phenomena in Eastern Europe.

Despite these challenges the three countries have achieved significant successes. The region has negotiated the difficult transition years without suffering violent conflict of the kind that paralyzed the Balkans, the Caucasus, and Central Asia. Eastern Europe gained much sympathy by deciding not to preserve military nuclear capacity and transferring weapons inherited from the Soviet Union to Russia. Furthermore, disagreements between Russia and Ukraine regarding the status of the Soviet Black Sea fleet have been satisfactorily managed and largely resolved, sparing Europe a major security risk.

However, there are plenty of regional security issues reaching beyond the borders of Eastern Europe to feature on the security agenda of the whole continent. The Transnistrian conflict in Moldova is one example. There are also difficult issues of supply and transit of Russian fuel. The key challenge for the three countries is still to strengthen contemporary state institutions to fully address economic, social, demographic, environmental, and security problems.

The legacy of the Chernobyl disaster—synonymous to the outside world with environmental problems in Eastern Europe—epitomizes the difficulties involved in dealing with all these problems at the same time. In the early hours of 26 April 1986 a violent explosion at the Chernobyl nuclear power plant, near the Ukrainian-Belarusian border, destroyed the reactor and started a large fire that lasted ten days. During the explosion and the fire a huge amount of radioactivity was released into the environment, spreading over hundreds of kilometers into Belarus, Ukraine, and beyond.

For the last twenty-one years, millions of Ukrainians and Belarusians have been living on contaminated land. Compulsory resettlement out of the more dangerous areas shattered the lives of hundreds of thousands. Many more chose to voluntarily abandon the environmentally unsafe and economically depressed region. Its mounting health problems and a catastrophic demographic situation were compounded by accelerating outward migration by young and able-bodied people. Prohibitions pervade the everyday lives of a whole generation of people still living in the contaminated areas. They can never again graze their cattle in meadows, pick berries and mushrooms in surrounding forests, or till their own fields.

Chernobyl affected one-fifth of Belarusan territory and a quarter of its population. In the early 1990s as much as 20 percent of the national budget was spent on remediation efforts, which would result in economic meltdown even in a stable, healthy economy. The economic, social, and environmental burden of Chernobyl was no lighter in Ukraine, which had to deal with the safety of the destroyed reactor as well. The disaster also clearly demonstrated that an accident in one country may threaten human lives and health all over a continent. Twenty-one years after the disaster the influential Blacksmith Institute still lists Chernobyl among the ten most polluted places in the world.

Given this legacy, the recent announcements of plans by the governments of Belarus and Ukraine to expand the use of nuclear power reflect the dramatic challenges facing these countries. Their current dependence on energy imports is seen as a key security concern. The region does not have sufficient energy resources of its own, but energy is critically important for both social stability and economic development, particularly with such energy-intense economies. The energy issue is all the more important because Eastern Europe stands at the crossroads of east-west and north-south energy corridors linking Russia to Western Europe, and the Black Sea to the Baltic.

The quest for secure energy supplies by whatever available means may have serious implications for the environment in Eastern Europe, already facing severe against acute problems. While some of these are inherited from the Soviet era, others are caused by the decline in state control during the transition years. A third category is related to the recent economic upturn and newly spurring industrial activities. Serious environmental issues facing the region include pollution in industrial and mining regions, accumulation of toxic waste, land degradation, and scarcity of safe drinking water. But at the same time the region has significant natural resources which, if used wisely, may support its long-term economic prosperity.

2.1. THE GEOPOLITICAL POSITION

Despite common borders and many similarities, the three countries of Eastern Europe do not constitute a region in the sense of political community. Belarus, Moldova, and Ukraine have not yet developed visible capacity and projects for regional integration. On the contrary, Eastern Europe is a zone of geopolitical attraction among major powers, including the Russian Federation to the east and the European Union to the west. Eastern Europe's pivotal location at the intersection of strategic transport corridors, such as between Russian and Caspian producers of fuel and European energy consumers, further amplifies such influence.

After expanding eastward over the last decade, the EU seems to be experiencing "enlargement fatigue." Its capacity to absorb additional members was compromised, in particular, by the failure in 2005 to ratify a new European Constitution. The EU is also the most important trade partner for all three countries. It is therefore still important for the EU to have friendly, politically stable, and economically prosperous countries on its doorstep, forming a solid bulwark against unwanted migration, terrorism, and other threats such as drug, arms, and human trafficking. Ukraine and Moldova are the only two European countries among the "top ten" sources of illegal migrants to the EU. EU's most comprehensive attempt to deal with Eastern Europe is through its Neighborhood Policy, which aims at strengthening stability in the region and cross-border cooperation.

On the eastern side, Eastern European countries must forge new relations with Russia, with which they share strong historic, cultural, and social ties. Russia is keen to maintain secure transit routes through Eastern Europe while retaining the ties of the past and developing political and economic cooperation. Travel to and from Russia is still visa free. Simplified border regulations and cultural affinity facilitate the transfer of several million Eastern European migrant workers in Russia, and other economic ties. Russia remains a key market for Eastern European products and the most important energy supplier for all three countries. As is the case with the EU, this economic cooperation makes relations with Russia extremely important and political disagreements—for example regarding the settlement of the Transnistrian conflict in Moldova—very painful. Russian security interests are also related to the presence of its military facilities in Moldova (Transnistria) and Ukraine (Crimea).

Since the disintegration of the USSR, various international bodies involving part of post-Soviet states have been set up. The first of these, the Commonwealth of Independent States (CIS) was established in 1991. The CIS currently includes twelve former Soviet republics. Among further initiatives the most notable was the Collective Security Treaty signed in Tashkent in May 1992 between all CIS countries excluding Moldova and Ukraine. An economic integration initiative, the Eurasian Economic Community (EurAsEC), was started in 2000 and currently involves six former Soviet republics (including Russia and Belarus) as members, and Ukraine and Moldova as observers. EurAsEC aims to offer free trade, a common customs policy, and, in the long term, monetary union. Also notable in the region is the Organization for Democracy and Economic Development—GUAM, which includes Georgia, Ukraine, Azerbaijan, and Moldova.

2.2. INTERNAL SECURITY CHALLENGES

Internal problems and tensions are no less important than geopolitical challenges. Not only may they weaken young states and increase their vulnerability to external factors, but they may also present security challenges in their own right. Not surprisingly, such internal security factors feature prominently in the national security doctrines of all three countries.

Many of the internal developments are common to other post-Soviet states. Though expanding, the region's economies still lag behind most of their neighbors, with Moldova one of the poorest European countries in terms of per capita GDP. All the countries suffered economic decline in the 1990s, followed by some recovery over the last five years. However, this recovery has gone hand in hand with painful economic restructuring. In the past Belarus, Ukraine, and Moldova were intricately linked to the rest of the Soviet economy. The collapse of the USSR and economic liberalization opened up local markets, increased competition, and severed some of the ties with former Soviet republics. However, access to Western markets, especially in the EU, has been very limited and often conditional on political or further economic reform. Moreover, the new patterns of trade with Europe have increasingly consisted of exports of raw materials in exchange for imports of manufactured goods. Finally, it has proven difficult to restructure the old heavy industry that was often the mainstay of the Soviet-era economy.

Economic restructuring has consequently not delivered on its promise of universally higher living standards and political stability. The decline in agricultural production contributed to increased poverty and further deterioration in the basic infrastructure of rural areas in all three countries. Social problems have also become more acute in some heavily industrialized regions. In certain cases this has coincided with tension and conflict. Here again, the most striking example is Transnistria, home to almost all Moldovan industry with traditionally strong ties to the former Soviet economic space. Another example of a region suffering from economic restructuring is the heavily industrialized Donbas region in Ukraine, where economic and social problems mesh with issues of environmental and energy security.

The economic and social problems of rural and heavily industrialized areas are aggravated by demographic trends, severely affected by the declining birth rates that are now below the replacement level in all three countries. The populations of Ukraine and Belarus will shrink significantly, with Ukraine expected to lose nine to fifteen million people over the next fifty years. Outgoing labor migration makes the situation even worse, hitting Moldova particularly hard, with an estimated 1,000,000 Moldavians (i.e. 40 percent of the active population) working abroad.

Other serious, in some cases severe, problems include the spread of HIV/AIDS and tuberculosis. The rate of increase in HIV/AIDS infections in the region is among the highest in the world, though significant differences among the countries have been reported. Ukraine, with an adult infection rate of 1.4 percent, is the hardest-hit country in Europe. The governments of the three countries are making a considerable effort to attract international attention and obtain assistance in addressing this serious problem.

Coping with these difficulties requires effective, resourceful, and committed state government. However, government bodies in the region are not always able to implement reform of social welfare, health care, and education. They themselves are often in need of reform, to effectively deal with public sector corruption, for example.

As already pointed out, internal and external security challenges are closely linked. On the one hand, internal weaknesses increase vulnerability to external threats; on the other hand, external pressures often shape economic and political reforms with their social, environmental, and other security repercussions. Energy, among other issues, is at the core of both internal and external security challenges in the region.

2.3. THE ENERGY DILEMMA AND CHERNOBYL LEGACY

Given the tragic Chernobyl legacy, why are both Belarus and Ukraine currently considering expanding their nuclear energy generating capability? The answer lies in the special role played by energy and energy security in Eastern Europe.

Energy is vital for the internal and external security of all three countries. A secure, affordable domestic energy supply is critical to economic development, particularly in energy-hungry industrial sectors. It is also essential to meet social needs (heating, transportation, etc.), especially for vulnerable groups. Since the region's own energy resources and production capacities, especially in Moldova and Belarus, are insufficient, a significant proportion of energy has to be imported, primarily from Russia. This is, in turn, a major factor in the external security of Eastern Europe. Another factor is the location of the region at the crossroads of major energy transport corridors linking producers in Russia and the Caspian region with consumers in Central, Western, and Northern Europe. In the context of rising global demand for energy and higher hydrocarbon prices, the stability of oil and gas transportation routes is becoming increasingly important for Russia, the EU, the United States, and other countries.

A good illustration of the external aspect of energy security was the heated debate over arrangements for the supply of Russian natural gas to

Belarus and Ukraine, tariffs for transporting gas across these countries, and ownership of gas transportation facilities. Belarus, traditionally a Russian ally, was purchasing Russian gas at US$47 per cubic meter until the end of 2006. From 2007, the price of the gas was increased to more than US$100 per cubic meter. In the context of price negotiations, Belarus also agreed to sell 50 percent of shares of Beltransgaz—the Belarus national gas distribution and transportation company—to Russia's state-owned Gazprom. The dispute between Russia and Ukraine over gas prices in early 2006 resulted in disruption of gas supplies to Western Europe and sparked a strong reaction from the EU that had worldwide resonance. While most observers considered that Russia was exerting political pressure by increasing gas prices, others pointed out that before the 2006 deal Gazprom had been supplying Ukraine at a fifth of the market price, equivalent to Russia subsidizing the Ukrainian economy by $3–5 billion a year. A similar dispute over tariffs on the export of Russian oil and its products to and through Belarus resulted in a brief disruption of oil supplies.

Imported energy is important to fuel economic development, particularly energy-hungry heavy industry, such as machine building and steel production in Ukraine and fertilizer and chemical production in Belarus. Refining of oil products in Mozyr and Novopolotsk used to be a key sector in the Belarus economy, but profits may drop substantially after Russia imposed tariffs on the export of oil to Belarus in January 2007. The survival of much of the metallurgy and machine-building industry in the Donbas depends directly on a cheap, secure supply of natural gas currently imported from Russia, or on finding an alternative such as electricity from Ukraine's domestic power sources. Most of Ukraine's heavy industries were inherited from the Soviet Union and are often located in environmentally and socially stressed areas, while forming the mainstay of the existing economy. It may not be economically feasible to restructure them to improve energy security. Moreover, these industries are socially (and politically) important, as they constitute the main source of employment in densely populated areas with a poorly diversified economy.

There are many other ways in which energy is linked to social and ultimately political issues. Even with current tariffs often below cost-recovery levels, heat and electricity bills are a burden for poor people. In 2006 utility bills (primarily electricity) represented 40 percent of an average pensioner's income in Moldova. Raising tariffs to cost-recovery levels may render heat and electricity virtually unaffordable for many.

Throughout the difficult 1990s, the energy supply in Eastern Europe remained relatively secure due to the slowdown in industrial activity and substantially underpriced imports of oil and gas from Russia and Central Asia. Recently energy demand in the region has reached and surpassed the 1991 level at the same time as world oil prices have increased dramatically.

Russia, for its part, has started a reappraisal of the political and economic costs and benefits of providing indirect energy subsidies. These factors are forcing the three countries to urgently rethink their energy supply options.

The need is so pressing that Belarus and Ukraine are turning to nuclear power to solve their energy problems. Belarus plans to build a domestic nuclear power plant by 2015, while the Energy Strategy adopted by Ukraine proposes building new nuclear reactors and extending the service life of existing ones. This raises obvious technological challenges locating reactors and finding adequate water resources for cooling, particularly in Ukraine, which is already short of water in many areas. But the deployment of nuclear power is also associated with various security challenges ranging from enforcement of nonproliferation to concerns about terrorism, the operation of reactors, and radioactive waste disposal. In addition, it may aggravate social and political tensions, already reflected in the hostile response by Ukrainian NGOs, opposed to plans to expand the nuclear power base.

On the other hand, Ukraine and Belarus are determined to increase energy efficiency and implement cleaner energy technologies. The need to increase energy independence has focused fresh attention on the coal sector, which currently provides up to half of all energy and fuels up to a quarter of the electricity production in Ukraine. Belarus also has substantial deposits of brown coal. The importance of coal to the region could potentially increase, but would require major capital investment. Much as nuclear power, it could result in significant environmental risks, although new technologies may ensure cleaner (albeit more expensive) coal-based energy generation. Other domestic energy supply options, such as hydropower or using wood and other bio-fuels, are associated with environmental, social, and security impacts as well (e.g., the impact of newly-built hydropower facilities on downstream areas).

Whatever the strategic choices, restructuring of the energy sector in Eastern Europe will continue, and will have a major impact on the economy and social stability as well as the state of the environment. As long as the key lessons of Chernobyl remain on the agenda, these impacts need to be fully understood and integrated into policy making.

2.4. ENVIRONMENTAL CHALLENGES FACING THE REGION

The Chernobyl disaster is the foremost, though by no means the only, example of the region's major environmental problems, largely associated with past disregard for the environment and the rapid industrialization and modernization of the USSR. Much of this legacy did not receive sufficient attention during the difficult transition years, when declining living standards and political and economic instability took precedence over environmental issues. The transition and recent economic recovery created new environmental challenges, many of which interact with security issues at the local, regional, and national level (Figure 1).

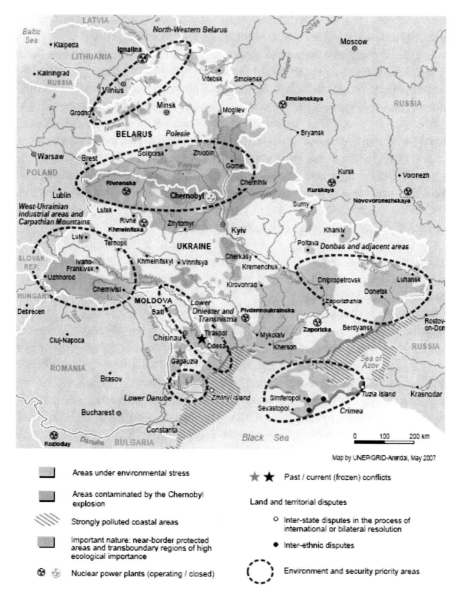

Figure 1. Environment and Security Priority Areas in Eastern Europe.

Major environmental problems inherited from the Soviet era are often located in and around large industrial centers. This is a result of intensive industrialization in compact areas, inefficient use of energy and natural resources, and disregard for local environmental concerns. Air and water pollution, accompanied by degradation of the landscape and ecosystems, is acute in the industrial zones in Ukraine and Belarus. The wetland areas of Polesie in southern Belarus are another type of territory under stress, intensive drainage and deforestation carried out to recover land for farming having damaged ecosystems and ultimately caused a drop in agricultural productivity. Serious environmental degradation also threatens the ecosystems of the Carpathian Mountains and the Azov and Black seas.

Environmental degradation often goes hand in hand with the declining health of local people. This overlaps with more recent economic and social problems that have often hit hardest the very same heavily industrialized areas that have the most serious environmental problems. In turn, social and economic difficulties shift attention and resources away from the environment, further aggravating the situation and creating a vicious circle that poses an additional threat to social stability. It is interesting to note that in the USSR, environmentally degraded areas often adjoined large, relatively untouched ecosystems with rich biodiversity.

The Soviet command economy's ability to restrict economic development to designated areas resulted in a specific patchwork of environmental degradation. The remaining wilderness areas have significant potential for nature conservation and tourism.

While some forms of environmental damage were reduced during the transition, others became much worse. The positive effects of transition included improved resource efficiency resulting in more realistic pricing of natural resources, new foreign and domestic investment in cleaner technologies, and a cutback in subsidies for heavy (particularly military) industry. On the down side, deregulation associated with market liberalization resulted in laxer environmental controls. The economic and political difficulties distracted the attention of the public and policy-makers from environmental issues. The increasing focus of business on profit-making encouraged more intensive exploitation of natural resources. Environmental degradation around large industrial facilities was often made worse by chronic underinvestment in their maintenance. In addition, trade liberalization in some cases resulted in shifts toward more pollution and resource-intensive industries. The all-pervading commercial propaganda that accompanied the rise of market economies strengthened consumerist behavior among those fortunate enough to be able to consume.

Strong, dynamically adaptive environmental protection agencies are needed to tackle this legacy and meet new challenges. Substantial progress in this field has been achieved in all three countries, particularly in view of the fact that at the time of independence even the ministries in charge of environmental protection were barely functional. In addition to progress at home, the three countries have played a remarkable part in international agreements and European processes, such as Environment for Europe, with Kyiv hosting the Fifth Ministerial Conference in 2003. Progress in drafting modern environmental legislation has been boosted by the countries' commitment to bringing environmental norms in line with EU directives.

At the same time, environmental bodies in the region are still generally weak compared to their Western and Central European counterparts (reflected, in particular, in the relatively low Environmental Sustainability Index scores of all three countries). Institutional development is particularly hampered by the insufficient priority given to the environment by the political agenda and mass media. Global environmental issues such as climate change, biodiversity conservation, and unsustainable consumption attract little public attention. At the same time, environmental problems causing direct health, social, or economic impacts (contamination by hazardous substances, safety of water, or land degradation) continue to generate significant public interest.

3. Conclusions

Over the next decade, and perhaps longer, the region will continue to face tough challenges modernizing its economy and radically reforming its energy systems, while building sustainable democratic societies. Such simultaneous political and economic transformation is only possible with strong external stimuli and support of the type provided by the EU to its candidate countries in Central Europe and the Baltic States. Yet the EU, with its current "enlargement fatigue," has certain constraints in helping in a substantial way.

The countries consequently have a long way to go before state institutions mature and a culture of dialogue and democratic representation is firmly established—a necessary precondition for developing long-term solutions to strategic challenges, including those related to the environment. Unless they are at least partly resolved, for example the tensions such as those found in Moldova's Transnistrian region and security-linked social issues in Crimea, they will continue to work against stabilization and democratic transition.

The most dramatic internal factor in the long term is probably the demographic situation. Ailing, aging, and shrinking populations will be increasingly unable to shoulder the burden of social transformation and economic modernization. Under the circumstances, the most active part of the population will go on looking for a better future outside the region, further restricting the potential of the countries. While these processes are difficult to stop, efficient, legitimate, and capable national elites could limit the damage by modernizing education and health care and boosting family-friendly social security measures. Solving environmental problems in each country, and particularly in socially stressed areas, though perhaps not decisive, would certainly contribute to this process.

The single most important external factor shaping the future security of Eastern Europe is the interplay of political and economic interests in the pan-European region. Many in Eastern Europe are attracted by Western models, but drawn East by historic, cultural, and linguistic affinities, and, last but not least, by close trade and energy links. Most probably the three countries of Eastern Europe will continue to search for a balance between the two poles. However, the three states themselves are not passive objects in a geopolitical game, but active players, and much of the regional security architecture will depend on the ability of Chisinau, Kyiv, Minsk, and other capitals to seek mutual understanding and reach compromises.

How does this affect interaction between environment and security? We still do not know whether Eastern European economies will stagnate, decline, or grow, and in what way, or whether growth will be based on resource- and energy-intensive industries or on technology- and labor-intensive activities and services. It is not clear how the continuing transition will define the political landscape of the three countries. But these factors will certainly define the agenda in the region, influenced by environmental and security limitations.

There is obviously an urgent need to mitigate threats and strengthen cooperation on Eastern Europe's external and internal borders. International institutions can make a meaningful contribution by easing tension, solving environmental problems, supporting energy security, boosting regional stability, and promoting stewardship of global ecosystems—but to do so they must cooperate with one another systematically in a drive to untangle the complex relationships between energy, security, and the environment.

References

Emerging Risks in the 21st century—An agenda for action, 2003, Paris.

Energy Strategy of Ukraine until 2030, Order No–145-p, accepted by Cabinet of Ministers of Ukraine 05.03.2006.

Esty, D., M. Levy, and T. Srebotnjak, 2005, *Environmental Sustainability Index: Benchmarking Environmental Stewardship,* Yale Center for Environmental Law and Policy, New Haven, CT.

Europe, Russia and in-between, 2006, *Economist,* October 26.

European Neighbourhood Policy Strategy Paper, European Commission, COM, 2004, 373 final, Brussels.

German Advisory Council on Global Change, 2000, *World in Transition—Strategies for Managing Global Environmental Risks,* Berlin.

Global security, 2006, Ukraine Special Weapon.

Hopquin, B., 2006, Tatars fight for recognition in Crimea, *Guardian Weekly,* March.

Kinley, D., ed., 2005, *Chernobyl Legacy: Health, Environmental and Socio-economic Impacts and Recommendations to the Governments of Belarus, the Russian Federation and Ukraine,* International Atomic Energy Agency, Vienna.

Kupchinsky, R., 2005, *Problems in Ukraine's Coal Industry Run Deep,* RFE/RL, Prague.

Lieven, A., 2006, The West's Ukraine illusion, *International Herald Tribune*, January 5.

Living with Risk—A Global Review of Disaster Reduction Initiatives, 2004, UN/ISDR, New York and Geneva.

McFaul, M., 2001, *Russia's Unfinished Revolution: Political Change from Gorbachev to Putin,* Cornell University Press, Ithaca, NY.

Meacher, M., 2005, One for oil and oil for one, *The Spectator,* 5 March.

Mulvey, S., 2006, Ukraine's strange love for nuclear power, BBC News, April 26.

OSCE, 2005, *Scoping Study on the Elimination of Rocket Fuel Component Stocks in Ukraine,* 28 September.

Polyakov, L., 2004, New security threats in Black Sea region, NATO *Summer Academy Reader,* June.

Rosenkranz, G., 2006, Nuclear power—Myth and reality, Nuclear Issues Paper 1, Bern.

Transparency International, 2007, Corruption Perception Index 2006.

UNAIDS (United Nations Programme on HIV/AIDS), 2005, *2005 AIDS Epidemic Update.*

UNDP (United Nations Development Programme), 2007, Beyond scarcity: Power, poverty and the global water crisis, *Human Development Report 2006,* Palgrave Macmillan, New York.

UNPD (United Nations Population Division), 2007, *World Population Statistics.*

Vasylevska, O., 2006, The cost of recycling: Ukraine to build a nuclear dump, *The Day,* 6 June.

Weinthal, E., 2004, *From Environmental Peacemaking to Environmental Peacekeeping,* Woodrow Wilson Center, Washington.

World Economic Outlook Database, 2007.

Yeremienko and Vozniuk, 2005, *Ukraine: Environmental Overview,* Kiev, Ukraine, 26–29 July.

ENVIRONMENTAL ISSUES OF THE KYRGYZ REPUBLIC AND CENTRAL ASIA

A. K. TYNYBEKOV[1]* ,V. M. LELEVKIN[2] AND
J. E. KULENBEKOV[3]

[1-3] *Kyrgyz Russian Slavic University, Kyrgyz Republic*

Abstract: The region of Central Asia suffers from ecological disasters on a vast scale with many more occurrences threatened. The extent of anthropogenic factors of influence on environmental degradation directly affects the scale of the damage. Further, the more extensive the territory impacted, the greater the ecological threat, even encroaching upon adjacent states. The regions of the Tien-Shan mountain range are among those areas especially endangered by ecological threats. Using advanced methods of measurement, scientists have investigated numerous issues: water ecosystems, the impact of radioactive wastes on the environment, pollution of rivers, the hydropower energy system, and glacial degradation.

Keywords: Environmental issues; anthropogenic; hydropower stations; Central Asia; Aral Sea; Kyrgyz Republic; Issyk-Kul Lake

1. Introduction

Human history in Central Asia dates tens of thousands of years into the distant past. In world history Central Asia is best known for the legendary *Silk Road*. Dating from the second century BC, merchandise was transported along a network of trails that connected the eastern and western ends of the Eurasian land mass. Central Asia was always seen as one land, historically and geographically. Then in 1924 the Soviets divided Central

---

*Address correspondence to A. K. Tynybekov, Kyrgyz Russian Slavic University, Kyrgyz Republic; e-mail: azamattynybekov@mail.ru

P.H. Liotta et al. (eds.), Environmental Change and Human Security, 407–432.

Asia into five nations, more or less along linguistic lines. Also in 1924, the Soviet Union closed the borders of Central Asia to the rest of the world. For the seventy years of Soviet-imposed isolation, Central Asia ironically functioned both as the breadbasket for the rest of the Soviet Union and as its military research center.

Central Asia covers four million square kilometers in area, stretching from the Caspian Sea in the west to Mongolia and China in the east, and from the southern Urals in Siberia to the north of Iran (see Figure 1). The terrain varies from steppes to mountains and deserts. The population of Central Asia in the countries of Kazakhstan, Kyrgyzstan, Tajikistan, Turkmenistan, and Uzbekistan is about sixty million people. The largest ethnic groups are Turkic (Kazakh, Kara Kalpak, Kyrgyz, Tatar, Turkmen, Uyghur, and Uzbek) and Persian (Tajik). Numerous other nationalities are represented, especially Russian and some Korean.

The Kyrgyz Republic, a sovereign state located in the northeast part of Central Asia, was formed in 1991 following the collapse of the Soviet Union. The territory covers 900 km from the west to the east, and 410 km from the north to the south; an area of $198.5 \, km^2$. The population is 5.1 million people. Kyrgyzstan is a country of striking contrasts: of blue lakes and green valleys, of powerful mountains covered with eternal snow and lush alpine meadows, of wild raging rivers and virgin walnut and fruit-tree forests.

The countries of Central Asia have many common environmental problems. The entire region suffers from ecological disasters on a vast scale with many more occurrences threatened. Further, the more extensive the territory impacted, the greater the ecological threat, even encroaching upon adjacent states. The extent of anthropogenic factors of influence on environmental degradation directly affects the scale of the damage. We can see in Figure 2 the transboundary nature of the Syr-Darya River water basin.

In the wake of a UN report that found that Central Asia is particularly vulnerable to desertification, experts are saying that all countries in the region urgently need to develop a joint strategy to combat the threat. The report warns that climate change, overexploitation of land, and unsustainable irrigation practices are degrading the soil. As a result, some fifty million people could be displaced in the next ten years.[2]

The basic sources of environmental issues are:

1. Impact of the mining industry: uncontrolled dumping and storage of toxic waste products and radioactive material from mining operations, resulting in pollution of air and water throughout Central Asia
2. Radioactive and chemical pollution
3. Manmade threats to the environment: the threat of eco-terrorism
4. Natural and manmade cataclysms (earthquakes, floods, mudflows, fires)

Figure 1. Kyrgyz Republic in Central Asia.

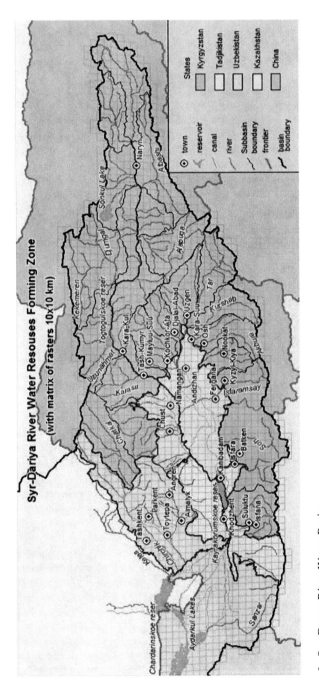

Figure 2. Syr-Darya River Water Basin.

5. The Aral ecological crisis and its consequences for the natural environment in the Central Asia region

6. Water resource management policy, including irrational use, the influence of artificial water basins and dams, and hydroelectric power stations' impact on the environment

7. Use of regional water and power resources

8. Ecological and radiological monitoring of the southeast coastal zone of Issyk-Kul Lake

9. Lidar Station (Teplokluchenka) for satellite imaging

10. Phytoplankton in Issyk-Kul Lake

11. Assessment of ecological risk with use of GIS technologies

12. Glaciers in Tien-Shan Mountains and

13. Kumtor gold mining impact on the environment

2. The Basic Sources of Environmental Issues

2.1. IMPACT OF THE MINING INDUSTRY—URANIUM EXTRACTION

Radioactive tailing dumps from Soviet-era extraction of uranium located in the mountain regions at Mailuu-Suu in the Kyrgyz Republic, in the Chkalovsk and Taboshar regions in Tajikistan, and at a number of closed uranium mines in the Ferghana Valley, which spans the Kyrgyz Republic, Tajikistan, and Uzbekistan, exist as chemical and biological hazards (see Figure 3).

Since 1946, waste products from uranium pit mines of the former Soviet Union, former Czechoslovakia, and former East Germany have been stored near the Mailuu-Suu River, which flows through the hillsides of the Tien-Shan Mountains and forms the border between the Kyrgyz Republic and Uzbekistan. The surface area of this nuclear burial ground covers 1,000 ha. By rough estimates, more than fifty years of accumulation of radioactive substances deposited in Central Asia today totals about two million cubic meters of fatally dangerous waste products. These nuclear waste products, which were hastily buried using the technology of the first half of the last century, have been measured recently to produce up to 4,000 μR/h, which exceeds all maximum permissible norms.

Numerous burial sites of waste products from processing uranium ore, and twenty-three of the fifty tailing dumps in the Kyrgyz Republic, are located in areas of known landslides and earth flow activity. If these radioactive waste products seep into and pollute the Mailuu-Suu, Kara-Darya, and Syr-Darya Rivers, the waters of these rivers, which flow through densely

Figure 3. Radioactive, Chemical and Biological Hazards in Central Asia.

populated areas throughout Central Asia, would certainly cause irreparable damage to both the environment and the populace.

The issue of radioactive pollution at Mailuu-Suu is not the only topical issue for the Kyrgyz Republic and Uzbekistan. In 1995 the State Committee of Mountain Engineering Supervision of the Republic of Uzbekistan took up the question of the threat of radioactive pollution entering the waters of the Kara-Darya River, which flows into the Syr-Darya River.

In April 1996, under the aegis of the International Foundation for Ecology and Health (ECOSAN), the first meeting was held at its offices in Tashkent, attended by representatives and experts from Uzbekistan, the Kyrgyz Republic, Kazakhstan, the Russian Federation, and other interested foreign states to resolve the Kara-Darya River issue.

On the international level, at the second session of the preparatory committee, held in Geneva in 1998 to consider a 2000 Nonproliferation Treaty of Nuclear Weapons (NYNW) conference, the delegations from the Kyrgyz Republic and Uzbekistan raised the questions of ecological rehabilitation of exploited places and burial sites of radioactive waste products, the development of effective measures to implement the required rehabilitation, and the need for assistance. These questions were submitted to the United Nations and the International Agency on Atomic Energy (IAAE). The IAAE has been monitoring burial sites of radioactive waste products to estimate the potential threat of the radioactive deposits. The World Bank has allocated one million dollars for implementation of this program.

Natural cataclysmic threats to the environment also exist. Today natural phenomena such as earthquakes, floods, mudflows, landslides, and earth flows are a great danger for the Central Asia countries. Landslides and earth movement are common annual occurrences when warm temperatures melt the heavy snowpack in the mountains of the Kyrgyz Republic, and each incidence results in mud with radioactive substances being washed into the Mailuu-Suu River. The German Foundation for International Cooperation gave a system that warns of landslide danger. The warning system sounded an alarm in 2006 when heavy rains in the spring caused land movement on hillsides, destroying the highway and a dam.

2.2. RADIOACTIVE AND CHEMICAL POLLUTION

In arid areas of Central Asia, potable water requirements are met using underground water resources, which surface as a result of natural underground pressure or with the help of deep pumps. Significant changes in the mineral content of the artesian water have recently been observed; even radioactive infections have been noted. According to definitions by hydrogeologists, artesian water is formed in the deep underground lakes called water lenses; the diameter of the lenses can reach as much as some tens or hundreds of kilometers.

In Uzbekistan and Kyrgyzstan, uranium mining today widely uses the method called "underground leaching." At the uranium deposit sites, a network of deep drainage holes are drilled, a system of pumps injects water into these holes to a certain depth, and then the pumps are reversed to suck the water solution that now contains the uranium components to the surface. Although this underground leaching system is economically effective, its fatal impact on the ecology of artesian waters is obvious.

It is impractical for governments to try to control access to underground water resources, and those who use underground leaching to extract uranium do not always follow ecologically sound practices. The resulting scale of pollution to natural underground water resources reaches a menacing level.

2.3. MANMADE THREATS TO THE ENVIRONMENT: ECOLOGICAL
TERRORISM

Religious extremists have the capability to implement their terrorist acts in the territory of Uzbekistan, Kazakhstan, the Kyrgyz Republic, and Tajikistan in order to achieve their political or economic purposes. Potential targets creating the greatest danger to the life and health of the people of Central Asia include the natural dams in the high mountain lakes of the Kyrgyz Republic and Tajikistan and the radioactive tailing dumps near one of the major rivers that flow through those same areas. If radioactive and toxic waste products of uranium manufacturing get into the waterways of the Kyrgyz Republic and Tajikistan, polluted water would be circulated throughout the territory of Uzbekistan, Kazakhstan, and Turkmenistan. Such a situation could bring irreversible environmental changes on local and regional scales, leading to political tension, economic destabilization, and uselessness of extensive agricultural lands for many decades due to radioactive nuclides and toxic impurity.

Ecological threats include technological terrorism. Western experts recognize the technological terrorism threat of using nuclear, chemical, and bacteriological weapons or radioactive and hard toxic chemical and biological substances. A second type of technological threat is an attempt of extremists to acquire nuclear facilities and other strategic industrial targets.

The western experts emphasize that availability of the technological equipment to produce poison substances and the possibility of self-manufacturing increase the probability of a crime that is a greater threat than nuclear terrorism, because acts of chemical and bacteriological terrorism can be prepared and carried out by one person. The threat of ecological terrorism for Kyrgyzstan and its neighbor states is a real and present danger. In some of the countries surrounding Central Asia all the prerequisites for nuclear and technological terrorism have matured; furthermore, the worsening transborder sociopolitical situation and impoverishment of vast masses might support the potential threat.

All these are reasons that the special attention of the world community is required to reduce and control the potential ecological threats from targets that are located in the transboundary zone of the Kyrgyz Republic and Tajikistan.

2.4. THE ARAL ECOLOGICAL CRISIS AND ITS CONSEQUENCES
FOR THE ENVIRONMENT OF THE CENTRAL ASIA REGION

Surface level reduction of Aral Sea water has led to the formation of large areas of drained sea bottom that are occurrence centers for dust storms that send aerosols of minerals into the atmosphere. The mineralization of an atmospheric precipitation decreases with distance from the territory of Aral Sea (see Figure 4). The mineralization results in changed radiation and microphysical properties of air mass because of increasing intensity of the absorption of sunlight.[2]

Most of Central Asia lies in an arid climatic zone with high temperatures, seasonal droughts, and low annual rainfall. Analysts agree that improper land use, deforestation, excessive grazing, and poor coordination of water use is putting Central Asia at massive risk. Areas near the Aral Sea, which suffer from soil salinization as well as desertification, are at particular risk. Hikmatullo Ahmadov, a leading scientific fellow at the Tajik Academy of Agricultural Sciences, writes that Central Asia can only tackle the problem if all of the countries work together. Each state should introduce proper land-use methods and regulate grazing while cooperating on farming technology, forestry, and water management.[1]

2.5. ECOLOGICAL RISKS TO POWER STATIONS IN THE KYRGYZ REPUBLIC

In 1934 the Soviet government began building a power supply system in the mountain territory of Kyrgyzstan, resulting in a network of lines supplying 13.2 KV. The modern power supply system has twenty power stations with the common established capacity of 3.4 million kilowatts; a 5,800 km nationwide system of bulk power supply fit with a voltage of 110–500 KV to the main transmission lines. An electricity distribution network more than 63,000 km long delivers 0.4–35 KV. The main nationwide thermal network is 490 km long (see Figure 5). From these power stations the Republic annually develops about 12–14 billion kilowatt hours of electric power. The Soviet-built Kyrgyz power supply system does not have analogues in world practice of network construction.

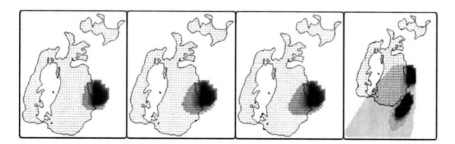

Figure 4. Satellite Photos Illustrate the Results of Aral Sea Drainage, 18 May 1998 and 18 September 1998.

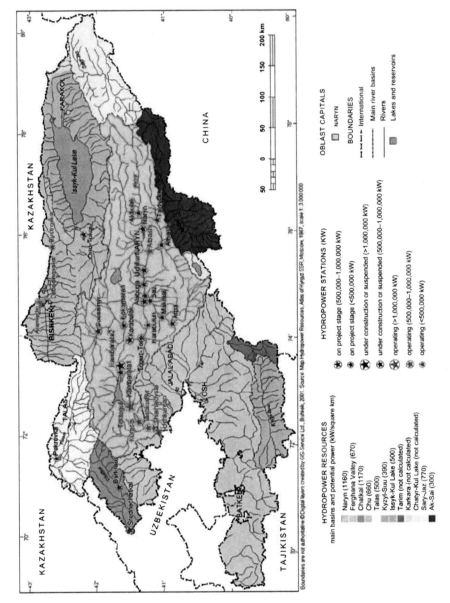

Figure 5. Hydropower Stations of the Kyrgyz Republic.

For hydro resources the Kyrgyz Republic rates third among CIS countries (after Russia and Tajikistan). The rivers of the Republic have an extremely high concentration of potential capacity per 1 km section. Specifically, the power density of the Naryn River with inflows exceeds such mighty rivers as Volga and Angara.

Today barely 9 percent of the energy potential is captured, creating 142.5 billion kilowatt hours. Using modern hydro-energy technology it is possible to create 72.9 billion kilowatt hours. However, at present converting only forty-eight billion kilowatt hours would be economically effective for the Republic. The Naryn River basin has the potential for 56.9 billion kilowatt hours. It is possible to construct twenty-two hydroelectric power stations for the development of electric power in the order of thirty billion kilowatt hours on the Naryn River and its inflows.[6]

The main reasons for pollution of water resources with radioactive nuclides and other toxic substances are imperfection of hydraulic engineering constructions at tailing dumps, degeneration of their protective constructions, and the lack of storage site filtration systems. From the centers of pollution begins the migration of the radioactive nuclides in a hydrographic network in underground waters; a halo of pollution extending tens of kilometers is formed.

At present, the situation is aggravated because, with the collapse of the USSR, Russian Atomic Society (RAS) sites in Central Asia have been ownerless and neglected for sixteen years, leading to degradation of their protective construction. In addition, the adverse impact of uranium-mining sites on the environment and increased probability of transboundary accidents exists because the storage sites of the RAS were placed near channels, marshes, and floodlands of the Syr-Darya River basin. Degradation of the environment of uranium mining sites in the upper Syr-Darya River is the result of the nuclear program development and extraction of uranium by the former USSR from 1944 to 1967.

Many of the tailing dumps are located in areas of dangerous geological activity, including earthquakes, tectonic activity, and landslides. The Naryn, Mailuu-Suu, and Susamyr Rivers all flow through this geologically active region into the Syr-Darya river basin. Calculations show that at full destruction 1.2 million cubic meters of the listed storage sites and radioactive tailings would be carried into the river cone, covering an area of 300 km^2 with radioactive nuclides of 10,000 Ci.

For example, we shall consider a geo-ecological situation near the Mailuu-Suu town that illustrates mistakes at the initial stage of uranium mining. Radioactive waste products were stored in twenty-three tailing dumps and thirteen off-balance ore slag-heaps in a valley of the Mailuu-Suu River. The total volume of radioactive tailings consists of two million cubic meters,

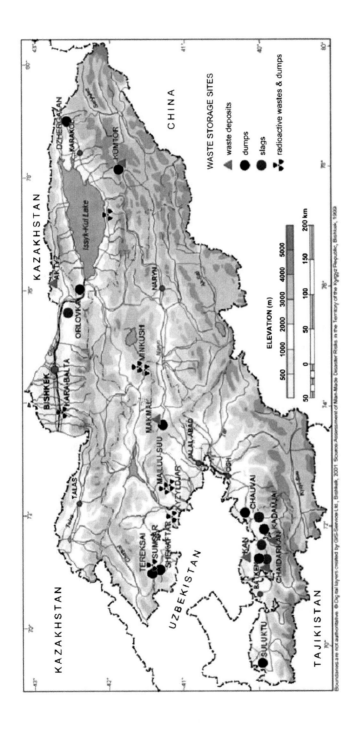

Figure 6. Radioactive Wastes, Dumps, Tailings, and Slags.

and with total radioactive nuclides of 50,000 Ci. The distance from Mailuu-Suu sites to the Uzbekistan border is 30 km.

Kyrgyz researchers have shown that many of tailing dumps including No. 3, 5, 7, 9, 10, and 18 and slag-heaps are located near the riverbed and inflows of the Mailuu-Suu River became sources of regular radioactive pollution of the river waters because of imperfectly designed dams, the water saturation of tailings, and unreliable waterproofing (see Figure 6).[8] Since the 1990s many more landslides have been observed in the seismically active Mailuu-Suu area (see section 2.1 regarding the landslide warning system in the Mailuu-Suu area).

2.6. USE OF REGIONAL WATER AND POWER RESOURCES

At a summit of the Central Asia Energy Cooperation (CAEC) attending heads of CA states agreed upon the "Mutually beneficial and effective utilization of water and power resources"; however, issues of payment among states remain unresolved. Requests by the Kyrgyz government at CAEC for compensation of some part of the interstate hydraulic engineering operation expenses were not satisfied. Because the water and power resources originate in the Kyrgyz Republic, members of the Kyrgyz Parliament are creating a legislative base for regulation of rights, priority needs, and conditions of water resources division among the CAEC states. However, requiring payment for water supplied to Central Asian states deficient in natural water resources could provoke conflict among states. This issue must take into account both interstate economic interests and maintaining the ecological balance of Syr-Darya water basins, especially considering the Aral Sea problem.

An example of interstate economical conflict occurred in 2004. Uzbekistan sells natural gas to Kyrgyzstan and Kyrgyzstan sells electricity to Uzbekistan. In the winter of 2004, Uzbekistan cut off the natural gas supply to Kyrgyzstan because of outstanding debts. In response, the Kyrgyz government, under the pretext that this move was necessary in order to generate more electric power for the heating system of the country, began to increase the volume of water flowing through the Toktogul hydropower station, which, in turn, reduced the water level of Toktogul reservoir. A sharp increase in water levels was recorded in all reservoirs lying downstream of the Toktogul hydropower station on the Syr-Darya River. In the spring, overflow from the too-full reservoirs on the Syr-Darya flooded agricultural areas in Uzbekistan and Kazakhstan. That summer when Uzbekistan, the largest consumer of water in the region for its vast cotton cultivation, needed water for irrigation, the Kyrgyz Republic did not have the resources in the capacity of Toktogul reservoir, the largest in Kyrgyz Republic.[4]

3. Environmental Investigation of the Issyk-Kul Region

3.1. RADON AIR POLLUTION

In the northeastern corner of Kyrgyzstan lies emerald-blue Issyk-Kul Lake, an alpine lake comparable in its beauty to the lakes of Switzerland. The waters of 134 rivers flow from the mountains that surround Issyk-Kul, many of them originating in eternal glaciers, and feeding into the lake. The Issyk-Kul basin is surrounded by 834 glaciers; in warm weather more than 60 percent of the river water comes from glacial melt.[9]

As Issyk-Kul has no outlets, mineral substances deposited in the lake by river water accumulate in the lake. Radiological research conducted by the International Science Center of the Kyrgyz Republic under the framework of project INTAS from 1998 to 2005 defined the radiological conditions in the southeastern section of Issyk-Kul lake.

The research project "Radiological Monitoring of the Issyk-Kul Region" financed by FAST (USA) studied the radiological conditions in the Issyk-Kul region over a two-year period, 1996 to 1998. Data was received about the level of background radiation of the Issyk-Kul region. A cartographic map of indexes of the background radiation level of the Issyk-Kul region was drawn up.[10] For example, near Kadji-Sai village uranium deposits were mined during Soviet times. Mining was suspended, but the tailing dams remain. A river flows near the tailing dumps, possibly polluting this river, which empties into Issyk-Kul Lake. This village is in the territory that has been researched to measure the background radiation.

According to the medical statistics of the Kyrgyz Republic, the death rate from malignant tumors in the Kyrgyz Republic amounts to 61.7 percent of deaths for the entire population. However, in the Issyk-Kul region the rate is 81.6 out of 100 patients, a rate of more than 1.5 times higher than for the general population of the Republic. Among the malignant diseases recorded in the Issyk-Kul region the most recorded cause of death is stomach cancer (10 percent), second is lung cancer (8.6 percent), and third is malignant tumors of the lymphatic system (3.9 percent).[11]

For people living on the southern coast of Issyk-Kul in the Dzhety-Oguz district (78.5 percent) and in Karakol town (86.2 percent), the death rate among the population exceeds the data for the general Issyk-Kul region (62.9 percent), and exceeds the rate for all of the Kyrgyz Republic (61.7 percent).

In the Issyk-Kul basin drilling into natural water resources taps into warm water containing normal levels of radon. Large settlements along the shores of the lake and popular recreational resorts have been established. For decades local villagers and guests at Issyk-Kul's many health resorts have bathed in this warm water, believing it to be healthful for them.

3.2. PHYTOPLANKTON IN ISSYK-KUL LAKE

Lake Issyk-Kul is a large, nonfreezing, slightly saline, deep water reservoir with an area of $6,200\,km^2$—making it the twenty-third largest lake in the world. The lake is an isolated pool located at an elevation of 1,608 m and possesses special hydrological characteristics. The great bulk of the primary organic substance determining the biological efficiency of a reservoir is synthesized by phytoplankton, so studying the ecological conditions of photosynthetic phytoplankton in Issyk-Kul is of great scientific and practical interest. The action of various ecological factors and human pollution can result in a change in the concentration and photosynthetic activity of algae. Therefore, the measurement of the productive characteristics of the phytoplankton allows for the estimation of the condition of the water environment as a whole.

Recently, the distribution was determined using methods to measure the fluorescence of the chlorophyll. These methods are both highly sensitive and allow for the quick estimation of the characteristics of the phytoplankton without affecting its physiological state. It is known that the output of fluorescence (Fo) at low light exposure correlates with the concentration of chlorophyll and the biomass of algae in reservoirs. This parameter depends primarily on the concentration of all light-harvesting pigments in the cells of algae and can serve as an indicator of an abundance of phytoplankton.

For the first time we carried out in situ research of the ecological condition of natural phytoplankton, its abundance, and the photochemical activity of its cells in Issyk-Kul and its inlets using the submersible fluorometer. The method of measurement with submersible fluorometer is pump and probe (see Figure 7).[12]

Our research on the distribution of fluorescence in the waters of Issyk-Kul traced the relationship between fluorescence of phytoplankton and the concentration of biogens. The maps that were drawn up from this data show the distribution of values of Fo (Chl*), Fv/Fm, and in the eastern part of Issyk-Kul the concentration of inorganic nitrogen on the horizontal axis, and on balance show that the distribution of fluorescence parameters are correlated with the concentration of inorganic nitrogen in the water, which is one of the most important components of mineral nutrients.

The greatest concentration of this element was present in the coastal areas along a line from Karakol to Tamga, which corresponds to less hilly terrain in this area and accordingly has greater enrichment of inflows by soil particles. Along this coastline we have also found a connection between the contents of mineral substances and the intensity of the fluorescence of phytoplankton. The greatest values of both parameters of fluorescence fell in an inlet at Karakol where a significantly higher concentration of mineral substances was observed.

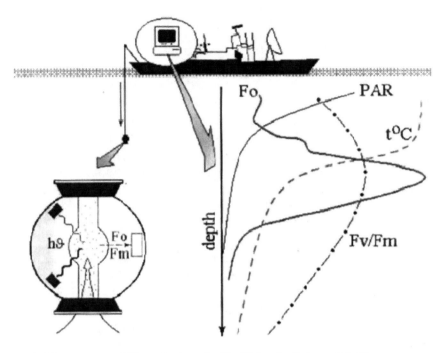

Figure 7. Measurement of Phytoplankton In Situ Using the Submersible Fluorometer.

Thus, the measurement of phytoplankton in situ with the use of the submersible fluorometer has allowed us to study, at a number of stations around Issyk-Kul, the distribution of the production parameters of algae, their abundance and photochemical activity, and which factors of the environment, including pollution, influence these variables.

Comparing the parameters for illumination, temperature, and concentration of salts of nitrogen and phosphorus in pelagian and arctic lake was shown that the concentration of biogens and water stratification has the greatest value for abundance and photosynthetic activity of phytoplankton. This comparison corroborates the oligotrophic character of the waters of Issyk-Kul Lake and that the optimum state for growth and developments phytoplankton was at depth of 25–55 m. Further at this depth, cells of layers have a higher sensitivity to occurrence of biogens.

A great quantity and activity of the reactionary centers of photosynthesis was found in areas where water brought terrigenous sediments to the river.

Chlorophyll of fluorescent phytoplankton was measured by way of survey vessel, using the submersible fluorometer with automatic registration of geographical coordinates at the stations where the probe was taken. Spatial-time maps of Issyk-Kul were drawn up.

3.3. ASSESSMENT OF ECOLOGICAL RISK WITH THE USE OF GEOGRAPHICAL INFORMATION SYSTEM (GIS) TECHNOLOGIES

In 2000 GIS technologies were used to analyze the mudflow risk in the area of the southern coast of Issyk-Kul, from Ton village in the west up to Kyzyl-Suu village in the east, to study the Ton river valley flood danger. Tujuk-Ter, Koltor, and Korumdu are moraine-glacial lakes in the upper ward of Terskei Ala-Too mountain range, which belong to the first category of breakthrough mudflow danger.

Similar mudflows could occur from mountain lakes in the valleys of the following rivers: Tosor, Tamga, Barskaun, and Chon-Kyzyl-Suu in the upper ward of the Terskei Ala-Too mountain range. These mountain lakes are also of the first and second categories of breakthrough danger. The period of annual mudflow danger runs from June to October. July is the most dangerous month for mudflow; 39 percent of cases of mudflow occurred then because of hydro-meteorological activity. The annual heavy snowfall in the region melts at its highest levels in July, increasing water levels in glacial lakes. The rates of mudflow cases for other months are June: 28 percent; May: 13 percent; August: 11 percent; and September: 9 percent.

This whole area has a strong flood danger. In order to compare the danger in the area, GIS technology was used to create a map showing an overlay of three layers of possible risk. We showed generalized characteristics of various kinds of danger (see Figure 8).[13]

Zoning of territory is determined by the number of points and categories of risk intensity. This map shows an overlay of three layers of possible risk. Differentiation of risk categories is made in the following gradation:

- Risk of 1–2 points: first category of risk. The risk is acceptable, but the first category of risk in this case is not visible because it is situated in the upper region of the mountain range.

- Risk of 3–4 points: second category of risk. The risk is partially acceptable.

- Risk of 5–7 points: third category of risk. The risk is unacceptable.

Recommendations are based on ecological risk assessment, and are directed at reduction or elimination of risk. Recommendations depend on the nature and degree of the danger:

- Carrying out the prevention measures, organizing a system of notification for the population, and training people for rescue action.

- Full evacuation of people, removal of everything that can be removed from the predicted lesion zone, construction of dams and mudflow protecting installations.

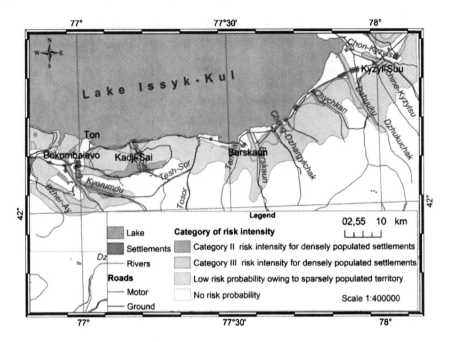

Figure 8. Synthetic Map to Show the Danger Using GIS Technology.

Application in the Kyrgyz Republic of multilevel GIS technology is focused on flood danger assessment and monitoring mudflow cases to minimize loss of life and property. GIS technology could be used for geo-information mapping of mountain areas and for making operative decisions of analytical tasks such as spatial modeling and engineering calculations. New methods of display would increase the speed and quality of analysis.

3.4. GLACIERS IN THE TIEN-SHAN MOUNTAINS

Glaciation is one of the important elements of the underlying surface in the Ton river basin. The glacier of Tien-Shan Mountains has great influence on quantity of glacial melt (see Figure 9).

The authors of this paper undertook a more detailed study to define an area of modern glaciation using two data sources for the various time periods:

• Scale of topographical map is 1:250,000, and the map was drawn up on the basis of aerial photography data made in 1963.

• Space image of NASA from June 2001.

Map layers were digitized with use of Arc GIS 8.3 software; geo-referencing of space images was carried out using of ENVI 3.5 software.

Figure 9. Glacier of Tien-Shan Mountains.

Great attention was paid to the accuracy at delimitation of the glaciation borders because a small percent of cloudiness on a space image could lead to a mistake in definition of the real sizes of researched glaciation. One digitized layer shows indistinct borders on twenty-three glaciers because of the presence of insignificant cloudiness on the space image. Layers of twenty-three researched glaciers are shown in Figure 10.

Scientists reported the reduction of glacial areas in the world at the beginning of the 1970s and especially their sharp reduction at the beginning of the 1980s. Similar reduction has occurred on glaciers of Terskei Ala-Too ridge. According to our research on one of the glaciers of the Kyrgyz Ala-Too ridge, which forms the Sokuluk River basin, the speed of glacial area reduction nearly doubled: 0.6 percent for the period from 1963 to 1986 to 1.3 percent for the period from 1986 to 2000. Figure 11 shows the diagram of changes of the glaciation area observed from 1963 to 2000.

For the last thirty-eight years the area of glaciers in Kyrgyzstan has decreased more than 28 percent. From 1963 to 1986 the area of glaciers decreased 13.3 percent, and from 1986 to 2000 17.1 percent. Eight glaciers had completely disappeared within the period from 1963 to 2000. They belonged to the first class ($<0.5\,km^2$). In spite of the fact that first-class glaciers represent only one-forth of all areas that glaciation occupy, they account for 40 percent of the total area of disappeared glaciers. While for the period from 1963 to 1986, 9.1 percent of first-class glaciers disappeared, for the period

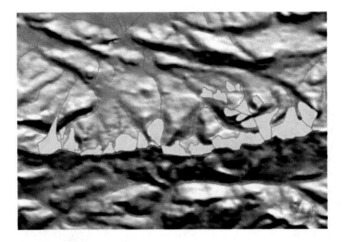

Figure 10. Layer of Glaciers of Ton River Basin.

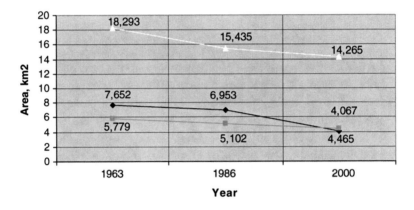

Figure 11. The Changing Diagram of the Glacier Area for the Period 1963–2000.

from 1986 to 2000 41.5 percent of glaciers disappeared. This testifies to the fact that glaciers of an area less than 0.5 km are melting more intensively than glaciers of other classes.[1]

Accordingly, GIS technology is the most suitable tool for the processing, analysis, and storage of observed results. One of the most important advantages of GIS technology is an opportunity for multilevel synthesis and analysis. Combining the thematic data of the various contents and analyzing results of such synthesis, new information is received and complex analysis could be carried out on a wide spectrum of factors and flood formation states.

3.4.1. Lidar Station (Teplokluchenka) for Satellite Imaging

The Kyrgyz Republic borders China in the southeast and conducts complex geo-monitoring of that territory because Kyrgyzstan's environment has been affected by the radioactive dust from the LobNor nuclear test polygon, which moves in Kyrgyzstan's direction, probably causing glacial degradation, polluting glacial runoff, and ultimately contaminating Issyk-Kul Lake (see Figures 12–13).

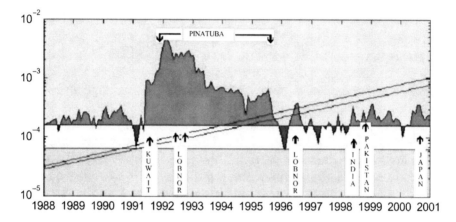

Figure 12. Change of Concentration of Stratospheric Aerosol at Heights of 15–30 km.

Figure 13. Atmospheric Brown Cloud Above Central East China and Dust Storm Above China. *Source*: NASA satellite.

Lidar Station (Teplokluchenka) (LST) of the Kyrgyz-Russian Slavic University is located in the Kyrgyz Republic at a height of 2,000 m above sea level in the southeastern part of the Issyk-Kul high mountain region (Central Tien-Shan within geo-coordinate: 42,50N; 78,40E), LST was established in 1987 for complex monitoring the tropospheric and stratospheric aerosol of Central Asia.[3] Since 2004 LST has been part of the Lidar network of the Commonwealth of Independent States—GIS-LiNet (see arrow on Figure 14).

3.5. KUMTOR GOLD MINING IMPACT ON THE ENVIRONMENT

The Barskaun River, one of the ten of major rivers of the Issyk-Kul basin, on average supplies 4 m³/s water to local irrigation. The Barskaun River basin's catchment area, located 6 km above Barskaun village, is measured on a hydrological post at Sasyk River Creek, which is 346 km² under the 3,480 m average altitude of catchment.[5]

The Barskaun River belongs to the river type fed by glacial snow melt, with maximum flow in August; almost half of its annual flow comes from significant levels of glacier melt from July to September—more than 60 percent in the case of the Barskaun River. On 20 May 1998 an accident occurred involving an automobile transporting cyanide to the Kumtor mining site. On one of the bridges crossing the Barskaun River a cyanide container fell into the river. Some cyanide spilled into the Barskaun River and flowed to the village downstream. Hydrological investigation of the Barskaun River

Figure 14. Location of Lidar Station (Teplokluchenka) on Satellite Image.

was carried out; comparison of the Barskaun River basin to other regional river basins showed no differences. However, the accident increased public interest in this region as to the contents of river water with special attention to indexes that affect the stability of cyanide in water.

The increased concentration of chlorine and the common water rigidity of the Barskaun River Creek can be explained by water leaching from nearby agricultural lands into the creek. The anomalous presence of sulfate-ion in natural waters, especially in a mountain landscape typical of the Barskaun River basin, has two possible explanations: (1) leaching containing plaster, and (2) other easily soluble sulfates of strata and oxides of sulfides, first of all pyrite, as the most widespread among them.[7] Thus, the most probable explanation for the anomalous water content of Barskaun River with a higher concentration of sulfate-ion relative to other regional streams is the oxidation process of sulfate, more specifically sulfur ore.

The Canadian company Kumtor, one of largest gold-mining companies worldwide, has been operating for many years in the watershed of the Barskaun River of the Terskei Ala-Too range. Ore deposit levels of gold-mining sites are often associated with high levels of sulfur ore in the strata. Therefore, it is possible that the hydro-geochemical anomaly of the Barskaun River could be linked to operations at the gold-mining site, probably resulting from the spreading of sulfur ore beyond the gold deposit area.

4. Recommendations

Establishment of an "Interstate Ecological Committee for Central Asia" (IECCA), which should:

- Be an independent elective organization, be familiar with the legal aspects of environmental preservation, and be vested with the appropriate legal authority.
- Have the authority to perform inspections at local sites, control performance norms for ecological safety, and report infringements of IECCA standards.
- Hold periodic and emergency conferences on environmental issues.
- Provide a forum for complex interstate decision making regarding transnational ecological issues and their solutions.
- Function as the Interstate Coordination Water-Economic Commission (ICWEC) to regulate water distribution, to maintain accounting and control of water consumption, and to enforce application of water economy in irrigation.

- Maintain an information network for both collecting information relevant to Central Asia's ecological issues and distributing information to interested scientists, agencies, and other parties.

- Maintain interaction with international organizations interested in environmental issues.

- Promote nature conservation policies and public education.

- Referee as disinterested party in dispute arbitration.

- Forecast possible future environmental damage, including possible acts of sabotage or terrorism.

- Establish an Emergency Response Center for cases of ecological emergency situations. Not all ecological damage is slow, observable, and predictable. A massive earthquake in the mountains of Kyrgyzstan would cause major flooding in Uzbekistan requiring immediate binational action.

5. Conclusions

Results of the 1996–1998 research show that on the southern coast of Issyk-Kul the usual level of environmental radiation is within the limits of normal, but the index of the level of internal radiation (inside homes and industrial buildings located near tailing dams) sometimes exceeds the natural norm. Clearly, it is necessary to carry out further detailed research to find out the reasons for the higher content of radon in the air at certain locations around Issyk-Kul. The alarming statistical data indicates some urgency for research into the causes of the unusually high death rate from malignancy among the population living near Issyk-Kul and recommendations for solutions to this problem (see section 3.1).

Thus, phytoplankton can serve as the indicator of cleanliness of water in Issyk-Kul Lake, which has important values for the tourism industry and recreational resorts (see section 3.2).

Our researches testify that using the submersible fluorometer is perspective to studying photochemical activity of cells and productional parameters of natural phytoplankton. The spatial-time maps of Issyk-Kul drawn up using GIS technologies for research of various hydro-biological parameters of the lake with geographical coordinates are perspective as well.

GIS technology is providing valuable information today in the Kyrgyz Republic regarding flood and mudflow danger assessment to minimize loss of life and property. For mountainous countries such as the Kyrgyz Republic, future applications of GIS technology are limitless. Synthesizing thematic data and analyzing results of information from different scales and time periods becomes possible (see sections 3.3, 3.4).

Thus, the conclusion of the anomalous water content of Barskaun River is in fortified concentration of sulfate-ion relatively to other regional streams, the most probable source of which is the oxidation process of sulfate, likely sulfur ore from the Kumtor gold mining site (see section 3.5).

The immense scale of ecological deterioration in Central Asia is in proportion to the scale of this vast region whose terrain and issues include mountain ranges dotted with ancient glaciers now melting to the point of disappearing; steppe lands tilled for productive agriculture now irrigated by increasingly polluted rivers; deserts, both natural and now manmade by encroaching desertification; lush ancient forests of virgin walnut and fruit trees now increasingly diminished by mindless human destruction (see sections 2.1–2.8).

The transboundary nature of many ecological problems in Central Asia requires multinational solutions. The five states of Central Asia, Kazakhstan, Kyrgyzstan, Tajikistan, Turkmenistan, and Uzbekistan, as representatives of their collective sixty million people, need a common ecological policy for Central Asia. Such a policy would be the basis for interstate cooperation toward the common goal of environmental preservation: what is the ecological status quo of Central Asia and how can we work together to preserve our natural environment and resources?

References

1. Belenko, V., and Ershova, N., Tracing glacier wastage of North Tien-Shan (Kyrgyzstan) on Sokuluk river basin.
2. Babajanova, R., Kadam, A., and Valiev, B., Central Asia threatened with desertification, *Times of Central Asia;* www.timesca.com (July 13, 2007), p. 6.
3. Chen, B. B., and Lelevkin, V. M., Complex monitoring the tropospheric and stratospheric an aerosol of Central Asia.
4. Muhamedzianov, A., Eurasia; http://centrasia.org/newsA.php4?st=1156136880 (2006).
5. Resources of superficial waters of the USSR—the Basic hydrological characteristics (Leningrad, Hydrometeoizdat, 1973), p. 308.
6. Substantive provisions (concept) of national water strategy of the Kyrgyz Republic (Bishkek, 2001); The project of water strategy of the Kyrgyz Republic (Bishkek, 2003).
7. Shvartsev, S. L., exec. editor, Bases of hydrogeology, Hydrogeochemistry—Science (WITH, Novosibirsk, 1982).
8. Torgoev, I. A., and Charskii, V. P., Ecological consequence of extraction of radioactive ores in Kyrgyzstan (Bishkek, *оцэи,* 1998), p. 55.
9. Tynybekov, A. K., and Hamby, D. M., The radiological characteristic of southern coast Issyk-Kul lake, Works of Institute of Management, Business and Tourism, 2nd edition (Bishkek, 1999), pp. 9–17.
10. Tynybekov, A. K., Radiological researches in Ton and Dzhety-Oguz areas of Issyk-Kul, Studying of mountains and a life in mountains: the International conference (KSMI, Bishkek, 2000), pp. 336–342.

11. Tynybekov, A. K., The radiological characteristic of coastal zones Issyk-Kul lake; the Environment and health of the people, Proceedings, T. VII (Bishkek, 1999), pp.78–86.
12. Tynybekov, A. K., and Matorin, V. M., Research of natural phytoplankton on Lake Issyk-Kul with use of the submersible fluorometer, The bulletin of the Moscow State University, 16th edition, *Biology* 1, 22–23 (2002).
13. Tynybekov, A. K., Torgoev, I. A., and Alyoshin, U. G., Estimation of risks of natural processes of southern coast of lake Issyk-Kul, the International conference on human health and strategy of an environment, the Program of a new millennium (Bishkek, May 2001) pp. 14–16.

ENVIRONMENTAL CHANGE IN THE ARAL SEA REGION

New Approaches to Water Treatment

RASHID KHAYDAROV[1] AND RENAT KHAYDAROV[2]*

[1,2] *Institute of Nuclear Physics, Ulugbek, Uzbekistan*

Abstract: This paper deals with novel approaches to solution of the important freshwater problems facing the Aral Sea Region. These approaches are based on using novel water treatment techniques developed taking into account local climatic and economic conditions. The local problem of removing water hardness and inorganic and organic contaminants can be solved by using proposed fibrous sorbents on the basis of Polyacrylonitrile (PAN). A special oligodynamic method that is particularly effective against typical types of pathogens in the Aral Sea region is proposed to solve a drinking water disinfection problem. As for the proposed solar powered water desalination technique based on a direct osmosis process, the separation there is driven by natural osmosis, which does not require external pumping energy as in the reverse osmosis process. The specific power consumption of the direct osmosis desalination process is less than $1\,kWh/m^3$ for sea water. On the basis of the findings the pilot device with productivity of $1\,m^3/h$ has been constructed. It consists of solar batteries with the capacity of $500\,W$ for pumping various fluids (feed, brine, product, and working solution) of the desalination device, solar energy heat exchangers for the recovery of working solution, water pretreatment unit on the basis of fibrous sorbents, and a water disinfection device with a very low energy consumption of $0.1\,W$-h. Due to the financial support of UNESCO in 2005, the device was installed in a village in the Aral Sea Region to remove salts with total concentration of about $17\,g/l$. Two hundred fifty water disinfection devices have been installed in manual artesian well water pumps through the support of JDA International (Colo., USA). More than twenty water purification systems consisting of filters on the

---

*Address correspondence to Institute of Nuclear Physics, Ulugbek, 100214, Tashkent, Uzbekistan

433

P.H. Liotta et al. (eds.), Environmental Change and Human Security, 433–447.

basis of developed fibrous sorbents and water disinfection units were installed in different villages of the Aral Sea Region.

Keywords: Direct osmosis; solar desalination; water hardness; water disinfection

1. Introduction

Irrigation for agriculture has been used in Central Asian countries for more than five thousand years, but during the last fifty years it has become unsustainable. It has resulted in an ecological crisis and significant environmental problems in the Aral Sea, which is located in the Republics of Uzbekistan and Kazakhstan (see Figure 1). The main problems faced in the Aral Sea basin can be summarized as follows.

- The level of water has dropped by 15 m since 1960.
- Only 5 km^3 of water reaches the Aral Sea, compared to 50–60 km^3 in 1960.
- The Aral Sea has shrunk to less than half its size during the last forty years.
- The amount of drinking and irrigation water in this region is insufficient and its quality is poor.
- Water and more than 50 percent of irrigated land have high salinity, and their salinization continues to increase.
- Biodiversity has been drastically reduced as a result of desiccation and shrinking of the Aral Sea, moreover, the loss of river habitat in the deltas has occurred.
- Desertification of the Amu-Darya and Syr-Darya river deltas has changed the climate in the Aral Sea region.
- All of these factors have had an adverse socioeconomic impact on the population in the region such as health risks, poor nutrition, and unemployment.

The Aral Sea disaster, which threatens the life, health, and habitat of the population, is well known all over the world due to the efforts of the region's countries as well as public and international organizations. Although measures have been taken with support from international and regional institutions, all these efforts have failed to mitigate the ecological problems, improve water quality, correct the imbalance in water use/consumption, and ensure the well-being of the population and the viability of

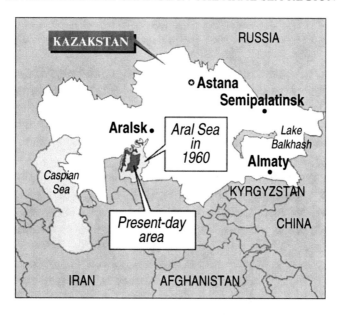

Figure 1. Map of the Aral Sea Region.

the natural ecosystems in the region. Water that is unsuitable for agriculture and municipal needs with a mineralization of 1.5–1.8 g/l and hardness of twice the Maximum Acceptable Concentrations (MAC) continues to be used by water consumers. Due to a shortage of drinking water in the Aral Sea Region, 86 percent of the population drinks and otherwise uses water that does not meet international quality standards recommended by the World Health Organization (WHO).[1]

As most of the rural areas of the Aral Sea Region do not have permanent access to electricity, usage of solar energy or batteries for powering various water treatment devices seems to be a very perceptive trend.

Three major types of contaminations of drinking water in the Aral Sea region can be identified as follows: (1) brackish underground and well water with heightened mineralization; (2) fresh artesian and surface water contaminated by bacteria; (3) surface and well water with increased concentrations of hardness ions, heavy metal ions, and pesticides.

Many methods, such as distillation, electrodialysis, reverse osmosis, and freezing are used for water desalination all over the world.[2–8] Unfortunately they are not widespread in Uzbekistan due to high power consumption and/or material cost (distillation method), low productivity (solar desalination), low efficiency of desalination of water with high concentration of salts (electrodialysis), and requirements for periodically supplying chemicals (ion exchange method) or electricity (reverse osmosis method). Therefore it is very important to develop a new low-energy-consuming technique for water desalination.

Chlorination, ozonation, ultraviolet irradiation, and oligodynamic methods are used for water disinfection all over the world. But they cannot be widely used in remote regions of developing countries due to high energy consumption or necessity of periodically supplying chemicals. Oligodynamic methods can be recommended for these countries, but they must be modified to meet local water regulation limits for disinfectants.[9]

Different types of resins and fibrous sorbents are used for water purification from radionuclides, heavy metal ions, and organic contaminants. An advantage of the fibrous ion-exchange sorbents over resin is the high rate of the sorption process, effective regeneration, and a small value of pressure drop of the sorbent layer for purified water. The specific surface of fibrous sorbents is $(2-3)~10^4$ m^2/kg, i.e., about 10^2 times greater than that of resin (10^2 m^2/kg). Therefore, the rate of the sorption process on fibrous sorbents is much greater than that of resin. The disadvantage of the known fibrous sorbents is a low exchange capacity, and it is necessary to improve this.

The purpose of this work was to develop new low-energy-consuming, cheap, and effective methods and equipment to treat all three types of contaminated water.

2. Removing Water Hardness and Inorganic and Organic Contaminants

Fibrous sorbents on the basis of Polyacrylonitrile (PAN) have been developed for water purification.

2.1. MATERIALS AND METHODS

Polyacrylonitrile (PAN) cloth with a surface density of 1.0 kg/m^2 and thickness of 10 mm was utilized as the raw material for making ion-exchange sorbents.

One- to ten-percent solutions of NaOH, 5–40 percent solution of hydrazine hydrate $NH_2NH_2~H_2O$ and 0.5–5 percent solutions of polyethylenimine $(-NHCH_2CH_2-)_x[-(CH_2CH_2NH_2)CH_2CH_2-]_y$ were used for the treatment of PAN cloth to make ion-exchange sorbents. Solutions of metal salts labeled by radionuclides (^{32}P, ^{51}Cr, ^{60}Co, ^{65}Ni, ^{64}Cu, ^{65}Zn, ^{82}Br, ^{89}Sr, ^{99}Mo+^{99m}Tc, ^{115}Cd, ^{124}Sb, ^{131}I) were used to optimize and reveal the best technology of making cation- and anion- exchange sorbents with respect to the high exchange capacity and good mechanical properties to remove inorganic and organic contaminants.

2.2. RESULTS AND DISCUSSION

Kinetics of saponification of the PAN fibers treated by a 5–40 percent solution of $NH_2NH_2 \cdot H_2O$ at 40–95°C were studied in the range of NaOH solution concentration from 1 to 10 percent at 25–70°C with temperature increments of 5°C. For example, the kinetics of saponification by 5 percent NaOH solution of samples treated by 20 percent $NH_2NH_2 \cdot H_2O$ solution at 70°C are given in Figure 2.

Increasing the temperature of the treating solution and duration of treatment causes filter capacity increase, but the strength of the fibers degrades. Experiments[10] have shown that optimum condition is in treatment by 20 percent $NH_2NH_2 \cdot H_2O$ solution at 70°C for 30 min and by 5 percent solution of NaOH at 25–30°C for 1 h. Exchange capacity (Cu^{2+}) of the sorbents is 3.5–4.0 meq/g.

Anion-exchange sorbents are made by treatment of cation-exchange filters in H-form by water solution of polyethylenimine. Amine groups attach to carboxy groups by electrostatic forces. Kinetics of anion-exchange groups' formation at concentrations of polyethylenimine from 0.5 to 5 percent and the temperature range from 20°C to 70°C with increments of 5°C were studied. Figure 3 demonstrates the kinetics curve at 25°C and concentration of polyethylenimine of 1 percent and Figure 4 shows the dependence of exchange capacities against the concentration of polyethylenimine at 25°C and treatment time of 1 h.

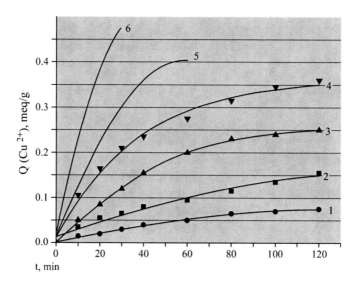

Figure 2. Kinetics of Saponification of the Fibers in 5 Percent NaOH Solution at 25°C (1), 30°C (2), 40°C (3), 50°C (4), 70°C (5) and 90°C (6).

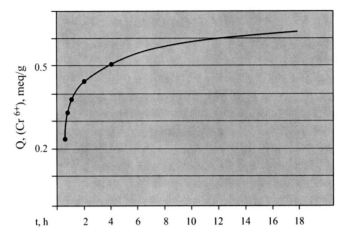

Figure 3. Kinetics of Anion-Exchange Groups Formation at 25°C in 1 Percent Polyethylenimine Solution.

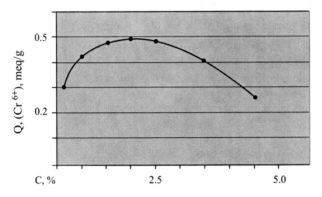

Figure 4. Dependence of Exchange Capacities Against the Concentration of Polyethylenimine at 25°C.

The treatment of the cation-exchange sorbents by a 1 percent solution of polyethylenimine at 25°C for 1 h was selected as the optimal condition for the anion-exchange sorbents production. Capacity (Cr^{6+}) of the sorbents is 3.0 meq/g.

Dependence of the distribution coefficient K_d for different ions against pH of the solutions is presented in Table 1.

Specific behavior of K_d of Co(II), Ni(II) and Cu(II) is explained by dependence of relation between M^{n+} form/hydrolyzed forms in the solution with the value of pH.[11] These results show high drinking water purification capability of the sorbents at pH from 6 to 8. Created fibrous sorbents have a high exchange capacity comparable with that of resin. An advantage of

TABLE 1. Distribution Coefficient Kd (ml/g) for Different Ions and Organic Substances (C0 = 10mg/l, V = 50ml, W = 0.5g).

Elements	Exchanger	pH of solution							
		2	3	4	5	6	7	8	9
Co(II)	Cationic	2,600	2,300	2,000	1,700	1,000	126	138	150
Ni (II)		600	870	920	990	750	430	510	780
Cu(II)		400	600	480	400	560	650	560	460
Zn(II)		2,000	4,000	5,000	4,000	1,900	1,700	1,400	900
Sr(II)		25	45	100	300	1,000	1,900	8,000	6,000
Cd(II)		830	680	520	380	240	97	75	46
Sb(II)		190	150	130	120	120	115	90	70
Cs(I)		100	200	900	1,900	3,200	4,000	4,000	1,500
Cr(VI)		150	100						
M-I131	Anionic	3,100	2,800	2,600	2,300	2,100	1,900	500	150
M-P32		3,200	3,000	2,700	2,500	1,900	1,100	300	150

Figure 5. Water Purification System Developed for Rural Areas of the Aral Sea Region (Height of the Largest Column Is 2,400mm).

the fibrous ion-exchange sorbents over resin is the high rate of the sorption process (about 10^2 times greater)[10] and small value of pressure drop of the sorbent layer for purified water.

At the present day newly developed sorbents are used in drinking water filters and systems for removing organic and inorganic contaminants. Figure 5 demonstrates the water purification system used in the Aral Sea Region to treat artesian water for organic and inorganic contaminants. Productivity of the system is $5\,m^3/h$, total pressure drop is very small (approximately 0.02 MPa), and purification and regeneration processes do not consume any additional energy. More than twenty systems were installed in different villages of the Aral Sea Region.

3. Water Disinfection

The objective of this investigation was to improve existing oligodynamic methods and to find the optimal alloy composition of electrodes that are most effective for water disinfection from typical types of pathogens in the Aral Sea region.

3.1. MATERIALS AND METHODS

Experiments were conducted with following bacterial cultures: *E. coli* (272, 33218, NCTC 10538), *Legionella pneumophila* serogroup 1, *Salmonella typhimurium* (251), *Salmonella typhi* (495), *Salmonella paratyphi* (8138), *Salmonella anatum* (4370) and *Vibrio cholerae asiaticea* (157).

The cultivation, culture enrichment, and testing of bacteria were performed following the Standard Methods for Testing Chemical Procedures[12] for the evaluation of disinfection, which are analogous of U.S. Standard methods for the examination of water and wastewater.[13]

Electrodes made of Ag, Cu, Au, and their alloys with different concentrations of the metals were prepared to find the optimal alloy composition for water disinfection from different pathogens.

3.2. RESULTS AND DISCUSSION

The oligodynamic disinfection process can be described by the following empirical formula[1]:

$$N(t) = N_0 \exp (-Ct/K)$$

where N_0 is the initial concentration of bacteria; N is the final concentration of bacteria; C is the effective concentration of metal ion composition in the water for a given ion collection, mg/l; K is the coefficient of resistance for a given ion collection, depending on bacteria types and concentrations of cations and anions in water (which can react with disinfecting metal ions); t is the time after introducing the metal ions into the water, i.e. disinfecting time. This formula shows that the concentration of metal ions C and time t necessary for water disinfection depend logarithmically on the initial concentration of bacteria, N_0.

Our experimental results[14-16] presented in Table 2 demonstrate the synergetic disinfecting effects obtained by using alloys with various silver/copper/gold combinations. In this table coefficient K is given for each composition of electrodes. Standard deviations of the coefficient K are given for p = 0.68. Concentrations of disinfecting ions in the water can be recalculated by using the value of concentration of Ag ions for each composition of electrodes.

Table 2 shows that the best disinfection is obtained by using an alloy of silver/copper/gold composition with concentrations of metals in the ratio 70–90 percent/10–30 percent/0.1–0.2 percent, respectively. These particular compositions allow reaching the minimum value of the coefficient K, and provide considerable reduction of metal ion concentrations in the water in comparison with other combinations.

Table 3 summarizes coefficients K for different pathogens and disinfecting times when employing the electrodes described above. These data correspond to concentrations of silver, copper, and gold of 30 µg/l, 7.5 µg/l, 0.075 µg/l, respectively.

The portable devices shown in Figure 6 have been tested at the Moscow Medical Academy (Russia), Institute of Medical Parasitology and Tropical Medicine (Russia), Envirotech Laboratories Inc. (Massachusetts, USA),

TABLE 2. Minimal Concentration of Disinfecting Metal Ions for Killing E. Coli with Initial Concentration 1,000 1/l (Final Concentration ≤1 1/l) within 30 Min.

Composition of electrodes, percent				
Ag	Cu	Au	Concentration of Ag$^+$ (mg/l)	K (mg·s/l)
100	0	0	1.5	390 ± 120
99.9	0	0.1	1.5	390 ± 120
99.8	0	0.2	1.5	390 ± 120
90	10	0	0.1	26 ± 8
89.9	10	0.1	0.05	13 ± 4
89.8	10	0.2	0.05	13 ± 4
80	20	0	0.05	13 ± 4
79.9	20	0.1	0.03	8 ± 3
79.8	20	0.2	0.03	8 ± 3
70	30	0	0.05	13 ± 4
69.9	30	0.1	0.04	10 ± 3
69.8	30	0.2	0.04	10 ± 3
50	50	0	0.5	130 ± 40
49.9	50	0.1	0.3	80 ± 25
49.8	50	0.2	0.3	80 ± 25
40	60	0	1.0	261 ± 80
39.9	60	0.1	0.8	210 ± 60
39.8	60	0.2	0.8	210 ± 60

TABLE 3. Coefficient K and Disinfecting Time
(Initial Concentration 1,000 CFU/l).

Pathogen	K (mg·s/l)	t (min)
Typhoid-Paratyphoid, Legionella pneumophila	15–30	60
Salmonella	8–15	30
V. cholera	8–15	30

Figure 6. Usage of Pen-Sized Water Disinfection Device.

Rogers Associates (Colorado, USA), EN Technology Inc. (Korea), the Ministry of Public Health (Uzbekistan), and the Department of Pathology of the University Sains of Malaysia. Typical test results are shown in Table 4. The pen-sized devices with resource of $3\,m^3$ of treated water can be used by population to prevent bacterial contamination in life conditions and emergency situations.

The devices with productivity of $5–50\,m^3/h$ have small sizes, very low energy consumption of 0.5–5 W-h and can be powered by accumulators, solar batteries, or 110–220 V AC. They can be used for disinfection of potable water systems, water storage systems, cooling water for air ventilation systems, and swimming pools, and to prevent the consequences of bioterrorism and any natural cataclysms. For example, due to the financial support of JDA International (Colorado, USA) 250 devices with productivity of $1\,m^3/h$ have been installed in manual artesian well water pumps in the Aral Sea region of Uzbekistan for disinfection of underground water. The devices,

TABLE 4. Test Results for the Portable Device.

	Initial concentration 10^9 CFU/l E. coli			Initial concentration 10^8 CFU/l E. coli		
Disinfecting time	3 min insertion	5 min insertion	10 min insertion	3 min insertion	5 min insertion	10 min insertion
30 min	100–150	100–150	100–150	90–100	50	45
2 h	100–120	30	3	70	6	5
2 days	15	<1	<1	2	<1	<1
3–20 days	<1	<1	<1	<1	<1	<1

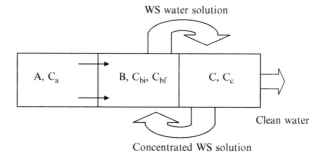

Figure 7. Principle of Operation of Direct Osmosis Desalination Method.

with a productivity of 50 m³/h, have been installed in more than twenty settlements of the region.

4. Direct Osmosis Water Desalination

4.1. PRINCIPLE OF THE METHOD

To describe the developed water desalination method based on the direct osmosis process (see Figure 7), let us suppose that there is salted water (the feed) containing contaminants with a concentration of dissociated ions C_a in section A. In section B there is a water solution of a certain substance with initial concentration C_{bi}. Let us call them a working fluid (WF) with a working substance (WS), respectively. Sections A and B are separated by a semipermeable membrane that is permeable to water molecules and impermeable for substances dissolved in section A and to the working substance. Let us also suppose that the condition $C_a << C_{bi}$ is met. Then in order to equilibrate the difference in chemical potential, osmosis movement starts, i.e., water begins to diffuse through the membrane from section A to section B. In other words in section B WF is diluted by clean water pulled out from section A. The water solution of WS with the final molar concentration C_{bf} from section B is transported to section C, where water and WS are separated or water is removed from a part of diluted WF.

The main problem with this water treatment method is the selection of WS. The simplest and the most accessible methods of separation of water from WS in section C are evaporation, freezing, and forming crystalline hydrates.

4.2. MATERIALS AND EQUIPMENT

In our experiments we used diethyl ether $(C_2H_5)_2O$ as the working substance and RE-1812-LP reverse osmotic membrane produced by SAEHAN Inc. (South Korea) as the semipermeable membrane for direct osmosis process. In a pilot device, solar batteries of 12 V with a total power of 500 W were used as an energy source and sorbents described above were used to remove salts of hardness and ions of iron. In the pilot device the water disinfection apparatus described above was used to kill bacteria to prevent destruction of semipermeable membrane.

4.3. RESULTS AND DISCUSSION

Test results of the device for water with different salinities are given in Table 5. Total electric energy consumption of the desalination process can be essentially

TABLE 5. Test Results of the Laboratory Experimental Model.

C_a of NaCl (g/l)	Recovery (relation of permeate and feed flow rates)	Flow rate of WF (l/h)	Part of WF for water separation	C_{bi} (g/l)	C_{bf} (g/l)	E_h (kJ)	E_s (kJ)	P_r (W)	C_c of NaCl (mg/l)
1	0.1	60	0.5	5	3.0	1,600	29	500	<10
	0.5	300	0.1	5	4.6	1,600	44	500	
		60	0.5	7	4.1	1,600	40	500	
		300	0.1	7	6.4	1,600	62	500	
4	0.1	60	0.5	20	12	1,600	112	500	<10
	0.5	300	0.1	20	18	1,600	176	550	
		60	0.5	25	15	1,600	140	500	
		300	0.1	25	23	1,600	220	550	
20	0.1	60	0.5	40	23	1,600	224	600	<20
	0.4	300	0.1	40	37	1,600	352	600	
		60	0.5	60	35	1,600	336	600	
		300	0.1	60	55	1,600	528	600	

decreased by using a solar heater instead of the electric heater used in the laboratory experimental device. Through support of UNESCO in March 2005, the pilot device with productivity of $1 m^3/h$ was installed in Turtkul village in the Aral Sea Region. The total concentration of salts in initial water was about 17 g/l. Test results of the device are shown in Table 6.

Our experiments have shown the following advantages of the method: (a) the method does not require a high pressure pump to push water through the membrane; (b) the electric energy consumption of the method is very low ($0.5 kWh/m^3$ in comparison with $2–5 kWh/m^3$ for RO and $2–4 kWh/m^3$ for MSF and MED) when solar energy or heat from ambient air are used to heat the water to 36–40°C; (c) the surface area of the solar heat exchanger of the device is 3–30 times less than that of the solar thermal desalination process because of lower solar energy consumption ($80 MJ/m^3$ in comparison with $2,500 MJ/m^3$ for the solar thermal desalination process without energy recovery and $250–500 MJ/m^3$ with energy recovery); (d) the method can be used for desalination of water with salt concentrations up to 40 g/l; and (e) since the disinfecting device is used in the water pretreatment process the water keeps disinfecting properties for a long time (not less than one year) and water produced in hot weather can be stored in water storage systems for cold seasons.

5. Conclusions

The Aral Sea crisis is one of the worst human-made environmental catastrophes of the twentieth century. The drying of the Aral Sea due to human misuse and overuse of its water clearly shows us all the human costs of climate change. This paper deals with novel approaches to solution of the important freshwater problems facing the Aral Sea Region. Test results have shown the efficacy of newly developed types of water treatment devices,

TABLE 6. Purification of Water Sample Taken from the Aral Sea Region.

Items	Influent water (mg/l)	Effluent water (mg/l)
pH (at 25°C)	6.7	6.7
Ca	560	1.2
Mg	400	1.6
Cl	4,590	7.1
NO_3	300	<0.02
SO_4	3,825	1.8
Na	3,300	2.3
K	1,200	1.6

taking into account the climatic and economic conditions of the Aral Sea Region. The developed desalination devices based on the direct osmosis technique used solar batteries as an energy source and can be utilized in remote regions for desalination of water with concentrations of salts up to 40 g/l. They have low energy consumption ($0.5\,kWh/m^3$ in comparison with 2–$5\,kWh/m^3$ for RO and 2–$4\,kWh/m^3$ for MSF and MED) when solar energy or heat from ambient air are used to heat the water to 36–40°C. The surface area of the solar heat exchanger of the device is 3–30 times less than that of solar thermal desalination device.

The developed water disinfection devices have small sizes, very low energy consumption of 0.5–5 Wh for productivity of 5–$50\,m^3/h$, and can be powered by accumulators, solar batteries, or 110–220 V AC. They can be used for disinfection of potable water systems, water storage systems, cooling water in air conditioners, and to prevent the consequences of bioterrorism and any natural cataclysms. The presented results show that the capacity of developed cation-exchange sorbents is 3.5–$4.0\,meq/g$ (Cu^{2+}) and that of the anion-exchange is $3.0\,meq/g$ (Cr^{6+}), which is comparable with the capacity of resin. An advantage of the fibrous ion-exchange sorbents over resin is the high rate of the sorption process (about 100 times greater) and a small value of pressure drop of the sorbent layer for purified water. Newly developed sorbents have been used successfully in water treatment systems for removing organic and inorganic contaminants in different villages of the Aral Sea Region.

References

1. UNDP Energy and Environment Unit, *Water—Critical Resource for Uzbekistan's Future* (United Nations, New York and Geneva, 2007).
2. B. Nicolaisen, Developments in membrane technology for water treatment, *Desalination* 153(1), 355–360 (2002).
3. K. Wangnick, 2004 IDA Worldwide Desalting Plants Inventory, Report No.18, 2004 (Wangnick Consulting, Gnarrenburg, Germany).
4. E. Tzen and R. Morris, Renewable energy sources for desalination, *Sol. Energy* 75(5), 375–370 (2003).
5. L. Garcia-Rodriguez, Renewable energy applications in desalination: State of the art, *Sol. Energy* 75(5), 381–393 (2003).
6. L. Garcia-Rodriguez, Seawater desalination driven by renewable energies: A review, *Desalination* 143(2), 103–113 (2002).
7. L. Huanmin, J. C. Walton, and A. H. P. Swift, Desalination coupled with salinity gradient solar ponds, *Desalination* 136(1–3), 13–23 (2001).
8. E. Zarza and M. Blanco, Advanced M.E.D. solar desalination plant: Seven years of experience at the Plataforma Solar de Almería, *Proceedings of the Mediterranean Conference on Renewable Energy Sources for Water Production* (Santorini, Greece, 1996), pp. 45–49.
9. U. Rohr et al., Impact of silver and copper on the survival of amoeba and ciliated protozoa in vitro, *Int. J. Hyg. Environ. Heal.* 203(1), 87–89 (2000).

10. R. A. Khaydarov and R. R. Khaydarov, Purification of Drinking Water from [134,137]Cs, [89,90]Sr, ^{60}Co and [129]I, in: *Medical Treatment of Intoxication and Decontamination of Chemical Agents in the Area of Terrorist Attack*, edited by C. Dishovsky (Springer, The Netherlands, 2006), pp. 171–181.

11. J. Kragten, *Atlas of Metal-Ligand Equilibria in Aqueous Solution* (Ellis Horwood, Chichester, 1978).

12. Drinking Water. Hygiene requirements and quality control, GOST 2874–82 of Russian Federation (1982).

13. American Public Health Association, *Standard Methods for the Examination of Water and Wastewater*, 19th ed. (American Public Health Association, Washington, 1995).

14. R. A. Khaydarov and R. R. Khaydarov, Water disinfection using electrolytically generated silver, copper and gold ions, *J. Water Supply Res. T.—AQUA* (53), 567–572 (2004).

15. R. A. Khaydarov and S. B. Malyshev, Patent of Russian Federation: Water Disinfecting Device, N 2163571, 27. 02. 2001.

16. R. A. Khaydarov, B. Yuldashev, S. Korovin, and Sh. Iskandarova, Patent of Republic of Uzbekistan: Water disinfectant device, N5031, 08.09.1997.

ENVIRONMENTAL CHANGE OF THE SEMIPALATINSK TEST SITE BY NUCLEAR FALLOUT CONTAMINATION

GULZHAN OSPANOVA[1]*, GULNAR MAILIBAYEVA[2], MANAT TLEBAYEV[3] AND MAYRA MUKUSHEVA[4]

[1,4] *Satpayev Kazakh National Technical University, Kazakhstan*

Abstract: This paper presents an evaluation of the current nuclear fallout contamination of the Semipalatinsk Nuclear Polygon's test sites and the effects of secondary contamination of ecosystems through water and food chain migration of radionuclides, and an overview of health hazards to the population of the affected areas.

Keywords: Environment; Semey nuclear polygon; explosions; radionuclides; health

1. Introduction

Atomic test polygons that were actively operated in the territory of Kazakhstan for four decades have caused considerable damage to the environment in the republic. Approximately 70 percent of total nuclear tests of the former USSR were carried out in Kazakhstan. Most of them, including air and surface explosions, were conducted at the Semey Nuclear Polygon (SNP) near the city of Semipalatinsk. Nuclear tests were repeatedly conducted in the Semipalatinsk Nuclear Test Site since 1949. A total of 456 nuclear explosions occured between 1949 and 1989, as well as tests of the first atomic (1949) and hydrogen (1953) bombs.[1] $9 \cdot 10^{16}$ Bq ^{137}Cs were entered into the environment as a result of nuclear explosions at the SNP compared to the Chernobyl incident in Ukraine, when the amount of fallout was $6 \cdot 10^{16}$ Bq. Thus, the SNP is one of the main sources of radiation hazard for the population of Kazakhstan. The significance of the problems facing the Semipalatinsk Test Site has been confirmed by the UN General Assembly.[2]

---

*Address correspondence to Gulzhan Ospanova, Chemical Technology of Inorganic Materials Department, Kazakh National Technical University, 22 Satpayev str., Almaty 050013, Kazakhstan; e-mail: ogulzhan@yahoo.com, osp_gul@mail.ru

P.H. Liotta et al. (eds.), Environmental Change and Human Security, 449–458.

2. Overview of Semipalatinsk Test Site History

The Semipalatinsk Test Site (STS) was the primary testing venue for the former USSR's nuclear weapons. It is located in the steppes of northeast Kazakhstan, south of the valley of the Irtysh River, and it covers an area of 18,000 km² (Figure 1). The scientific buildings for the test site were located about 150 km west of the town of Semipalatinsk (later renamed Semey).

The site was selected in 1947 by the Soviet government, probably due to its remote location and vast expanse of sparsely populated steppes. The first Soviet atomic bomb, Operation First Lightning (nicknamed *Joe One* by the Americans) was deployed in 1949 from a tower at the Semipalatinsk Test Site, scattering nuclear fallout on nearby villages of Kazakh nomads. The same area ("Experimental Field," a region 40 miles west of Kurchatov City) was used for more than 100 subsequent above-ground weapons tests.

There were multiple facilities that conducted nuclear tests within the STS. Once atmospheric tests were banned, testing was transferred to underground locations at Chagan and Murzhik in the west, and at the Degelen Mountain complex in the south, which is riddled with boreholes and drifts from both subcritical and supercritical tests. The Chagan River complex and nearby "Balapan" in the easternmost part of the STS includes the site of the extremely powerful Chagan test, which resulted in the formation of radioactive Lake Chagan (a.k.a. Atomic Lake). This artificial nuclear lake, formed as a result of nuclear exposure excavation on January 15, 1965, covers the area of 7×8 km². Overall, between 1949 and the cessation of atomic testing in 1989, 340 underground shots and 116 atmospheric explosions were conducted at

Figure 1. Semipalatinsk Test Site.

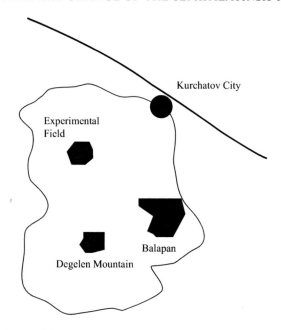

Figure 2. Semey Nuclear Polygon.

the STS. The last nuclear test conducted at Semey Nuclear Polygon took place in Balapan in November 1989. The SNP is shown in Figure 2.

As a result, a part of the population of the region was subjected to living in a highly radioactive area. The testing period can be categorized into two parts according to the level of radiation these people were exposed to. The first period is from 1949 to 1963, when atmospheric tests and tests erupting and blowing off soil above ground were conducted. The second is from 1963 to 1989, when all tests were conducted underground. The atmospheric tests from 1949 to 1953 caused the greatest radiation damage to people living in the region. During this period, an enormous amount of radiation accumulated in the area, which later brought about grave consequences. The ancient steppes, the nomadic way of life, and numerous animals and plants all became hostages of the warfare of powerful forces.

The changes in the political system during the second half of the 1980s allowed the beginning of a movement to shut down the testing site. In 1988 it grew into the international movement "Nevada-Semey" led by the well-known Kazakh poet and public figure Olzhas Suleimenov. This movement won the support and sympathy of people all over the world. Meetings, demonstrations, letters, appeals to governments, collections of thousands of signatures, and picketing at the testing sites took place.

The Semipalatinsk Test Site was officially closed by Kazakhstan President Nursultan Nazarbayev on August 29, 1991.

3. The Current Assessment of the Nuclear Fallout Contamination

Today, eighteen years after the last explosion, the level of biologically significant long-living radionuclides in the soil still exceeds the acceptable concentration. Therefore, long-living radionuclides migration is an urgent problem in the Semipalatinsk Test Site area.

Our research shows that effects of second contamination have begun to appear. These effects include radionuclides migration with underground waters, radioactivity transfer on the earth's surface by melted and downpour waters, wind transfer, and radionuclides migration through the food chain.

Long-living radionuclides are accumulating in agricultural produce and animals. Thus, the problem of transferring basic dose form radionuclides in the soil-vegetation-animal system is becoming of paramount importance.

3.1. RESEARCH METHODOLOGY

Our analysis of the current radiation situation at the test site consisted of the following: radiospectrometric and radiochemical examination of samples of soil, plants, water, and biological objects; statistical processing of sample examination results; and estimation of the complex of physical models for radionuclides migration. We carried out studies of the conduct of radionuclides in the food chain and water basin, prognosis of biological accessibility of this radionuclide in the soil, and analysis of radioactive accumulation in vegetation of the SNP territory.

The evaluation of the nuclear fallout contamination of the Semipalatinsk Test Site is based on the system analysis that includes the following factors: (1) calculation of radionuclides migration in the following systems: "soil-plant," "plant-animal" and "water-animal"; (2) simulation of ^{137}Cs, ^{90}Sr radionuclides distribution in the soil-plant-animal system of "critical" sites; and (3) calculation of ecological parameters in order to reveal the population radiation doses for the purpose of economic usage of critical sites.

3.2. SOIL AND WATER CONTAMINATION

About 30 percent of the test site is contaminated with Cesium-137 (^{137}Cs) and Strontium-90 (^{90}Sr). These elements are the basic long-living products of radiation division, and their content determines the contamination level. Within the vast territory of the SNP, areas called "Experimental Field," "Balapan," and "Degelen" are highly contaminated with radiation. However, the capacity of external radiation doses in the soil is relatively similar in the sites where nuclear tests were not carried out directly. Exceptions like the "Experimental Field" and "Balapan" regions were the epicenters of surface and atmospheric nuclear explosions. Their content of ^{137}Cs and ^{90}Sr are $2 \cdot 10^{16}$ Bq and $1.3 \cdot 10^{16}$ Bq respectively. ^{137}Cs contamination occupies a territory of 10–12 km, which equals to approximately 3 percent of the total test zone. Other areas in the test site show a notable concentration of transuranic elements (i.e., plutonium and americium), and this represents a serious danger to the people.

Degelen Massif, another contaminated site, is situated in the southern part of the SNP. The contaminated territory has a configuration of two "spots" of irregular shape. In 1992 water effusion was registered in twenty-seven galleries of Degelen Massif. As a result of this accident, radioactive compounds contaminated twenty-four mouth sites with an exposure dose of 1–5 mR/h. The maximum contamination density of Atomic Lake is 3,700 kBq/m^2 (100 Ci/km^2).

3.3. FOOD CHAIN CONTAMINATION

For the first time, models of ^{137}Cs and ^{90}Sr migration by food chains were developed for the "Experimental Field" and "Balapan" sites of the Semipalatinsk Test Site. The mathematical models are parameterized based on the experimental information characterizing the features of grassland ecosystem components. The main purpose of these models is to predict ^{137}Cs and ^{90}Sr content in the components of ecosystem.

Prediction assessment carried out using the developed models showed that radionuclides concentration in horse and sheep milk from animals that grazed at the "Experimental Field" and "Balapan" may continue to exceed the standards established for food products for a long time. ^{137}Cs content in horse milk and sheep milk at "Balapan" site is respectively four times and two times higher than allowable norms for food products in the Republic of Kazakhstan. According to our prognostication, the radionuclides concentration in the milk of horses and sheep will be exceeding the standards for the next twenty-five to thirty years.

In the "Experimental Field" ^{137}Cs content in milk of animals 1.7 times higher than the acceptable allowable norm of 100 Bq/l. ^{137}Cs content in meat is at 800 Bq/kg, which exceeds the allowable norm of 160 Bq/kg by 400 percent. Our models estimate that the content of ^{137}Cs in meat will be 570 Bq/kg in 2010.

3.4. AGRICULTURAL AND ECONOMIC IMPACT

About two million hectares of agricultural land were contaminated with radioactive substances. Land contaminated with the maximum radioactive level extends to as much as seven million hectares. Villages around the test site are situated at 15–50 km from the test site. In comparison, the nearest village to the Nevada test site in the USA was 150 km from the test site. In the Semipalatinsk area nuclear fallout covered a territory of 340,000 km² with the population of 1,700,000 people.

As a result of farm animals grazing at the most contaminated areas (in a region of Atomic Lake), local shepherds received the highest exposure doses. They are the most at-risk group of the STS population with the total annual exposure dose for the population category of about 1.13–1.47 mSv. The most exposure is received by shepherds who graze animals year-round.

Additional exposure doses to the population from radioactive contamination of the Semipalatinsk Test Site from nuclear tests and the radiation contamination of the lands adjacent to the STS resulted in significant damage to the local economy. The investigation of the economic impact is worth further investigation.

The prediction results can be used to develop recommendations for the former Semipalatinsk Test Site land economic use. A complex of mathematical models can be used for different scenario calculation, particularly to develop the agrochemical protection measures aimed at reduction of radionuclides accumulation in agricultural products. This research could be used as a basis for the ecological and economic justification of balanced environmental management at the STS.

4. Radiation Hazard for the STS Population

Nuclear tests caused irreparable damage to health and the environment. The testing site is closed, but its trace still remains on our land. Nowadays access to the territory of the former SNP is open. There are no fences around the perimeter; people and animals roam freely across the former test site. Active industrial mining of coal and salt on its territory has begun. In addition to raw material extraction sites, there are cattle pastures and settlements present in the area. Sites of permanent living for herdsmen and their families (both winter and summer camps) have appeared on the polygon territory. Also herds of horses inhabit the area.

The region has the highest level of cancer diseases, changes in the genetic structure of people and animals, the shortest life span, and the highest mortality rate in Kazakhstan. More than one million people were exposed to periodic and chronic ray radiation during the operation of the SNP. However, in several regions the cumulative dose of radiation exposure was much higher and reached the dangerous level of 2–4 Gy.

The nuclear tests and all other activities of the Semipalatinsk Test Site were carried out in total secrecy, and the local population was not informed of the testing; hardly anyone was aware of the danger and hazards of these nuclear tests and their consequences. Because of the lack of sufficient safety measures, soil, water sources, and food were heavily contaminated. No medical survey with a clear purpose of investigating the health impact of exposure to radiation was conducted until the 1960s.

In 1962 the Semipalatinsk Radiation Clinic started a medical study that covered 10,000 people living in fourteen residential areas of the three districts: Abaiski, Beskaragaiski, and Zhana-Semeisky. The radiation exposure impact on the population in the Semipalatinsk region manifested in both acute and chronic symptoms. Long-term medical observation of the surveyed people's health condition provided evidence of increased incidents of disorders of the circulatory and neurological systems. The study also revealed that due to the nuclear tests, the subjects' lifespan decreased by ten years on average and they showed earlier signs of aging.

Both morbidity and mortality rates caused by tumors among people living near the test site increased dramatically. The frequency distribution shows the highest increases first in the period from the fourth to the fifteenth years and then from the twenty-third to the twenty-seventh years of operation of the testing site. The average annual morbidity of tumors in the area is 40 percent higher than tumor morbidity of the population that was not exposed to radiation.

The Regional Clinic of Oncology showed that the number of leukemia patients in the Semipalatinsk region grew sevenfold from 1975 to 1985 compared with the previous decade. Long-term environmental damage has affected the gene structure and caused increased incidents of congenital deformities and developmental abnormalities.

The incidence of chromosomal aberrations in leukocytes among people living in the Semipalatinsk region is 4.5–5 times higher than in those not exposed to radiation. Congenital deformities accounted for 6.4 percent of all diseases and 2.3–7.3 percent of deaths in the period between 1986 and 1988. The number of children with diseases of the nervous system and mental disorders is increasing. The birth of babies with mental and/or physical disabilities is increasing.

During the nuclear testing period, radioactive clouds passed over the residential areas and the people suffered from external gamma ray exposure as well as chronic internal radiation exposure by absorbing radioactive substances contained in food, water, and air. After nuclear tests, in certain residential areas the radioactivity sometimes reached levels higher than 500 R/h. After underground tests, radioactive gases leaked to the surface and people were repeatedly exposed to low-level radiation. Overall, people living in the radiation-affected areas received external and internal gamma ray radiation of 5–100 rems.

Due to the natural and climatic conditions around the test site not only the people who lived in the areas neighboring the test site, but also those living in places like Semipalatinsk City, which is more than 100 km from the epicenter, could see the "nuclear mushrooms." The radioactive clouds carried by strong winds sometimes spread over areas hundreds of kilometers away from the test site. As a result of this, the people living in the regions of Semipalatinsk, Pavlodar, Karaganda, and Eastern Kazakhstan as well as the Altai Region of the Russian Federation were affected by the tests.

During the forty years of nuclear testing conducted at the SNP in the Semipalatinsk Region, radioactive fallout fell several times on different residential areas, including the villages of Dolon, Cheremyshki, Mostik, Kanonerka, Znamenka, Sarzhal, Karaul, and Kainar, and local residents were exposed to massive amounts of radiation. The ground and atmospheric tests conducted in the Degelen Mountains were accompanied with radioactive fallout over the districts of Abaisk, Abralinsk, and Zhana-Semeisk in the Semipalatinsk Region. The residential areas such as Abraly and Kainar were under particularly unfavorable conditions.

On September 24, 1951, a ground test was conducted in the Degelen district. It was identical to the atomic explosion made in 1949. The radioactive fallout affected mostly the population of Kainar village. In August of 1953 the residents of Kainar village were again affected when a nuclear bomb exploded (with a yield of 470 kt). Many of them had not been evacuated to the safety zone. By 1963, the population of the above-mentioned areas was affected by radioactive fallout at least ten times.

The unique demographic condition of these areas, which have extremely low population mobility, created a situation where people who experienced virtually all the nuclear tests made from 1949 to 1989 still live in the affected area. Unsurprisingly, this exposure has drastically harmed the health of the local population.

The oncological morbidity and mortality rates of the people of these areas radically increased during radiation exposure period from 1960 to 1965. In 1965, for example, the oncological mortality rate reached 300–350 for a population of 100,000. The major causes of death were tumors of stomach, intestines, and lungs, as well as leukemia. The second peak in mortality due to tumors was observed between 1970 and 1975, many years after initial exposure. During this period, the oncological mortality rate of Kainar's population again reached 300 per 100,000. In 1983 the mortality rate from tumors caused by radiation was 184 per 100,000.

Genetic radiation aftereffects on the people of Kainar village was the cause of the increase in the incidence of growth disorders observed among newborn babies. During several years the congenital disorder incidence among newborns reached 1.9–2.6 percent. The growth disorders caused by radiation consisted in large part of microcephaly and growth disorders of the skull. There are 3,200 children in the Abralinsk District; 149 of them are handicapped and have serious congenital abnormalities. Immuno-deficiency, leukemia, and lymphatic tumors have almost continuously appeared among minor and adult members of Kainar village. These diseases have weakened the defense mechanism of the human body and caused an increase in the incidence of infectious diseases, especially among children.

In response to the considerable increase of the general prevalence of diseases and the death rate of the population, the Republic of Kazakhstan issued the law "About social protection of population who suffered due to the nuclear explosions" to offer social benefits to the population affected by the SNP.[3]

It is impossible to see the radiation, feel it, or touch it, but its consequences are terrible. In the Semipalatinsk region alone, there are about 850,000 registered sufferers.

5. Conclusions

The Semey nuclear polygon, which actively operated in the territory of Kazakhstan for four decades, has caused considerable damage to the health and environment in the republic. There was significant radioactive contamination of the ecosystem during the operation of the testing site. Even nowadays, effects of second contamination have begun to appear: migration of radionuclides with underground waters, radioactivity transfer on the earth surface by melted and downpour waters, wind transfer, and radionuclides migration through the food chain.

It is necessary to limit the access of the general population to the "critical," still highly contaminated polygon sites ("Experimental Field," "Balapan," "Degelen," and Atomic Lake area). The concentration of ^{37}Cs and ^{90}Sr, as well as other transuranic elements, represents a serious danger to people in these areas.

Other sites can be used with strong radioecological monitoring of agroecosystems. There are increased levels of radionuclides in agricultural products and our models predict the effects to last at least another quarter of a century.

Health hazards of radiation exposure still manifest themselves among the local population. The prevalence of genetic mutations, birth defects, and tumor mortality is the price that the local population pays for the years of exposure to gamma ray radiation. Kazakhstan cannot solve this problem single-handedly. It is an issue of global importance: the Semipalatinsk region needs international support to eliminate the damage of nuclear testing, especially in the areas of medical and social rehabilitation.

References

1. Mikhailov, V. N. et al., USSR nuclear tests (1996).
2. Resolution adopted by the UN General Assembly (71st plenary meeting, November 27, 2000), International cooperation and coordination for the human and ecological rehabilitation and economic development of the Semipalatinsk region of Kazakhstan.
3. Republic of Kazakhstan law No. 1787-XII (December 18, 1992), About social protection of population who suffered due to the nuclear explosions at the Semipalatinsk Test Site.

SECTION V

ENVIRONMENTAL CHANGE AND HUMAN IMPACT LINKAGES

SUMMARY, CONCLUSIONS, AND RECOMMENDATIONS FOR POLICY AND RESEARCH

DAVID A. MOUAT[1]* AND WILLIAM G. KEPNER[2]

[1] Desert Research Institute
Division of Earth and Ecosystem Development
2215 Raggio Parkway
Reno, Nevada 89512 USA
[2] U.S. Environmental Protection Agency
Office of Research and Development
944 E. Harmon Avenue
Las Vegas, Nevada 89119 USA

1. Conclusion: Environmental Change and Human Impact Linkages

What are the human impacts of environmental change? How might land be used and what would be the potential benefits or consequences? Numerous questions arise as the world we know becomes smaller in our perception and the human population it supports becomes more dependent on the circumstances of a globalized economy. Increasingly, technology makes information instant and accessible to many. A first premise of the concept of security is that protection of human life from environmental, economic, food, health, personal, and political threats is a vital core value. A second argument is that actions that guard against environmental degradation represent a key element to security by ensuring sustainable resources and the continuation of providing ecosystem services that sustain human well-being. The human element of these premises, often partitioned as *human security*, and the ecosystem element is typically thought of as *environmental security*; both concepts, nonetheless, are so intrinsically interlinked that one term does not usually occur without the other.

One goal of this scientific volume is to examine both human and environmental security and their linkages in the context of both natural and social sciences, in order to suggest strategies for reducing our collective vulnerability to pervasive threats. New global challenges await us in the near future, with

---

*Address correspondence to D. A. Mouat, Desert Research Institute, 2215 Raggio Parkway, Reno, Nevada 89512 USA; tel: +001-775-673-7402; fax: +001-775-674-7557; e-mail: dmouat@dri.edu

461

P.H. Liotta et al. (eds.), Environmental Change and Human Security, 461–469.
© 2008 *Springer Science + Business Media B.V.*

the potential to affect the security of multiple nations and regions. What is new is that these challenges lie outside the "traditional" concept of threat and represent events of unimaginable scale and proportion with potential dire consequences. Some examples include mass uncontrolled trans-border migration, disease pandemics such as bird flu, international crime, regional climate variation, increased intensity and frequency of severe weather events, and large-scale environmental degradation such as desertification. Accordingly, this volume includes sections on recognizing environmental change, on subsequently identifying the key security challenges that face us in the early twenty-first century, on exploring common mechanisms to respond to global environmental change, and on responding to hazard impacts. Last, as environmental and human security are often inextricably interlinked and their common problem sets are often equal among all societies, this forum endeavors to explore these linkages, identify challenges, and seek solutions in a collaborative fashion across multiple scientific disciplines and geographies.

2. Integrating Human Security with Environmental Security

Sustainable societies are dependent on the goods and services provided by ecosystems, including clean air and water, productive soils, and the production of food and fiber (Millennium Ecosystem Assessment, 2005). Ecosystem services have been defined in a variety of ways; however, in the end they reflect the basic outputs of ecological function or processes that directly or indirectly contribute to human well-being and a sense of security.

In 2000 the United Nations Secretary-General initiated the Millennium Ecosystem Assessment (MA) to assess the consequences of ecosystem change for human well-being and the scientific basis for action needed to enhance the conservation and sustainable use of those systems and their contribution to human well-being. From 2001 to 2005, the MA involved the work of more than 1,360 experts worldwide. Their findings provide a state-of-the-art scientific appraisal of the condition and trends in the world's ecosystems and the services they provide, as well as the scientific basis for action to conserve and use them sustainably (2005). Ecosystem services have generally been characterized into four key categories, which include supporting, regulating, provisioning, and cultural functionality (Farber et al., 2006: Table 1).

It is the provision of these basic services and their probability for continuation into the future that serve as core to the concepts of human and environmental security. One dilemma that stands before us is to recognize and anticipate change in ecosystem services in all of its forms and to understand the impact to human security. This change in paradigm as reflected by the

TABLE 1. Ecosystem Functions and Services (From Farber et al., 2006).

Ecosystem functions and services	Description	Examples
Supportive functions and structures	Ecological structures and functions that are essential to the delivery of ecosystem services	
Nutrient cycling	Storage, processing, and acquisition of nutrients within the biosphere	Nitrogen cycle; phosphorous cycle
Net primary production	Conversion of sunlight into biomass	Plant growth
Pollination and seed dispersal	Movement of plant genes	Insect pollination; seed dispersal by animals
Habitat	The physical place where organisms reside	Refugium for resident and migratory species; spawning and nursery grounds
Hydrological cycle	Movement and storage of water through the biosphere	Evapotransporation; stream runoff; groundwater retention
Regulating services	Maintenance of essential ecological processes and life support systems for human well-being	
Gas regulation	Regulation of the chemical composition of the atmosphere and oceans	Biotic sequestration of carbon dioxide and release of oxygen; vegetative absorption of volatile organic compounds
Climate regulation	Regulation of local to global climate processes	Direct influence of land cover on temperature, precipitation, wind, and humidity
Disturbance regulation	Dampening of environmental fluctuations and disturbance	Storm surge protection; flood protection
Biological regulation	Species interactions	Control of pests and diseases; reduction of herb ivory (crop damage)
Water regulation	Flow of water across the planet surface	Modulation of the drought-flood cycle; purification of water
Soil retention	Erosion control and sediment retention	Prevention of soil loss by wind and runoff; avoiding buildup of silt in lakes and wetlands
Waste regulation	Removal or breakdown of nonnutrient compounds and materials	Pollution detoxification; abatement of noise pollution

(continued)

D.A. MOUAT AND W.G. KEPNER

TABLE 1. Ecosystem Functions and Services (From Farber et al., 2006)

Ecosystem functions and services	Description	Examples
Nutrient regulation	Maintenance of major nutrients within acceptable bounds	Prevention of premature eutrophication in lakes; maintenance of soil fertility
Provisioning services	Provisioning of natural resources and raw materials	
Water supply	Filtering, retention, and storage of fresh water	Provision of fresh water drinking; medium for transportation; irrigation
Food	Provisioning of edible plants and animals for human consumption	Hunting and gathering of fish, game, fruits, and other edible animals and plants; small-scale subsistence farming and aquaculture
Raw materials	Building and manufacturing Fuel and energy Soil and fertilizer	Lumber; skins; plant fibers; oils; dyes Fuel wood; organic matter (e.g., peat) Topsoil; frill; leaves; litter; excrement
Genetic resources	Genetic resources	Genes to improve crop resistance to pathogens and pests and other commercial applications
Medicinal resources	Biological and chemical substances for use in drugs and pharmaceuticals	Quinine; Pacific yew; echinacea
Ornamental resources	Resources for fashion, handicraft, jewelry, pets, worship, decoration, and souvenirs	Feathers used in decorative costumes; shells used as jewelry
Cultural services	Enhancing emotional, psychological, and cognitive well-being	
Recreation	Opportunities for rest, refreshment, and recreation	Ecotourism; bird-watching; outdoor sports
Aesthetic	Sensory enjoyment of functioning ecological systems	Proximity of houses to scenery; open space
Science and education	Use of natural areas for scientific and educational enhancement	A "natural field laboratory" and reference area
Spiritual and historic	Spiritual or historic information with significant religious values	Use of nature as national symbols; natural landscapes

MA has been driven by the integration of diverse disciplines associated with the natural and social sciences. Thus, an important challenge in the future is to develop research initiatives that cut across political boundaries and synthesize the unique contributions of all sciences to intelligently inform and support effective policy and decision making.

Particular foci for this activity may include stressor and threat identification and causal processes; mapping, ranking, and comparing risk vulnerability; addressing multiple stressors and cumulative impacts; and connecting socioeconomic and biophysical variability with policy action and social equity and justice. Last, threat assessment and its relation to human and environmental security transcend cultural and political-economic constructs. Inevitably, the concept of security becomes attached to either a defined human-environment system or a physical geography directly involving stakeholder and decision-making groups. At a minimum, the intent of assessing security and reducing risks to sustainable societies or human well-being includes the integration of best available science into policy with the purpose of translating theory into decision making and responsible action.

3. Summary of Contributions and Findings

The workshop and the contributions provided in these proceedings encompassed a broad spectrum of issues involved with environment and human security. An understanding of the linkages, conceptual approaches, and case studies are presented. The proceedings are divided into multiple sections: An Introduction, which describes why we need to consider the importance of environmental and human security; a section on environmental challenges with case studies in North Africa, the Balkans, and the Middle East; a section that emphasizes case studies involving human security; a section on human challenges with global case studies; and a concluding section on acting on hazard impacts with case studies in sub-Saharan Africa, the Black Sea Region, and Central Asia.

In the introductory section, "Approaches to Environmental and Human Security," Liotta and Shearer define environmental and human security through threats and vulnerabilities. Importantly, the critical uncertainties that face decision makers often impede efforts to deal with the issues. Liotta and Shearer argue that uncertainties must be folded into the decision-making process. In so doing, they suggest strategies for taking action. In critiquing traditional concepts of security, the authors use the metaphor "zombie concepts" of modernism, which emphasize the state and thereby fail to engage the multiple and interdependent processes of change that global societies face. In this context, new solutions beget increased risk and new knowledge yields greater uncertainty. The second metaphor the authors draw on is the "boomerang effect," which looks more narrowly at how policies intended to address some specific dimension of security can undermine other dimensions.

When these metaphors are considered as a set of related ideas, it becomes apparent that the world is confronted with socially produced and human-centered vulnerabilities. Further, the potential for local and localized risk has mutated into *systemic risk* that affects both the "developing" and "developed" parts of the world. Responses to climate change, in particular, must therefore accommodate thinking in terms of multiple facets of security. Owen follows Liotta and Shearer by carefully articulating the significance of the concept of human security as central to a variety of international actions including policy making. This paper first argues that measuring human security—despite its critics' concerns—is a worthy academic exercise with significant policy-relevant applications. Then, by analyzing the four existing methodologies for measuring human security, it is argued that all fall victim to a paradox of human security. After introducing the notions of space and scale to the measurement of human security, a new methodology is proposed for mapping and spatially analyzing threats at a subnational level.

The second section on "Environmental Challenges: Examples from North Africa, the Balkans, and the Middle East" consists of specific approaches at the case-study level as well as overviews of desertification. Mouat and Lancaster present a proactive approach that involves understanding and mitigating problems before they get to a crisis or even irreversible stage. They propose using alternative futures analysis to model the environmental effects of alternative human decisions on land use to allow the selection of desirable pathways of development and land use. Al-Alawi describes the status and causes of desertification in Jordan. The leading causes of desertification in Jordan are improper agricultural techniques such as failure to use contour plowing, overcultivation, conversion of rangelands to croplands in marginal areas, and uncontrolled human development. Water scarcity is one of the principal factors behind the improper agricultural practices. Strategies for controlling erosion, protecting potential agricultural lands, and water harvesting are given. Galatchi discusses global issues involving environmental change and international efforts to promote environmental conservation. Specifically discussed global issues include global warming, protection of the ozone layer, acid rain, pollution, and desertification. International efforts discussed include the Rio Earth Summit, the UNCCD, the UNFCCC and IPCC, the ITTA and ITTO, the UNCSD, and others. Bachev discusses environmental problems and associated management implications from years of inappropriate agricultural practices in Bulgaria and current attempts to change these practices. The paper goes into considerable detail regarding types of environmental degradation (pollution, erosion, and loss of biodiversity) and discusses the importance of national and international efforts at bringing about effective policies to deal with the problems. Nash et al. provide a regional assessment of land use and land use change in Morocco over a twenty-two-year period. They critically note that the ability to analyze and assess environmental change and to relate those changes to causative factors is essential in environmental decision making. Moreover, their paper is the only one in these proceedings that emphasizes the use of remote sensing for land cover characterization.

The third section, "Human Challenges: Case Studies," focuses on urbanization and security issues. Lawson et al. discuss human security in urban settings in the context of conflict, crime, and violence. They argue that the essential idea of human security "is that people rather than nation-states are the principal point of reference and that the security of states is a means to an end rather than an end in itself." They further state that "international peace and security is ultimately constructed on the foundation of people who are secure." Bobylev analyzes the importance of urban infrastructure on the sustainability of urban development. As urban infrastructure is impacted by environmental issues, the recognition of the linkages between urban infrastructure and environmental security provides a way forward to make cities sustainable. Hearne offers a compelling case for linking environmental security with regional conflict. He quotes General Anthony Zinni, who was asked to comment on the importance he placed on the environment as a security issue: "When environmental conditions, either natural or man-made, are destabilizing a region, a country, or have global implications, then there are major security implications." Those problems could affect stability and could result in conflict. Christie examines the concept of human security in the context of its use (and abuse) as a means of carrying out the dictates of the state in the guise of the human condition. He states that the use of the term "human security" is one of adoption, adaptation, and co-option. The options are either to remove the increasing practice of militarizing aid or "to continue to promote the ethical component of the framework, demanding that the lives and dignity of people trump that of state security, either our own or that of governments that nominally claim jurisdiction over people elsewhere." Vankovska and Mileski discuss theoretical and political tendencies of human and environmental ("in") security from "below" (that is, from the perspective of the "European periphery, particularly the Balkans"). They raise awareness that the people and societies concerned are outsiders or passive observers of the developments [and the debates], and often of their own fate. They close with the [rhetorical] comment: "Humans or environment—who has the priority when the resources are very limited, ambitions big and general patience low? Or better, where should one start with the survival policy?" Morrone concludes with a discussion of environmental justice and health disparities in Appalachia and how these local cases have global implications. She mentions that the individual case studies have set precedents that may lead to improvements of our understanding of the political, social, and environmental factors that lead to health disparities in poor rural populations worldwide.

The concluding section of this book perhaps best captures the diversity of both hazard impacts and geography addressed at the ARW Workshop. With significant contributions from Africa, the Black Sea region, and Central Asia, these six chapters address a spectrum of hazards, from those directly resulting from human activity such as nuclear fallout in Kazakhstan (the chapter by Ospanova et al.), contamination resulting from Chernobyl (Udovyk), and radioactive waste contamination in the Kyrgyz Republic (Tynybekov et al.), to those that are more clearly the results of interactions between human actions and environmental conditions such as desertification, poverty, and terrorism (Smith; Jebb et al.) and water salinization (Khaydarov and Khaydarov).

As stated by Jebb et al., "It is important to recognize that policy making and implementation will take time, intelligence, critical analysis, patience, and empathy," plus developing guidelines for action, such as those proposed by Smith, which may accelerate the process. Different factors dominate the situation in different regions, and affect the ability of individual countries to respond to hazards in a timely manner, and, as Udovyk's paper illustrates, the "ripple effect" cannot be ignored. Despite these challenges, the papers in this section offer a variety of actions for ameliorating the impact of hazards—for example, the innovative water purification systems installed in villages near the Aral Sea described by Khaydarov and Khaydarov and the use of GIS technologies to analyze the mud flow risk in an area of the Kyrgyz Republic as discussed by Tynybekov et al. Ospanova et al. suggest that the agricultural and economic impact of nuclear contamination could be assessed in several ways, including use of mathematical models results of different scenarios aimed toward developing agrochemical protection measures to reduce radionuclides accumulation.

4. Recommendations for Future Actions

The NATO SPS Advanced Research Workshop provided a number of perspectives on human and environmental security and included important case studies as examples. An opportunity exists to further develop the concepts of this workshop and the findings in this book to a variety of actions and collaborative international efforts. The first is that the participants of the ARW and hopefully the readers of this volume play a proactive role in bringing to the attention of policy makers and decision makers the observations made and conclusions drawn herein. Second, a follow-on workshop(s) targeted toward specific issues such as water and energy in the context of security would be most valuable. Finally, the true measure of our success will be not to produce volumes for bookshelves, but rather to translate our collaborative efforts into concrete actions with measurable results.

References

Farber, S., Costanza, R., Childers, D. L., Erickson, J., Gross, K., Grove, M., Hopkinson, C. S., Kahn, J., Pincetl, S., Troy, A., Warren, P., and Wilson M., 2006, Linking ecology and economics for ecosystem management, *Bioscience* **56**(2): 121–133.
Millennium Ecosystem Assessment, 2005, *Ecosystems and Human Well-Being*, Island, Washington, DC; www.MAweb.org.

INDEX

Printed in the United States
122951LV00001B/1-30/P